THE HEAVY TRANSITION ELEMENTS

A Macmillan Chemistry Text

Consulting Editor: Dr Peter Sykes, University of Cambridge

COMPLEXES AND FIRST-ROW TRANSITION ELEMENTS: David Nicholls

The Heavy Transition Elements

S. A. Cotton
School of Chemical Sciences
University of East Anglia

F. A. Hart
Department of Chemistry
Queen Mary College,
University of London

A HALSTED PRESS BOOK

JOHN WILEY & SONS
New York

First published in the United Kingdom 1975 by
The Macmillan Press Ltd

First published in the U.S.A. by
Halsted Press, a Division of
John Wiley & Sons, Inc.
New York

Printed in Great Britain

Library of Congress Cataloging in Publication Data

Cotton, S   A
   The heavy transition elements.

   "A Halsted Press book."
   1. Transition metals. 2. Transition metal
compounds. I. Cotton, S. A., joint author. II. Title.
QD172.T6C65 1975   546'.6   75-4950
ISBN 0 470-17681-4

Contents

Preface

The study of transition metals and their compounds occupies a prominent place in most first-degree chemistry courses. In many cases considerable attention is paid to the first-row transition series, much less attention to the second- and third-row metals, while the lanthanides and actinides are treated with the utmost brevity. The origin of this imbalance is probably to be found in a combination of two factors. The first is that the industrially important, well-known and abundant transition metals are the 3d metals such as copper, iron and nickel. Secondly, the quantitative aspects of ligand-field theory are more readily applied to the 3d metals than to the 4d, 5d, 4f and 5f metals because of the relative magnitudes of the physical parameters, such as the spin–orbit coupling constant and the crystal-field splitting parameter, that are involved.

This book, which is a sequel to a cognate volume[†] dealing with the 3d metals, gives an account of the 4d, 5d, 4f and 5f metals, which it is hoped will be adequate for any first-degree requirements and for postgraduate courses dealing with general aspects of transition-metal chemistry. The treatment is given in sufficient range and detail to allow considerable latitude to the course organiser and student in their choice of precise topic and level of approach. We have not hesitated to include a high proportion of descriptive chemistry, in the conviction that a sound knowledge of experimental facts forms the basis of any scientific discipline. This style of treatment may also be useful to research workers requiring a general view of some particular area of 4d, 5d, 4f and 5f chemistry; it is not intended, however, to provide a detailed introduction to research.

We would like to thank a number of our colleagues, particularly Professor D. C. Bradley, Dr D. M. P, Mingos and Dr P. Thornton for reading portions of the manuscript and making constructive comments. Any remaining errors are our own responsibility. We also wish to thank Mrs H. Matthewman and Mrs T. Gue for their very efficient typing of the manuscript. Finally, we both wish to thank Mrs Eileen Hart for preliminary typing and for much assistance and encouragement.

S. A. Cotton
F. A. Hart

[†] *Complexes and First-Row Transition Elements,* by David Nicholls

Abbreviations for Common Ligands

acac	acetylacetone anion
bipy	2,2'-bipyridyl
bzac	benzoylacetone anion
cp	cyclopentadienyl anion
diars	*o*-phenylenebisdimethylarsine
diglyme	2,2'-dimethoxydiethylether
diphos	1,2-diphenylphosphinoethane
dma	N,N-dimethylacetamide
dtpa	diethylenetriaminepenta-acetic acid anion
EDTA	ethylenediaminetetra-acetic acid anion
hal	halogen anion
hfac	hexafluoroacetylacetone anion
nta	nitrilotriacetic acid anion
oxine	8-hydroxyquinoline anion
phen	1,10-phenanthroline
py	pyridine
THF	tetrahydrofuran

TITANIUM

Metal: h.c.p.; m.p. 1680°; I_1: 6.83 eV; I_2: 13.57 eV; I_3: 27.47 eV
Oxides: TiO, Ti_2O_3, TiO_2
Halides: TiX_2 (X = Cl, Br, I), TiX_3 and TiX_4 (X = F, Cl, Br, I)

Typical donor atom/group	Oxidation State and Representative Compounds				
	0	1	2	3	4
			Hal	O,Hal,N	O,Hal,N,As,etc.
Co-ordination number					
3				√ $Ti\{N(SiMe_3)_2\}_3$	
4 tet					√√ $TiCl_4$
5 T.B.P.				√√ $TiBr_3(Me_3N)_2$	√ $TiOCl_2(Me_3N)_2$
5 S.P.					?
6	√† $Ti(bipy)_3$		√√ $TiCl_2$	√√√ $[Ti(urea)_6]^{3+}$	√√√ $TiCl_4(Cl_3PO)_2$
7					√ $TiCl(S_2CNMe_2)_3$
8					√ $TiCl_4(diars)_2$

† 'Suspect' ligand; ? Suspected; √ Known; √√ Several examples; √√√ Very common

1 Zirconium and Hafnium

The three pairs of metals Y, Lu; Zr, Hf; Nb, Ta show a striking resemblance between the lighter and the heavier metal of each pair, arising from the predominant stability of the highest, or group, oxidation state, together with the ionic nature of the bonding and the close similarity of ionic radii. Thus both zirconium and hafnium are rather poorly represented in oxidation states other than +4, and the ionic radii are $Zr^{4+} = 74$ pm and $Hf^{4+} = 75$ pm, leading to chemical properties that differ only in comparatively minor respects. However, hafnium has been investigated to a smaller extent than has zirconium, so the factual basis for the statement that their properties are similar is less complete than might be desirable. The chemistry is relatively straightforward, being mainly that of the 4+ ions. Since these are fairly large, high co-ordination numbers are frequent. There are no known carbonyls but numbers of σ- and π- bonded organometallics have been prepared.

The metals occur as zircon, $ZrSiO_4$, and baddeleyite, a form of ZrO_2. As would be expected on account of their similar properties, they always occur together but hafnium is much less abundant than zirconium and only one zirconium atom in fifty is on average isomorphously replaced by hafnium.

1.1 The Metals and their Aqueous Chemistry

Zirconium metal was isolated by Berzelius in 1824 by potassium reduction of a fluoride. Hafnium was not obtained until 1923, a lengthy fractional crystallisation of complex fluorides (as with niobium and tantalum) being necessary before the pure hafnium complex could be reduced with sodium. The hafnium had remained undetected by ordinary chemical methods and its presence was first demonstrated by X-ray spectroscopy.

Either metal is now prepared by reduction of the tetrahalide vapour.

$$ZrCl_4 + 2Mg \xrightarrow[Ar]{1150°} 2MgCl_2 + Zr$$

Excess Mg and $MgCl_2$ are removed by vacuum distillation; if necessary, the product may then be zone refined.

Both metals are high melting, having m.p.s 1852° (Zr) and 2150° (Hf). They have the hexagonal close-packed structure at ordinary temperatures. Zirconium metal is resistant to corrosion by air, most cold acids, and alkalis, but is attacked by hot aqua regia or hydrofluoric acid. Since it also has a low neutron absorption cross-section (weighted average of five isotopes, 0.18 barns†) it may be used for atomic-pile construction. Hafnium must be absent, however, since its average (six isotopes) is 105 barns. This separation is achieved by ion-exchange chromatography or by solvent extraction with tributylphosphate in ways essentially similar to those used for separations within the lanthanide and actinide series. The separation is, of course, carried out before preparation of the tetrahalide and reduction to the metal. Zirconium–niobium alloys are useful superconductors.

Because of hydrolysis, the hydrated ions $[M(H_2O)_x]^{4+}$ apparently do not exist in solution. Hydrous zirconium and hafnium oxides are soluble in aqueous HF, HCl, H_2SO_4 and HNO_3. Unlike the neighbouring metals, Nb and Ta, which form MO^{3+}, there is no evidence for ZrO^{2+} or HfO^{2+}. Thus an X-ray analysis of the compound $ZrOCl_2 . 8H_2O$, obtained from dilute hydrochloric acid solution, shows that it contains the polymeric cation $[Zr_4(OH)_8(H_2O)_{16}]^{8+}$. The Zr–(OH)–Zr

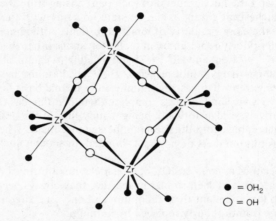

$\bullet = OH_2$
$\bigcirc = OH$

Figure 1.1 The tetrameric $[Zr_4(OH)_8(H_2O)_{16}]^{8+}$ cation found in $ZrOCl_2 . 8H_2O$
(after T. C. W. Mak, *Can. J. Chem.*, **46** (1968), 3491)

bridges and the eight-co-ordination, dodecahedral in this case, are as expected for a moderately large cation with a rather high charge number (see figure 1.1). It is uncertain whether this species predominates in aqueous solution; the degree of polymerisation is pH dependent, increasing with rise of pH. There is some evidence that a trimeric species is present in 2.8 M HCl. There is no true hydroxide, hydrated forms $Zr(OH)_4(H_2O)_x$ being obtained.

The fluoride and sulphate anions have greater affinities for Zr^{4+} than has Cl^- and uncharged or anionic species are readily formed at fairly low acid concentrations. Thus the hydrated sulphate $Zr(SO_4)_2 . 4H_2O$ crystallises from 6 M H_2SO_4; the structure involves square antiprismatic co-ordination of zirconium. Fluoro-complexes include $Zr_2F_8 . 6H_2O$ (dodecahedral co-ordination—see figure 1.2a) and $HfF_4(H_2O)_2 . H_2O$ (square antiprism with four bridging fluorines—see figure 1.2b).

† 1 barn = 10^{-28} m^2

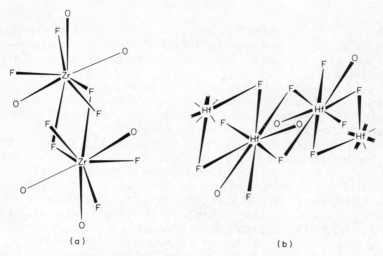

Figure 1.2 (a) The dimeric structure of $ZrF_4 . 3H_2O$; (b) the polymeric structure of $HfF_4 . 3H_2O$ (after D. Hall, C. E. F. Rickards and T. N. Waters, *Chem. Ind.* (1964), 713; *Nature, Lond.*, **207** (1965), 405)

Other salts of interest include the complex hydrated oxalate $Na_4[Zr(C_2O_4)_4] . 3H_2O$, which has dodecahedral co-ordination, and the hydrated nitrate $Zr(NO_3)_4 . 5H_2O$, obtained from cold concentrated nitric acid. The unsolvated nitrate $Zr(NO_3)_4$ may be obtained by

$$ZrCl_4 + 4N_2O_5 \rightarrow Zr(NO_3)_4 + 4NO_2Cl$$

It is volatile and the zirconium is doubtless eight-co-ordinated in a dodecahedral manner with bidentate nitrate groups; there is some spectroscopic evidence (infrared and Raman) for this. The hafnium compound firmly holds on to a molecule of N_2O_5 as $Hf(NO_3)_4 . N_2O_5$.

1.2 Oxides

Apart from a volatile unstable species, probably the monoxide, formed by heating zirconium-zirconium dioxide mixtures, the dioxides are the only oxides. At ordinary temperatures, monoclinic forms with irregular seven-co-ordination are stable; at very high temperatures the fluorite structure is adopted. The dioxides may be obtained by heating the hydrated hydroxides. Zirconium dioxide, being a rather inert substance after strong ignition (for example, it is then unattacked by hot aqueous HF) and being stable up to over 2000°, forms a useful refractory material and ceramic opacifier or insulator. For these purposes, addition of a little CaO gives a stable fluorite structure, thus avoiding the adverse mechanical consequences of phase changes on repeated heating and cooling.

1.3 Halides

As in the case of titanium, there is a considerable difference between the properties of the tetrafluorides and tetrachlorides. Thus the fluorides have higher melting points (ZrF_4, 932°; $ZrCl_4$, 438°) and the chlorides are readily hydrolysed whereas the fluorides are not.

The tetrafluorides may be obtained by the action of gaseous HF on the dioxides. In the solid state, ZrF_4 is polymeric with square antiprismatic co-ordination, but on heating to 800° the vapour gives an infrared spectrum consistent with a tetrahedral monomer. If this is a true interpretation, the fluorides YF_3, ZrF_4, NbF_5 and MoF_6 form a series interesting in its gradation of molecularity. As already mentioned, the fluorides form stable crystalline hydrates.

The tetrachlorides may be made from the oxides by treatment with CCl_4 vapour or thionyl chloride. Both chlorides are quite volatile and may be separated by fractional distillation or gas chromatography. In the vapour, electron diffraction and vibrational spectra (especially Raman) indicate tetrahedral structures for $ZrCl_4$ and $HfCl_4$. Solid $ZrCl_4$ is a linear polymer (zig-zag chains of $ZrCl_6$ octahedra joined by edges). Hydrolysis occurs instantaneously, giving the hydrated oxy-chloride. On heating $ZrOCl_2 . 8H_2O$ with thionyl chloride, the complex $(ZrCl_4 . SOCl_2)_2$ is obtained; this is a chlorine-bridged dimer, with oxygen-bonded thionyl chloride. This reaction illustrates the dehydrating (and chlorinating) ability of thionyl chloride, as well as its occasional tendency to co-ordinate.

The tetrabromides and tetraiodides may be made by such methods as

$$ZrCl_4 \xrightarrow[20°]{BBr_3} ZrBr_4$$

$$ZrO_2 \xrightarrow[400°]{AlI_3} ZrI_4$$

They are fairly low melting (bromides : Zr, 450°; Hf, 424°; iodides : Zr, 500°; Hf, 449°) and volatile. In the vapour phase, electron diffraction gives Zr–Br = 244 and Hf–Br = 243 pm.

Oxidation states other than +4 are represented by ZrF_2, MCl_2, MBr_2, ZrI_2; ZrF_3, MCl_3, MBr_3, MI_3 (M = Zr and Hf). There are certain anomalies connected with some of these compounds; thus the metal–halogen distances determined by X-ray diffraction for the trihalides are characteristic of the M^{4+} ions, and the low magnetic moment indicates metal–metal bonding. ZrX_3 (X = Cl, Br, I) are isomorphous; they have strings of octahedra sharing opposite faces. HfI_3 is similar. All are dark coloured solids having strong reducing properties. They are made by reactions such as

$$ZrCl_3 \xrightarrow[675°]{Zr} ZrCl_2$$

$$ZrI_3 \xrightarrow{380°} ZrI_2$$

$$ZrCl_4 \xrightarrow[Zr]{420°} ZrCl_3$$

The existence of certain lower halides is in doubt. Thus some reports state that reaction of Hf with $HfCl_4$ at elevated temperatures leads to the formation of $HfCl_2$ (which gives $HfCl_4$ and HfCl above 900 K), but others deny this.

One notable difference between the two metals lies in the iodides. While reduction of HfI_4 proceeds via $HfI_{3.3}$ to HfI_3, there being no evidence for a lower phase, the zirconium sequence is $ZrI_4 \rightarrow ZrI_3 \rightarrow ZrI_{2.86} \rightarrow ZrI_2$.

1.4 Other Binary Compounds

With hydrogen, the behaviour is rather similar to that of the lanthanides, the gas being absorbed reversibly and approaching a limiting concentration of MH_2; the structure is then distorted fluorite.

Nonstoichiometry is evident among the sulphides. These are made as follows

$$Zr \xrightarrow[700^\circ]{\text{excess S}} ZrS_3$$

$$ZrO_2 \xrightarrow{CS_2} ZrS_{1.8-2.0} \xrightarrow{\text{heat}} ZrS_{0.9-1.54} \text{ and } ZrS_{0.7}$$

The disulphide—violet and a semiconductor—has the cadmium iodide structure.

The metals both form refractory nitrides MN by direct combination; these have the sodium chloride structure. ZrN is a superconductor below 8.9 K. Carbides MC are formed at high temperatures by direct combination or by reaction of carbon with the dioxide. They too have the sodium chloride structure.

1.5 Complexes of Zirconium and Hafnium

1.5.1 Complex Halides

The general situation here is that many fluorozirconates and fluorohafnates, in which the co-ordination number tends to be eight, have been obtained. The octahedral chloro- and bromo-complex ions $[MX_6]^{2-}$ (M = Zr, Hf; X = Cl, Br) are also known. No complex iodides have yet been isolated.

Complex fluorides of the types AMF_8, AMF_7, AMF_6 and AMF_5 are known (M = Zr, Hf; A = different counter-ions). Of course, the stoichiometry gives no guidance concerning the co-ordination geometry, but some X-ray studies have been performed. Thus $[Cu(H_2O)_6]_2[ZrF_8]$ was found to have square antiprismatic co-ordination of the Zr^{4+} ion (Zr–F = 205–211 pm). In $[Cu(H_2O)_6]_3[Zr_2F_{14}]$ the co-ordination number of eight is maintained by dimerisation involving edge-sharing by two square antiprisms. Zirconium is seven-co-ordinate in Na_3ZrF_7. This compound contains pentagonal bipyramidal $[ZrF_7]^{3-}$ ions, but in the ammonium analogue the ion is distorted, probably as a consequence of N–H $\cdots$ F hydrogen bonding.

In the case of the $[MF_6]^{2-}$ complexes, six, seven and eight co-ordination have all been established. Thus Rb_2MF_6 contain the octahedral anion $[MF_6]^{2-}$, while K_2MF_6 contain M^{4+} in a polymeric eight-co-ordinated state and $(NH_4)_2ZrF_6$ has seven-co-ordinated zirconium. The hydrazinium salt $N_2H_6ZrF_6$ contains zig-zag chains composed of ZrF_8 bicapped trigonal prisms, four of these fluorines being shared between adjacent polyhedra. Clearly the lattice energy, which depends on the counter-ion, is the dominating factor rather than any inherent greater stability of a particular co-ordination polyhedron for the zirconium or hafnium ion. ^{19}F n.m.r. and Raman evidence indicate that octahedral $[MF_6]^{2-}$ species predominate in aqueous hydrofluoric acid solutions of zirconium and hafnium. No structures have been determined in the case of complexes $[MF_5]^-$ (see figure 1.3).

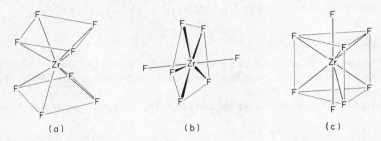

Figure 1.3 Zirconium fluoro-complexes: (a) square antiprismatic $[ZrF_8]^{4-}$ in $[Cu(OH_2)_6]_2 ZrF_8$; (b) distorted trigonal bipyramidal $[ZrF_7]^{3-}$ in $(NH_4)_3 ZrF_7$; (c) bicapped trigonal prismatic ZrF_8 (edge-shared) units in $(N_2H_6)ZrF_6$

Only one type of chloro- and bromo-complex is known and this is the octahedral anion $[MX_6]^{2-}$. This configuration has been established by X-ray diffraction for $Cs_2 ZrCl_6$, which has the $K_2 PtCl_6$ structure, and is strongly indicated by the infrared and Raman spectra of $Cs_2 MX_6$ (M = Zr, Hf; X = Cl, Br).

These complex halides may be made by wet or dry methods; for example

$$ZrF_4 + 2KF \xrightarrow{\text{HF aq}} K_2 ZrF_6$$

$$ZrF_4 + KF \xrightarrow{\text{fuse}} KZrF_5$$

$$HfOCl_2 + 2KCl \xrightarrow{\text{HCl conc.}} K_2 HfCl_6$$

As an example of oxohalide complexes of these metals, $(NEt_4)_4 [Zr_2 OCl_{10}]$ is thought to be analogous in structure to $Cl_5 MOMCl_5^{4-}$ (M = Re, Ru, Os); infrared evidence suggests that the stretching force constant for Zr—O—Zr vibration is only about 60 per cent of that of the other complexes.

1.5.2 Adducts of the Halides with Uncharged Ligands

The halides give complexes with a variety of uncharged ligands, including phosphine oxides, esters, ethers, sulphides, amines, phosphines and arsines. Thus both 'hard' and 'soft' ligands will co-ordinate. In many cases, however, physical investigation of the complexes has been necessarily limited to infared spectral analysis owing to lack of solubility or difficulty in obtaining a specimen suitable for X-ray diffraction. Typical complexes include $MCl_4 (OPCl_3)_2$, $MCl_4 (THF)_2$, $ZrCl_4 (SMe_2)_2$, $MCl_4 py_2$, $MCl_4 bipy$ and $MX_4 diars_2$ (M = Zr, Hf; X = Cl, Br). There is some infra-red evidence that many of these are octahedral (which is expected) and have the *cis* configuration (which is less usual). The diarsine complexes, like their titanium analogue, are eight-co-ordinate with dodecahedral co-ordination. It is interesting that on reaction of a mixture of MX_4 with o-phenylenebis(dimethylarsine) the

Zr complex precipitates immediately, while the hafnium analogue precipitates much later. Another example of differences between Zr and Hf lies in their complexes with tetrahydrothiophene where Hf behaves as a much better acceptor.

There are also a number of complexes arising from the trihalides, such as $ZrCl_3 bipy_3$ and $ZrX_3 py_2$ (X = Cl, Br, I), but structural investigation has been hampered by low solubility and lability. It is notable, however, that the magnetic moments for these complexes (0.3–1.3 B.M.) are considerably below the value expected for a $4d^1$ species; this may indicate some metal-metal interactions or be due to spin-orbit coupling.

1.5.3 Acetylacetonates and Related Compounds

The two metals form several different species containing β-diketonate ligands and these have proved amenable to fairly extensive investigation.

The tetrakis(acetylacetonates) are prepared in aqueous solution; for example

$$ZrOCl_2 \xrightarrow[Na_2CO_3]{Hacac} [Zr\ acac_4]$$

However, reaction in organic solvents, without base, gives a lower degree of substitution; for example

$$ZrCl_4 + 2Hacac \xrightarrow[boil]{Et_2O} ZrCl_2\ acac_2 + 2HCl$$

$$ZrCl_4 + 3Hacac \xrightarrow[boil]{benzene} ZrCl\ acac_3 + 3HCl$$

$Zr(acac)_4$ has been shown by single-crystal X-ray diffraction to be monomeric and eight-co-ordinate (see figure 1.4a). The molecular symmetry is D_2.

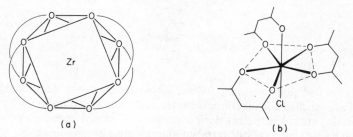

(a)　　　　　　　　　　　　(b)

Figure 1.4 (a) The antiprismatic geometry in $Zr(acac)_4$ (after J.V. Silverton and J. L. Hoard, *Inorg. Chem.,* **2** (1963), 243); (b) the structure of $Zr(acac)_3Cl$; distorted pentagonal bipyramidal.

The tris-species [M diket₃Cl] are monomeric and seven-co-ordinate (M = Zr, Hf; diket = acac, bzac, dbm). Crystallographic analysis of $Zr(acac)_3$ Cl shows that the configuration may be described as a distorted pentagonal pyramid with one oxygen and one chlorine occupying the 'axial' positions (see figure 1.4b). The halide can be replaced by π-C_5H_5 and an X-ray structure determination of one such compound, $[Zr(\pi$-$C_5\ H_5)hfac_3]$, again shows a distorted pentagonal bipyramid, with

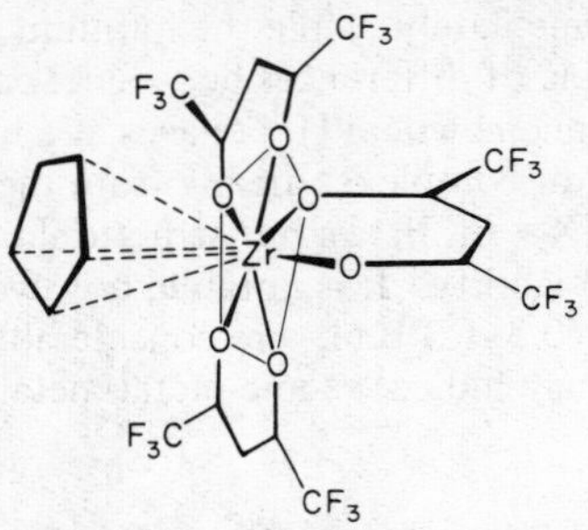

Figure 1.5 The structure of $Zr(CF_3.CO.CH.CO.CF_3)_3(\pi\text{-}C_5H_5)$ (after M. Elder, J. G. Evans and W. A. G. Graham, *J. Am. chem. Soc.*, **91** (1969), 1245)

the C_5H_5 at one apex (see figure 1.5). At room temperature the n.m.r. spectrum of this compound shows rapid configurational exchange, but at low temperatures the two types of diketonate ligand can be discerned.

Infrared and Raman spectra indicate that $[MX_2 acac_2]$ (M = Zr, Hf; X = Cl, Br) adopt the *cis* configuration in solution, but, except in the case of replacement of X by bulky alkoxide groups, the two environments of the methyl groups on the acetylacetonate ligands cannot be distinguished by n.m.r. even at $-130°$.

1.6 Alkoxides and Dialkylamides

These compounds, $M(OR)_4$ and $M(NR_2)_4$ (M = Zr, Hf) have been studied rather thoroughly. The co-ordination number of four is, of course, very low for Zr^{4+} and Hf^{4+} and intermolecular bridges of the type

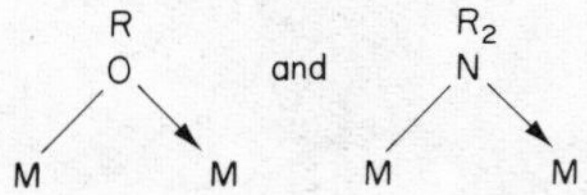

are readily formed giving polymers, whose molecularity may be determined in very dry solution by colligative measurements. However, the use of bulky alkyl groups such as $-CMe_3$ creates sufficient steric hindrance to prevent or reduce polymerisation, giving stable volatile tetrahedral monomers in suitable cases.

Methods of preparation are straightforward.

$$ZrCl_4 + 4EtOH \xrightarrow{\text{base}} Zr(OEt)_4 + 4HCl$$

$$ZrCl_4 + 4LiNEt_2 \longrightarrow Zr(NEt_2)_4 + 4LiCl$$

$$Zr(NEt_2)_4 + 4Bu^tOH \longrightarrow Zr(OBu^t)_4 + 4Et_2NH$$

Both classes of compound are very readily hydrolysed and must be prepared under rigorously anhydrous conditions. In boiling benzene $Zr(OEt)_4$ is 3.6-fold associated but $Zr(OCMe_3)_4$ and $Hf(OCMe_3)_4$ are monomeric. The latter two compounds are quite volatile, having vapour pressures of about 650 Pa (5 mm Hg) at 90°. The dialkylamide compounds, which are liquids or low-melting solids, give dithiocarbamates by CS_2 insertion

$$M(NR_2)_4 \xrightarrow{CS_2} M(S_2CNR_2)_4$$

The dithiocarbamates form a well-defined series of complexes, probably eight-co-ordinate.

1.7 The Borohydrides

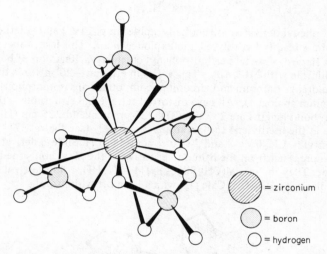

Figure 1.6 The structure of $Zr(BH_4)_4$ (after P.H. Bird and M.R. Churchill, *Chem. Commun.* (1967), 403)

The compounds $Zr(BH_4)_4$ and $Hf(BH_4)_4$ have been carefully studied (see figure 1.6). The zirconium compound, prepared by the reaction

$$ZrCl_4 + 4LiBH_4 \rightarrow Zr(BH_4)_4 + 4LiCl$$

has had its structure determined by X-ray diffraction at −160° (to reduce thermal motion and allow the H atoms to be 'seen'). It has interesting triple-hydrogen bridges and is as volatile as a metal carbonyl (m.p. 29°, extrap. b.p. 123°). The hafnium compound is similar. The proton n.m.r. spectra show only one type of proton, presumably due to rapid intramolecular exchange.

1.8 Compounds Involving Metal–Carbon Bonds

Zirconium and hafnium form just sufficiently effective σ-bonds with carbon for a few alkyls and phenyls to be obtained, although all are rapidly attacked by air.

Thus $ZrPh_4$ is believed to exist in ether solution at $-40°$ and $Zr(CH_3)_4$ has been prepared by the reaction

$$ZrCl_4 + 4LiMe \xrightarrow[-45°]{Et_2O/MePh} ZrMe_4 + 4LiCl$$

The compound, isolated as an etherate, is unstable, decomposing by $-15°$. $Zr(CH_3)(\pi\text{-}C_5H_5)_2Cl$ is much more stable, decomposing at $191°$. Other thermally stable σ-bonded derivatives may be prepared as follows

$$Zr(\pi\text{-}C_5H_5)_2Cl_2 \xrightarrow[\text{2. } H_2O]{\text{1. } LiC_6H_5} \{Zr(\pi\text{-}C_5H_5)_2C_6H_5\}_2O \text{ (m.p. } 250°)$$

$$Zr(\pi\text{-}C_5H_5)_2Cl_2 \xrightarrow{LiC_6F_5} Zr(\pi\text{-}C_5H_5)_2(C_6F_5)_2$$

The perfluorophenyl derivative is thermally stable (m.p. $219°$) and volatile (subl. $120°/1$ Pa) but is readily hydrolysed and explodes in air. This last property is a characteristic hazard of metal perfluorophenyl chemistry. Reaction of benzyl-magnesiumchloride with MCl_4 (M = Ti, Zr, Hf) in ether at $-20°$ affords the tetra-benzyls. The orange hafnium and zirconium compounds are isomorphous (but not with the titanium analogue). All have distorted tetrahedral structures — the mean metal–carbon bond lengths are 213 pm (Ti), 227 pm (Zr) and 225 pm (Hf). The average angle at the methylene carbon atoms is only $92°$ (see figure 1.7). A num-ber of neopentyl ($-CH_2CMe_3$) and 2-silaneopentyl ($-CH_2SiMe_3$) derivatives have recently been prepared from the lithium derivative of the alkyl concerned and the metal chloride. They include $M(CH_2SiMe_3)_4$ (M = Zr, Hf), colourless volatile liquids, which ignite in air, and $Zr(CH_2CMe_3)_4$, which is an off-white solid.

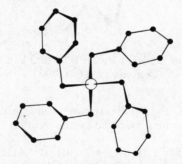

Figure 1.7 Tetrabenzylzirconium (after B.T. Kilbourn, G.R. Davies, J.A.J. Jarvis and A.J.P. Pioli, *Chem. Commun.* (1971), 677)

The cyclopentadienyls, as is generally the case, are much more stable than the alkyls and aryls. Very many compounds of the type $Zr(C_5H_5)_2X_2$ are known, (for example with X = Cl, NO_3, NCS, OBu, SPh, $SnPh_3$); and $Zr(C_5H_5)_4$ or $Hf(C_5H_5)_4$ may be made in a conventional way using NaC_5H_5; their detailed struc-ture (σ- or π-C_5H_5) is uncertain. The black pyrophoric $Zr(C_5H_5)_2$ may be obtain-ed by sodium naphthalenide reduction

$$Zr(C_5H_5)_2Cl_2 + 2NaC_{10}H_8 \xrightarrow{THF} Zr(C_5H_5)_2 + 2NaCl + 2C_{10}H_8$$

The tetra-allyl compounds $Zr(C_3H_5)_4$ and $Hf(C_3H_5)_4$ can be obtained by a low-temperature Grignard reaction; they are thermally unstable. N.M.R. evidence shows clearly that the allyl groups in $Zr(C_3H_5)_4$ are not symmetrically π-bonded. Above $-40°$, however, the characteristic AM_2X_2 pattern changes into AX_4 suggesting a rapid intramolecular rearrangement; the hafnium compound does not show this transition, being always AX_4. $(\pi\text{-}C_5H_5)_2ZrCl_2$ forms $(\pi\text{-}C_5H_5)_2Zr(allyl)_2$ on reaction with C_3H_5MgCl; infrared data suggest that one allyl is σ-bonded and the other π-bonded.

VANADIUM

Metal: b.c.c.; m.p. 1890°; I_1: 6.74 eV; I_2: 14.65 eV; I_3: 29.31 eV

Oxides: VO, V_2O_3, VO_2, V_2O_5

Halides: VX_2 and VX_3 (X = F, Cl, Br, I), VX_4 (X = F, Cl, Br) and VF_5

Oxidation State and Representative Compounds

Co-ordination number	0	1	2	3	4	5
Typical donor atom/group	CO	CO	O	N,O,Hal	O^{2-},O,N,Hal	O^{2-},F
3				$V\{N(SiMe_3)_2\}_3$ ✓		
4 tet				VCl_4^- ✓	VCl_4 ✓	$VOCl_3$ ✓
4 planar						
5 T.B.P.				$VCl_3(NMe_3)_2$ ✓✓	$VOCl_2(NMe_3)_2$ ✓	VF_5 (vapour) ✓
5 S.P.					$VO(acac)_2$ ✓✓✓	$[VOF_4]^-$ ✓✓✓
6	$V(CO)_6$ ✓	$[V(CN)_5NO]^{3-}$ ✓	$[V(OH_2)_6]^{2+}$ ✓✓	VF_6^{3-} ✓✓	$VO(acac)_2py$ ✓	$[VF_6]^-$ ✓
7				$[V(CN)_7]^{4-}$ ✓		
8					$VCl_4(diars)_2$ ✓	$VO(NO_3)_3.CH_3CN$ ✓

✓ Known; ✓✓ Several examples; ✓✓✓ Very common.

2 Niobium and Tantalum

The chemistry of this pair of metals, which have remarkably similar properties, is dominated by their highest possible oxidation state of +5, but not so completely as the +4 state dominates zirconium and hafnium chemistry. Thus well-defined lower oxidation states, particularly Nb(IV) and Ta(IV), are found but their chemistry is rather limited. This is also the first pair of heavy transition metals capable of forming typical low-valent π-bonded complexes.

The history of these metals extends back to the first decade of the nineteenth century, when C. Hatchett showed that a mineral brought from New England in the previous century contained a new element, which, having isolated its oxide, he appropriately named columbium. At the same time, A.G. Ekeberg isolated a similar oxide from minerals from the Baltic coast and called his new element tantalum, on account of its inability to enter solution (Tantalos was, in the Greek myth, unable to drink although tantalised by the sight of abundant water). After several decades during which these elements were believed to be identical and called tantalum, H. Rose showed that at least two different elements were involved and named the second element niobium (Niobe, Tantalos's daughter). The identities of the two metals were finally resolved around 1866 by several workers, particularly M.C. Marignac. Since then, niobium has also been called columbium, particularly in the U.S.A., until comparatively recently; the name 'niobium' was agreed internationally in 1949. It provides a rare example of the discoverer's name for his element not being ultimately adopted.

2.1 The Metals, their Occurrence and Extraction

As is usual in this part of the transition series, the metals have a considerable affinity for oxygen and occur as mixed oxides. Thus the isomorphous series columbite (Fe, Mn)Nb_2O_6-tantalite (Fe, Mn)Ta_2O_6 occurs fairly widely. An important source of niobium fairly free from tantalum is pyrochlorite CaNaNb_2O_6F.

The columbite or other ore may be treated by sodium hydroxide fusion followed by aqueous acid washing, leaving the mixed hydrated pentoxides. Although separation is possible by means of the greater readiness of tantalum to form the insoluble MF_7^{2-} complex (a method orginated by Marignac in 1866 and in commercial use until recently), the modern technique is to use a countercurrent liquid–liquid extraction of the metals into isobutylmethyl ketone from hydrofluoric acid.

The metallic elements may be isolated by reduction of the pentoxides (a) by alkali metals, (b) by carbon or (c) by electrolysis of fused complex fluorides.

Methods (a) and (c) are preferred for tantalum; (b) is good for niobium. The metals are high-melting (Nb, $2468°$; Ta, $2996°$) and tantalum is very dense ($\rho = 16.6$). They remain lustrous in air but react with oxygen or steam on strong heating. Niobium is a component of some steels, especially for high-temperature use, and Nb/Ti alloys are useful superconductors.

2.2 Oxides

The principal oxides are the pentoxides M_2O_5, which are produced by strongly heating the metals in oxygen or by heating the hydrous oxides. They are white, inert powders which dissolve in hydrofluoric acid or molten alkali. The inert pentoxides may be converted into more reactive hydrous oxides by fusion with potassium hydroxide, followed by dissolution in water and precipitation with acid.

Niobium forms a dioxide NbO_2 and a monoxide NbO in addition to the pentoxide; tantalum forms no well-defined lower oxides, although Ta_2O_5 will accept interstitial tantalum atoms to give compositions down to $Ta_{2.5}O_5$; the lighter metal is showing the expected greater stability of lower oxidation states compared with the heavier metal. Niobium dioxide has a fairly conventional structure based on that of rutile but having niobium atoms associated in pairs ($Nb-Nb = 280$ pm). This suggests metal–metal bonding, a view supported by the diamagnetism of this oxide. Niobium monoxide, a stoichiometric oxide, is reported to have a curious structure; the unit cell is that of NaCl with the eight corner atoms and the one central atom removed. Both the lower oxides are obtained by hydrogen reduction of the pentoxide at high temperatures.

The two pentoxides react with water at various pH values to give oxyanions, a rather complicated pH-dependent series of oxyanions being characteristic of this region of the transition series (Zr, Hf; V, Nb, Ta; Mo, W).

Fusion of niobium pentoxide with excess KOH, dissolution of the product in water and slow concentration ultimately produces potassium hexaniobate $K_8Nb_6O_{19}.16H_2O$. A similar tantalum species $K_8Ta_6O_{19}.16H_2O$ is isomorphous. In these compounds (see figure 2.1) an octahedron of metal ions contains a central

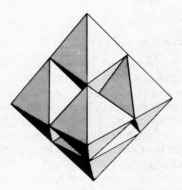

Figure 2.1 The hexameric $[M_6O_{19}]^{8-}$ ion (M = Nb, Ta); each octahedron represents an MO_6 unit (after M.T. Pope and B.W. Dale, *Q. Rev. chem. Soc.* (1968), **22**, 527)

oxide ion and has a bridging oxide ion along each edge. Each metal ion also has a terminal oxide ion. If the hexaniobate is crystallised under less alkaline conditions, a protonated salt $K_7HNb_6O_{19} \cdot 13H_2O$ may be obtained (niobium only). Further reduction of the pH value, however, leads to further polymerisation and eventual deposition of hydrous oxide. At this point, it is convenient to recall that the vanadium(V) solution species, in order of descending pH, are (principally): VO_4^{3-}; $V_2O_7^{4-}$ and HVO_4^{2-}; $H_2VO_4^-$ and $(VO_3^-)_n$ (for example, $n = 3$ or 4); $H_2V_{10}O_{28}^{4-}$ and VO_2^+. In the case of niobium and tantalum there is no definite evidence for MO_4^{3-}.

Some phases may be obtained in the solid state which are insoluble in water. These have no discrete complex anion like the hexaniobate but are mixed oxides. Thus fusion of potassium carbonate with tantalum pentoxide gives K_3TaO_4, $KTaO_3$, $K_2Ta_4O_{11}$ and KTa_5O_{13}. The corresponding niobium system gives rather similar results. The compounds $KNbO_3$ and $KTaO_3$ adopt the perovskite structure; they are known inaccurately — since there are no discrete MO_3^- anions — as potassium metaniobate and metatantalate, respectively. Some niobium and tantalum perovskites are ferroelectric and a number of these mixed oxides, for example $Ba(Na_{0.25}Ta_{0.75})O_3$, have been made and investigated.

A number of niobium and tantalum bronzes are known. This class of mixed oxide, well known for tungsten, has too many cationic charges for its stoichiometry, leading to delocalisation of metal electrons over the crystal with the corollary of metallic electrical conductivity and appearance. An example is $Sr_{0.8}NbO_3$.

2.3 Halides

The pentahalides and tetrahalides are, except for TaF_4, all known. The higher halides present few unusual features, but the lower halides introduce what is a characteristic feature of the chemistry of the group of metals Nb, Ta; Mo, W; Re, namely the ability to form metal clusters in compounds other than carbonyls.

All the pentahalides may be obtained by the action of the appropriate halogen on the heated metal. Most of them may be obtained by other methods also; thus $TaCl_5$ can be made in any of the following ways:

1) Treatment of tantalum metal with (a) Cl_2/Ar at $300°$ or (b) HCl gas at $400°$;
2) treatment of tantalum pentoxide with (a) carbon and Cl_2 at $500°$ or (b) CCl_4 at $300°$;
3) treatment of hydrous tantalum pentoxide with boiling $SOCl_2$ (best);
4) treatment of tantalum pentasulphide with Cl_2 at $300°$.

The pentafluorides are volatile white solids, m.p.s $80°$(Nb) and $95°$(Ta). A coordination of six is achieved in the solid state by tetramerisation using linear fluoride bridges (see figure 2.2), a type of structure characteristic of the pentafluorides M_4F_{20} (M = Nb, Ta, Mo, Ru, Os). On melting, the fluorides are quite viscous, doubtless because of the formation of $-MF_4-F-MF_4-F-MF_4-F-$ chains. In the gas phase polymers persist at low temperatures; there is Raman evidence that depolymerisation can take place at higher temperatures. Thus at $750°$ and 32.8 kPa (246 mm Hg) pressure only monomers appear to be present. The fluorides are only slowly hydrolysed in moist air; in general it is only the

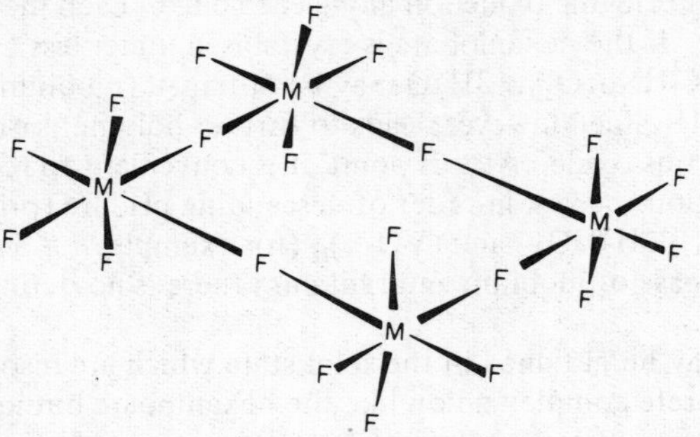

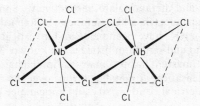

Figure 2.2 Structure of niobium and tantalum pentafluorides

Figure 2.3 The dimeric structure adopted by niobium and tantalum pentachlorides

transition-metal hexafluorides and higher fluorides that suffer instant hydrolysis.

The pentachlorides are dimeric, using bent chloride bridges (see figure 2.3).

It is an interesting feature of fluorine that in molecular compounds it bridges metals rather rarely, and usually by single linear links, while the other halogens bridge metals more commonly, and by double, angular bridges as here in $NbCl_5$. The pentachlorides remain dimeric in nonco-ordinating solvents but are monomeric in the vapour state, in which they are trigonal bipyramidal. The mixed halide $TaFCl_4$ is a fluorine-bridged tetramer. There is [93]Nb n.m.r. evidence ([93]Nb, 100 per cent, has $I = 9/2$) that all the species $NbCl_5$, $NbCl_4Br$, ... $NbBr_5$ exist, probably complexed, in acetonitrile solution. The relative proportions present agree with the amounts calculated on a statistical basis.

The pentabromides and pentaiodides are quite stable except for NbI_5, the compound of the most easily oxidised halide ion with the more easily reduced metal

$$2NbI_5 \xrightarrow[\text{vacuum}]{200°} NbI_4 + I_2$$

Niobium tetrafluoride has an interesting layer-type structure (see figure 2.4), which may be considered to be derived from the NbF_5 structure by additional fluoride bridging to give a planar polymer. It is made by reduction of the pentafluoride by niobium at 300°. Above 350°, it disproportionates

$$2NbF_4 \rightarrow NbF_5 + NbF_3$$

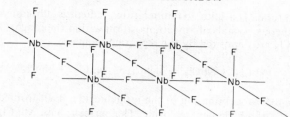

Figure 2.4 The polymeric structure of NbF_4

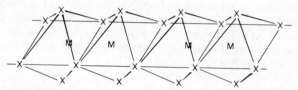

Figure 2.5 The structure of MX_4 (M = Nb, Ta; X = Cl, Br, I) showing the 'pairing' of the metal atoms that leads to diamagnetism

Tantalum tetrafluoride is as yet (1974) unknown.

Niobium and tantalum tetrachlorides are mutually isomorphous and are linear polymers showing pairing of the metal atoms (see figure 2.5); this metal–metal bonding doubtless causes the observed diamagnetism of these d^1 tetrachlorides (contrast the paramagnetic NbF_4). The tetrabromides are closely similar; all four compounds are prepared similarly to NbF_4. Niobium tetraiodide is made by the thermal decomposition of NbI_5 as previously mentioned; it has the structure of $NbCl_4$. Tantalum tetraiodide is made by

$$TaI_5 + Al \xrightarrow{350-500^\circ} TaI_4 + AlI_3$$

2.3.1 The Lower Halides of Niobium and Tantalum

These are of a rather complicated nature and have been the subject of considerable investigation. The main principles seem well established but one or two obscurities still remain. There are three main types: (a) conventional stoichiometric trihalides; (b) halides based on clusters of three metal atoms; (c) halides based on clusters of six metal atoms. Classes (b) and (c) cannot be expressed as $NbCl_n$ where n is integral, but can be expressed as, for example, Nb_3Cl_8. They also tend to be non-stoichiometric, the example just given existing with a little more or a little less chloride than the perfect phase Nb_3Cl_8.

Niobium and tantalum trifluorides are obtained by the action of HF gas on the metal at 225° under pressure. They have the ReO_3 structure. Niobium tri-iodide is also a stoichiometric trihalide and is formed by the reaction of the pentaiodide with niobium. The existence of tantalum tri-iodide is uncertain.

The bromides and iodides, however, all form compounds based on the perfect phase M_3X_8, but often with a deficiency of metal. Thus Nb_3Cl_8 is made by the reduction of $NbCl_5$ with niobium metal, the operation being carried out at about

450° in a sealed tube. If a suitable temperature gradient is subsequently maintained, homogeneous crystals of any composition within the range Nb_3Cl_8–$Nb_{2.55}Cl_8$ are deposited by the repeating sequence

$$\text{`NbCl}_3\text{' + NbCl}_5 \rightarrow 2NbCl_4 \rightarrow \text{`NbCl}_3\text{' + NbCl}_5$$

arising from the disproportionation of the tetrachloride, itself formed from the intentional presence of a little pentachloride. The perfect phase Nb_3Cl_8 has a structure based on the CdI_2 structure with a quarter of the Cd atoms omitted (see figure 2.6); the Nb atoms are in triangular groups of three, the short Nb-Nb distance indicating metal-metal bonding. There are three sorts of chloride; terminal, doubly bridging and triply bridging with respect to the Nb_3 cluster. The range of available compositions between the green Nb_3Cl_8 and the brown $Nb_{2.55}Cl_8$ may correspond with absences of a niobium atom in up to half the Nb_3 clusters, leaving a Nb—Nb grouping as in $NbCl_4$.

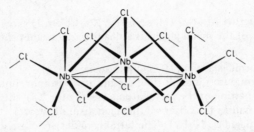

Figure 2.6 A trimeric Nb_3Cl_8 unit; the three types of chloride are indicated

The compounds '$TaCl_3$', '$NbBr_3$' and '$TaBr_3$' have the same structure as Nb_3Cl_8 and may be made in a similar way. They have regions of homogeneity each of which forms part of the range M_3X_8–M_5X_{16}. Finally, besides the NbI_3 mentioned above, there exists a Nb_3I_8 phase (from which NbI_3 is structurally distinct) having a structure related to that of Nb_3Cl_8 (and identical with that of a β-form of Nb_3Br_8) with Nb_3 triangles, in which the Nb—Nb distance is 300 pm.

The third class of niobium and tantalum halides have compositions approximating to $MX_{2.5}$ and are mainly based on the octahedral metal cluster M_6X_{12}. This may carry a triple positive charge and will then have six halides bridging to other octahedra, giving an overall formula M_6X_{15} or $MX_{2.5}$. Alternatively the M_6X_{12} octahedron may carry a double positive charge and have four coplanar bridging halides giving a sheet of octahedra of overall formula M_6X_{14}. As a third alternative, discrete divalent or tetravalent cations $M_6X_{12}^{2\text{ or }4+}$ are possible. We shall discuss only one example from each class.

On heating NbF_5 with niobium foil in a nickel tube with a temperature gradient from 900–400°, black crystals of Nb_6F_{15} are deposited. The structure is depicted in figure 2.7; the Nb—Nb distance is 280 pm. The compound is remarkably stable, being unattacked by acids, alkali or air. On heating Nb_3Cl_8 with niobium metal in a sealed quartz tube at 840°, black crystals of Nb_6Cl_{14} are obtained. These have a planar-bridged structure (see figure 2.8) involving slightly flattened octahedra

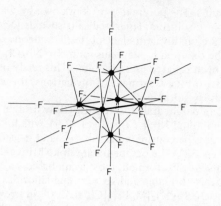

Figure 2.7 The structure of Nb_6F_{15}; of the 18 fluorine atoms shown, 12 form part of the Nb_6F_{12} unit – the remainder bridge adjacent octahedra

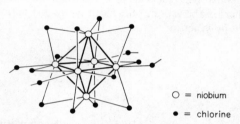

O = niobium

● = chlorine

Figure 2.8 The structure of Nb_6Cl_{14}; based on a Nb_6Cl_{12} core with four chlorines bridging in a sheet polymeric structure

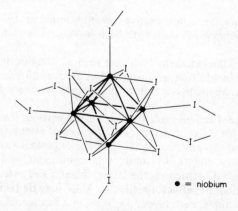

● = niobium

Figure 2.9 The structure of Nb_6I_{11}; there are eight face-bridging iodides

(Nb—Nb = 299 or 289 pm). The polymer is broken down by water, green $Nb_6Cl_{12}^{2+}$ ions passing into solution. Hydrated salts such as $[Nb_6Cl_{12}]Cl_2 \cdot 7H_2O$ may be obtained; they are uni-divalent electrolytes in aqueous solution. The $Nb_6Cl_{12}^{2+}$ cation may be oxidised by aqueous iodine to $Nb_6Cl_{12}^{4+}$; the corresponding Ta cation has been isolated as a sulphate. This ready oxidation has been explained by a molecular-orbital treatment of the cations, for which their high symmetry makes them very suitable.

As a further variant, an octahedral cluster with triply bridging iodine on each face is known in the compound Nb_6I_{11} (see figure 2.9), which is made by heating Nb_3I_8. The cluster can thus be considered as $Nb_6I_8^{3+}$ and its geometry is quite analogous to $Mo_6Cl_8^{4+}$ (p. 32). The octahedra are linked three-dimensionally, as in Nb_6F_{15}. Anionic clusters are also formed, most again based on M_6X_{12} units. $H_2Ta_6Cl_{18} \cdot 6H_2O$ has twelve bridging and six terminal chlorides, for example. On heating the pyridinium salts $(pyH)_2[M_6Br_{12}]Cl_6$ to $200°$ an irreversible transformation occurs with six chlorines exchanging positions with six bromines.

2.3.2 Oxohalides

A number of oxohalides such as MOX_3 (M = Nb or Ta, X = Cl or Br; M = Nb, X = I) may be prepared by controlled reactions of the pentahalide with oxygen. They appear to be monomers in the vapour phase, but X-ray diffraction shows solid $NbOCl_3$ to be a chain polymer of

units with bridging oxygens perpendicular to this plane completing the octahedral co-ordination.

2.4 Other Simple Binary Compounds

Niobium and tantalum are rather reactive toward nonmetals; for example, H, B, C, N, Si and P form binary compounds with both metals. A few of these will be briefly mentioned.

Interstitial nonstoichiometric hydrides are formed, the limiting compositions being about $MH_{0.8}$. In addition, an unstable but distinct dihydride NbH_2 with the fluorite structure can be made electrolytically; it rapidly reverts to NbH, especially in a vacuum.

A number of borides can be made by direct combination; for example, the diborides MB_2 have the AlB_2 structure. This consists of alternate layers of hexagonal nets of boron atoms and hexagonal close-packed metal atoms.

The hard, refractory, electrically conducting carbides MC and M_2C are also known. The former have the NaCl structure; the latter, the anti-CdI_2 structure.

From a large number of refractory silicides, MSi_2 may be taken as examples. They have layer structures, each layer consisting of a hexagonal net of silicon atoms with metal atoms at the centre of each hexagon ($CrSi_2$ structure).

2.5 Alkoxides and Dialkylamides

As is rather characteristic of the earlier transition metals, the groups $-OR$ and $-NR_2$ bond well with niobium and tantalum

$$MCl_5 + 5EtOH + 5NH_3 \rightarrow M(OEt)_5 + 5NH_4Cl$$

The pentaethoxides are readily hydrolysed, volatile liquids, which are associated in nonpolar solvents. They are prepared as indicated above and may be converted into other alkoxides or aryloxides by reaction with esters (Ta less readily than Nb)

$$Nb(OEt)_5 + 5CH_3COOR \xrightarrow{\text{boil}} Nb(OR)_5 + 5\,CH_3COOEt$$

$$(R = Bu^t, Ph, SiMe_3 \text{ and others})$$

The related dialkylamides may be prepared from $LiNR_2$

$$NbCl_5 + 5LiNR_2 \rightarrow Nb(NR_2)_5 + 5LiCl$$

They are volatile and monomeric; the co-ordination in $Nb(NMe_2)_5$ is shown in figure 2.10. The niobium(V) dialkylamides readily decompose to form $Nb(NR_2)_4$; magnetic and e.s.r. data indicate the latter to have a distorted tetrahedral structure. $Ta(NBu_2)_4$ is also known.

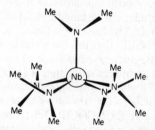

Figure 2.10 The structure of $Nb(NMe_2)_5$ (after C. Heath and M.B. Hursthouse, *Chem. Commun.* (1971), 143)

2.6 Complexes of Niobium and Tantalum

These are limited in variety, being mainly of three types each derived either from the penta- or tetrahalides. These are (a) MF_8^{3-} and MF_7^{2-} (b) MX_5L and (c) MX_4L_2. In addition the complex thiocyanates $M(NCS)_6^-$ are formed by reaction of the pentahalides with thiocyanate ion in acetonitrile.

In aqueous solutions of NbF_5 or TaF_5 which also contain fluoride salts and hydrogen fluoride, several species are in equilibrium. They are MF_8^{3-}, MF_7^{2-}, MF_6O^{3-}, MF_6^- and MOF_5^{2-}. Salts of all of them may be obtained under appropriate conditions. Thus $(N_2H_6)(TaF_6)_2$ and $(N_2H_6)TaF_7$ are obtained from anhydrous $HF, (N_2H_6)TaF_7 \cdot H_2O$ from 45 per cent HF and $(N_2H_5)_2TaF_7$ from 5 per cent HF. ^{19}F n.m.r. spectra of niobium(V) in aqueous HF indicate that $NbOF_5^{2-}$ predominates up to 30 per cent HF and NbF_6^- in stronger solution; solubility as well as concentration determines which species crystallises out. $NbOCl_5^{2-}$

and $TaCl_6^-$ appear to be the predominant species in strong aqueous HCl. The MF_6^- salts are octahedrally co-ordinated; MOF_5^{2-} salts are nearly so. Na_3TaF_8 shows the more usual square antiprismatic eight-co-ordination. $Cs_2[MCl_6]$ contains octahedral anions.

The pentahalides form monomeric adducts of the type MX_5L with oxygen, nitrogen and sulphur ligands. Thus the pentafluorides form volatile complexes $MF_5(Me_2O)$ and $MF_5(Me_2S)$ but they have a greater tendency than the other pentahalides to form $2:1$ complexes, exemplified by MF_5py_2, $MF_5(NH_3)_2$ and $MF_5(Me_2SO)_2$.

The pentachlorides, bromides and iodides all behave rather similarly. Thus six-co-ordinate monomeric complexes with Me_2O, Me_2S, $POCl_3$, Ph_3PO and C_4H_8S ligands are obtained. It is noteworthy that Me_2S is preferred to Me_2O, as shown by careful competitive experiments. There is a distinct tendency to form M=O linkages as shown by the following reactions

$$MCl_5 + 3Ph_3PO \rightarrow MOCl_3(Ph_3PO)_2 + Ph_3PCl_2$$

$$MCl_5(Et_2O) \xrightarrow{heat} MOCl_3 + 2EtCl$$

Reduction to the rather stable Nb(IV) and Ta(IV) complexes occurs readily. Thus at room temperature the seven-co-ordinate monomers MCl_5 diars are formed, but on heating with excess diarsine, reduction (of Nb only) occurs with the production of $NbCl_4(diars)_2$, an eight-co-ordinate complex similar to those formed by Ti, Zr, Hf; V. Pyridine in excess will reduce both pentahalides to MCl_4py_2.

The 4+ oxidation state includes the octahedral $Cs_2[MCl_6]$ and the orange dodecahedral $K_4[Nb(CN)_8].2H_2O$. The latter compound is made by the action of aqueous KCN on methanolic $NbCl_5$ that has been electrolytically reduced. E.S.R. measurements indicate that the dodecahedral structure is not maintained in solution, however; this behaviour compares with that of $[Mo(CN)_8]^{3-}$ and illustrates the configurational lability of higher co-ordination numbers.

Acetonitrile will reduce $TaCl_4$ to the tripositive state, and complexes $TaCl_3(MeCN)_2$, $TaCl_3(bipy)$ and $Ta(PhCOCHCOPh)_3$ may be isolated from the acetonitrile solution, the last two compounds being obtained on addition of the appropriate ligands.

In addition to the isoelectronic series of cyclopentadienyl hydrides $Ta(cpd)_2H_3$, $W(cpd)_2H_2$ and $Rf(cpd)_2H$ mentioned in chapter 9 (page 174), tantalum figures in a rather similar series of phosphine hydrides $TaH_5(Me_2PCH_2CH_2PMe_2)_2$, WH $WH_6(PMe_2Ph)_3$ and $ReH_7(Ph_2PCH_2CH_2PPh_2)$, in each of which the eighteen-electron rule is obeyed. $TaCl_5$ in benzene is treated with the diphosphine and n hydrogen at $115°$ and 10.3 MPa (1500 p.s.i.) pressure. The white crystalline product, m.p. $135°$, has a low Ta—H stretching frequency ($1544\ cm^{-1}$) presumably indicating a comparatively weak bond.

2.7 Organometallics

Compounds containing a σ-bond to carbon have been recently studied. The pentahalides react with dimethylmercury to give the monomethylated products $MeNbCl_4$, $MeTaCl_4$ or $MeNbBr_4$

$$2NbCl_5 + HgMe_2 \xrightarrow{-35°,\ CH_2Cl_2} 2MeNbCl_4 + HgCl_2$$

Reaction of MCl_5 with the more vigorous methylating agent dimethylzinc in pentane at low temperature affords compounds such as $NbMe_2Cl_3$ and $TaMe_3Cl_2$, which decompose slowly at room temperature. These halomethyls readily form six-co-ordinated adducts such as $NbMeCl_4(PPh_3)$ or $TaMe_2Cl_3(PPh_3)$. The bipyridyl adduct $TaMe_3Cl_2(bipy)$ has irregular seven-co-ordination. Pentamethyl tantalum has recently been synthesised by the reaction

$$TaMe_3Cl_2 + 2MeLi \xrightarrow{Et_2O} TaMe_5 + 2LiCl$$

It is a volatile yellow oil, decomposing at 25° to give methane. Phosphine adducts $MMe_5(Me_2PCH_2CH_2PMe_2)$ (M = Nb, Ta) have also been isolated.

A stable and well-characterised compound results from reaction of MCl_5 with the Grignard reagent Me_3SiCH_2MgCl; surprisingly the product is $[M_2(CSiMe_3)_2(CH_2SiMe_3)_4]$. Crystallographic analysis shows the co-ordination of each niobium atom (the tantalum compound is isomorphous) to be tetrahedral (see figure 2.11).

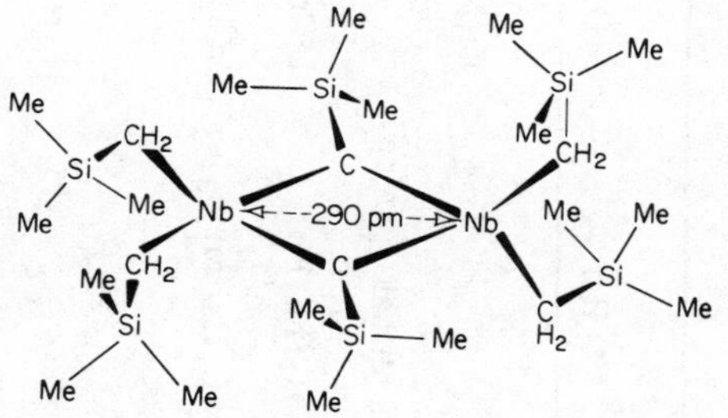

Figure 2.11 The dimeric structure of $[M_2(CSiMe_3)_2(CH_2SiMe_3)_4]$ (M = Nb, Ta) (after F. Huq, W. Mowat, A.C. Skapski and G. Wilkinson, *Chem. Commun.* (1971), 1477)

The Nb—Nb distance is 289.7 pm. The short Nb—C (bridge) distance (197 ± 2 pm compared to Nb—CH_2 216 pm) supports the presumed multiple character (about 1.5) of the bridge bonds. The deprotonation of the bridging carbon atoms is both well established and surprising. The ^1H n.m.r. spectrum has resonances at $\tau 8.5$ (CH_2, intensity 4) and $\tau 9.54, 9.79$ (CH_3, intensity 27). The compound is diamagnetic, as appropriate for Nb(V). In addition to the alkyls just discussed, there also exists an interesting completely phenylated anion $[Li(THF)_4]^+[TaPh_6]^-$.

To summarise this chapter, the chemistry of niobium and tantalum is notable for (a) its dependence on the carefully investigated halides rather than on co-ordination chemistry and (b) the great similarity of the two metals.

CHROMIUM

Metal: b.c.c.; m.p. 1890°; I_1: 6.76 eV; I_2: 16.49 eV; I_3: 30.95 eV
Oxides: Cr_2O_3, CrO_2, CrO_3
Halides: CrX_2 and CrX_3 (X = F, Cl, Br, I); CrF_4, CrF_5 and CrF_6

Oxidation State and Representative Compounds

	0	1	2	3	4	5	6
Typical donor atom/group	CO	CN,N	O,Hal	N,O,Hal,etc.	O,Hal	O^{2-},Hal	O^{2-}
Co-ordination number							
3				$Cr\{N(Pr^i{}_2)\}_3$ √			
4 planar			Cr(II) phthalocyanine √				
4 tet			$CrI_2(Ph_3PO)_2$ √	$LiCr(OBu^t)_4$ √√	$Cr(OBu^t)_4$ √√ †	$[CrO_4]^{3-}$ √√	$[CrO_4]^{2-}$ √√
5 T.B.P.			$[Cr\{N(C_2H_4NMe_2)_3\}Br]^+$ √	$CrCl_3(NMe_2)_2$ †			
5 S.P.						$[CrOCl_4]^-$ √?	
6	$Cr(CO)_6$ √√	$[Cr(CN)_5\,NO]^{3-}$ √	CrF_2 √√√‡	$[Cr(OH_2)_6]^{3+}$ √√√	$[CrF_6]^{2-}$ √√	$[CrOCl_5]^{2-}$ √√	
7		?					
8					$Cr(O_2)_2(NH_3)_3$ √	$[Cr(O_2)_4]^{3-}$ √	

† Not yet confirmed; ‡ Usually Jahn-Teller distorted; ? Suspected
√ Known; √√ Several examples; √√√ Very common

3 Molybdenum and Tungsten

The chemistry of these two industrially important metals bears a close general resemblance to that of niobium and tantalum. There is the same emphasis on the highest oxidation state, in this case six. Common to both pairs of metals are such important features as polymeric oxyanions, metal–metal bonding in lower halides, formation of bronzes, a certain paucity of classical co-ordination compounds with a resulting prominence of the binary halide chemistry, and lack of hydrated cations.

Tungstic acid has been known since 1779, and since both metals occur separately in ores and are obtained by carbon reduction of the heated trioxides, they were isolated soon afterwards. The ores are usually oxides $M^{II}MoO_4$ and $M^{II}WO_4$, except for molybdenite MoS_2, which is the main source of molybdenum. The main source of tungsten is wolframite $(Fe, Mn)WO_4$, and this element was for many years known as wolfram. Molybdenite is roasted to the trioxide, converted by ammonia into ammonium molybdate, which is purified and heated to reconvert to the trioxide; this is then reduced by hydrogen. The tungsten ores are fused with sodium hydroxide and the sodium tungstate is dissolved and acidified to give the hydrous trioxide, which is heated and reduced as before.

The metals are very high melting (Mo, $2610°$; W, $3410°$). They combine with oxygen, sulphur, fluorine and chlorine and are dissolved by the mixed acids HF/HNO_3 and by fused sodium peroxide but are rather inert otherwise. They are used as components of steels especially those for use at high temperature, and tungsten filaments for electric lamps are well known. Molybdenum disulphide is used as a graphite-like lubricant, since it has a layer structure. Molybdenum is important biologically in the nitrogen-fixing enzyme nitrogenase and exists in other proteins, such as xanthine oxidase. In many of these proteins it occurs along with iron.

3.1 Oxides

The principal oxides are the trioxides MO_3, which are made by heating the hydrous trioxides (themselves obtained by acidification of aqueous molybdates or tungstates) or by heating the metals in oxygen. The structure of the yellow WO_3 is a slightly distorted variant of the simple ReO_3 structure; but the white MoO_3, while also having 6:2 co-ordination, has MoO_6 octahedra arranged in layers, an arrangement that is unusual in oxide structures.

Molybdenum alone forms a violet pentoxide Mo_2O_5 on heating the trioxide with the metal, but both metals form dioxides on careful reduction of the trioxide with hydrogen. The dioxides are isomorphous and have a rutile structure distorted in

such a way that the metal atoms are grouped in pairs, with a close M—M distance
of 248 pm (Mo) or 249 (W) and a longer one of 306 (Mo) or 308 (W). The analogy
with NbO_2 is close, since MoO_2 and WO_2 are both diamagnetic. This example of
the instability in halide or oxide structures of a discrete metal ion relative to a metal
cluster of two, three or more metal ions illustrates a general tendency in this part of
the periodic table. Here the oxidation state is a normal one (that is, not low) and
the ligands are not π-acceptors; inter-metal bonds formed by metals in low oxidation
states with π-bonding ligands are more general. A simple example of the latter
class in the Cr, Mo, W triad is $[W_2(CO)_{10}]^{2-}$.

There are also several nonstoichiometric oxides. Many are based on the WO_3
structure but have an oxygen deficiency in a regular manner. They have definite
but fractional stoichiometries, for example $W_{18}O_{49}$. Reduction of the moist trioxides gives 'molybdenum blue' and 'tungsten blue', respectively. These are highly
coloured hydroxy-oxides, in which the metal is in a nonintegral oxidation state between +5 and +6, a typical stoichiometry being $MoO_{2.5}(OH)_{0.5}$.

The yellow aquo-ion $[Mo(H_2O)_6]^{3+}$, which is very unstable to air, is formed
from aquation of dilute $[MoCl_6]^{3-}$ solutions using an ion-exchange resin; there is
also evidence for dimeric oxo-bridged aquo-cations of Mo(IV) and Mo(V).

3.2 Oxy-anions

These compounds of Mo(VI) and W(VI) form a large and important class. They
may be conveniently classified as simple anions, isopolyanions or heteropolyanions.

The molybdate $[MoO_4]^{2-}$ and tungstate $[WO_4]^{2-}$ anions are tetrahedral and
discrete. Their salts may be made by dissolution of the trioxide in aqueous alkali
followed by crystallisation. In contrast to $[CrO_4]^{2-}$, they are not oxidising agents,
thus again confirming the greater stability of the heavier metals in the higher oxidation states.

When solutions of the molybdate or tungstate salts are acidified, polymerisation
occurs to give species that contain MO_6 octahedra joined corner-to-corner or edge-
to-edge. For molybdenum, the main species are the heptamolybdate $[Mo_7O_{24}]^{6-}$
and octamolybdate $[Mo_8O_{26}]^{4-}$ ions. For tungsten, principal species are the hexa-
tungstate $[W_6(OH)O_{20}]^{5-}$ in weakly acidic solution and the decatungstate
$[W_{10}O_{32}]^{4-}$ in strongly acidic conditions. The structure of the heptamolybdate

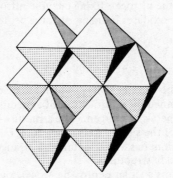

Figure 3.1 The heptamolybdate ion $[Mo_7O_{24}]^{6-}$ (after Lindquist, *Arkiv. Kemi,* **2**
(1951), 325

anion shows edge junctions only. It has been suggested that $[H_4W_4O_{16}]^{4-}$ is the primary aggregation product in isopolytungstate formation. As with other polymeric anions, while each individual structure is chemically perfectly reasonable and in accord with bonding theory, it is at present difficult in many cases to say why a particular degree of polymerisation or a particular structure is favoured over others.

The anions just discussed are known as isopolyanions, since they contain only one cationic element (Mo or W). There is a second class, the heteropolyanions, in which the molecule contains an additional cation, which is often a metalloid such as P or As, but may be a true metal such as Fe or Co. This heterocation, of which one or two may be present, is in a central position in the polyanion and is in tetrahedral or octahedral co-ordination with oxygen ions. A representative structure, that of the 12-molybdophosphate anion $[PMo_{12}O_{40}]^{3-}$, is shown in figure 3.2. The crystalline salts may be obtained from mixed solutions of their components,

Figure 3.2 The $[PMo_{12}O_{40}]^{3-}$ group: the structure is composed of four groups of three WO_6 octahedra around a central PO_4 tetrahedron

by, for example, acidification of a molybdate–phosphate mixture for $PMo_{12}O_{40}{}^{3-}$. Very many examples of heteropoly anions are known, the hetero-atom being Al, Cr or Fe(III); Ce, Th, Si, Ge or Sn(IV); P or As(V); Te(VI) or others. The structure of the heteropoly anion $[CeMo_{12}O_{42}]^{8-}$ is unusual in that it involves face-sharing between the MoO_6 octahedra; the cerium atom is twelve-co-ordinate.

An example of the complicated behaviour observable in these systems is afforded by $[V_2W_4O_{19}]^{4-}$ and $[VW_5O_{19}]^{3-}$, formed from V(V) and W(VI) in solution with appropriate pH adjustment. The former is stable in the pH range 4–7; above pH 7 it forms $VO_3{}^-$ and $WO_4{}^{2-}$ while below pH 4 it reversibly forms $[VW_5O_{19}]^{3-}$. These may possibly have the same structure as the isopolyanions $[M_6O_{19}]^{2-}$.

3.2.1 Tungsten Bronzes

This class of compound has already been mentioned in the discussion of niobium and tantalum. Reduction of Na_2WO_4 with molten zinc yields Na_xWO_3, which over the range $0.32 \langle x \langle 0.93$ has the perovskite structure with random vacancies in the larger cation position. Compared with WO_3 (which has a similar structure but, of course, with *no* sodium atoms), the extra electrons from the xNa^+ ions appear to be delocalised over the whole structure. It would be expected that they mainly reside on the W atoms, which would thus suffer a fractional reduction in formal oxidation state. The resulting solids are inert, yellow, red or violet substances with metallic lustre and electrical conductivity.

3.2.2 Other Binary Compounds

By direct combination, the metals form several sulphides. The most important are MoS_2 and WS_2 on account of their unique layer structure, which, taken together with their inertness, leads to useful lubricating properties. The structure, which affords only very weak binding between the layers with a consequent easy cleavage and greasy feel, is shown in figure 3.3. The co-ordination polyhedron is unusual, being a trigonal prism, which means that the sulphur atoms are in eclipsed pairs. There is considerable electron delocalisation, indicated by the metallic appearance.

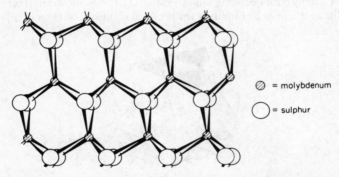

= molybdenum

= sulphur

Figure 3.3 The structure of MoS_2, showing the trigonal prismatic co-ordination of molybdenum (smaller circles) by sulphur (larger circles) (after A.F. Wells, *Structural Inorganic Chemistry*, Clarendon Press, Oxford (3rd edn, 1962)

3.3 Halides

The known halides are listed in table 3.1. There is a predominance of F and Cl in the higher oxidation states and lack of F in the lower, such as is observed for many other metals. As might also be expected, tungsten forms more higher halides than molybdenum.

Table 3.1

	VI	V	IV	III	II
Mo	F,Cl	F,Cl	F,Cl,Br,I	F,Cl,Br,I	Cl,Br,I
W	F,Cl,Br	F,Cl,Br	F,Cl,Br,I	Cl,Br,I	Cl,Br,I

The hexafluorides, prepared by direct combination, are colourless, very volatile liquids (b.p.: Mo, $35°$; W, $17°$). They are monomeric and octahedral and are very susceptible to hydrolysis. The dark blue WCl_6, made by direct combination, is a somewhat easily reducible solid; thus ethanolysis gives the trichlorodiethoxide $W^VCl_3(OEt)_2$. The hexachlorides also adopt monomeric octahedral structures. The hexabromide WBr_6 is unstable.

The pentahalides, if monomeric, would be co-ordinatively unsaturated and there are consequently tendencies towards polymerisation. Thus MoF_5 is tetrameric in the solid state, adopting the same structure as NbF_5 and TaF_5 (page 18). It is made by reduction of MoF_6 with molybdenum; the tetra- and trifluorides may also be obtained thus. WF_5 is slightly unstable; it tends to disproportionate to WF_4 and WF_6. The pentachloride $MoCl_5$, made by direct combination, is monomeric and trigonal bipyramidal in the vapour and is monomeric in benzene solution. However, it dimerises in the solid in the same way as do $NbCl_5$ and $TaCl_5$. The rather long Mo—Mo distance (384 pm) and a temperature-independent magnetic moment of 1.67 B.M. suggest that there is no direct metal–metal bond. WCl_5 is also a dimer.

The resemblance to Nb and Ta halides is also found in the tetrahalides, where the tungsten compounds are isomorphous with the Nb and Ta. The structures of the molybdenum tetrahalides are mainly unknown, but one form of $MoCl_4$ is isomorphous with the niobium and tantalum analogues. Another form adopts a structure based on edge-shared $MoCl_6$ octahedra; this displays 'normal' magnetic behaviour for two unpaired electrons (μ_{eff} = 2.42 B.M. at room temperature). The tetrahalides are made by routes suitable for halides of an intermediate oxidation state, such as

$$WF_6 \xrightarrow{\text{benzene, } 110^\circ} WF_4$$

$$MoCl_5 \xrightarrow{\text{benzene, } 80^\circ} MoCl_4$$

$$MoO_2 \xrightarrow{\text{CCl}_4} MoCl_4$$

$$MoBr_3 \xrightarrow{\text{Br}_2} MoBr_4$$

3.3.1 Lower Halides

The trihalides show an interesting difference between the two metals, those of molybdenum having conventional structures while those of tungsten are of the metal cluster type—a proclivity of the heavier metals. Thus $MoCl_3$, made by hydrogen reduction of $MoCl_5$, has a distorted $CrCl_3$ structure with metal atoms drawn together in pairs, 277 pm apart; $MoBr_3$ also involves octahedral co-ordination. WBr_3, made by direct combination, is W_6Br_{18}, while WCl_3 has a structure based on a cluster core of six tungstens, $W_6Cl_{12}^{6+}$. WBr_3 undergoes the following conversions

$$W_6Br_{18} \xrightarrow[\text{Br}_2]{140^\circ} W_6Br_{16} \xrightarrow[\text{vac}]{200^\circ} W_6Br_{14} \xrightarrow[\text{vac}]{320^\circ} W_6Br_{12}$$

These bromides almost certainly all contain W_6Br_8 clusters joined together by bromine atoms. This cluster is very typical of the dihalides of molybdenum and tungsten. All these halides are $(M_6X_8)X_4$ (M = Mo, W; X = Cl, Br, I) and contain the M_6X_8 cluster (figure 3.4). The cluster may be bridged to other clusters by four halide ions in positions 2, 3, 4, 5 and have two terminal halides in positions 1, 6. The 'dihalide' MX_2 (= M_6X_8 + ½ x 4X + 2X) then results. The noncluster halide ions are quite labile and many species such as $(Mo_6Cl_8)Br_6{}^{2-}$, $(Mo_6Cl_8)(OMe)_6{}^{2-}$, $(Mo_6Cl_8)(OH)_4(H_2O)_2$ or $(Mo_6Cl_8)Cl_4(Ph_3PO)_2$ may be prepared from the dihalides.

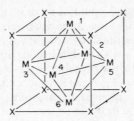

Figure 3.4 Geometry of the M_6X_8 cluster adopted by lower halides of molybdenum and tungsten

The metal–metal distance is quite short (Mo–Mo = 264 pm in Mo_6Cl_8) and the compounds are diamagnetic. This is simply explained by a M.O. approach to the bonding in the cluster. We first note, however, that there are five σ-bonds to halogen from each metal ion and we use five valence-bond d^2sp^3 hybrid orbitals to account for these bonds. It is the remaining unused orbitals that bond within the metal cluster, namely three d orbitals and one unused d^2sp^3 hybrid from each metal ion, 24 in all. These give 12 bonding and 12 antibonding M.O.s. We now account as follows, taking $(Mo_6Cl_8)Cl_6{}^{2-}$ as an example.

30 Mo–Cl σ-bonds, requiring 60 electrons

12 bonding M.O.s for cluster; 24 electrons

Total required, 84 electrons

2 terminal Cl^- ions provide 2 x 2 = 4 electrons

4 terminal Cl atoms provide 4 x 1 = 4 electrons

8 cluster Cl atoms provide 8 x (2 + 2 + 1) = 40 electrons

6 Mo atoms provide 6 x 6 = 36 electrons

Total available, 84 electrons

A second form of $MoCl_2$, obtained from $Mo_2(OAc)_4$ and HCl gas, may have the $CdCl_2$ structure.

A number of mixed halides exist; controlled fluorination of WCl_4 affords *cis*- and *trans*-WCl_4F_2 and in fact the whole series $WCl_{6-n}F_n$ is known. Reaction of $MoCl_2$ with AsF_3 yields $MoCl_2F_3$, while there is evidence for $MoBrF_4$ and $MoBr_4F_2$.

3.3.2 Polynuclear Halide Complexes

Not surprisingly, a number of these exist. $W_2Cl_9{}^{3-}$ is formed on reduction of $WO_4{}^{2-}$ in concentrated HCl; the structure adopted is that of two octahedra sharing a face. It is diamagnetic (W–W = 241 pm). Reaction with pyridine affords $W_2Cl_6py_4$, having a structure involving only two chlorine bridges. This compound is also diamagnetic.

$Mo_2X_9{}^{3-}$ (X = Cl, Br) have similar structures to that of $W_2Cl_9{}^{3-}$. They are virtually diamagnetic, owing to metal–metal interaction (Mo–Mo = 265 pm in the chloride). Reaction of $Mo_2(OAc)_4$ with ACl (A = alkali metal) gives two species, $Mo_2Cl_8{}^{4-}$ and $Mo_2Cl_8{}^{3-}$. The former is $(Cl_4MoMoCl_4)^{4-}$ (Mo–Mo = 214 pm), isoelectronic and isostructural with $Re_2Cl_8^{2-}$, while the latter is $(Cl_3MoCl_2MoCl_3)^{3-}$ (Mo—Mo = 238 pm), whose structure is derived from that of $W_2Cl_9^{3-}$ by removing a bridging chlorine. Other clusters $(Mo_3X_{13})^{7-}$, $(Mo_3Cl_{12})^{6-}$ and $(Mo_3X_{11})^{5-}$ (X = Cl, Br) have also been reported.

3.4 Oxohalides and Complex Oxohalides

When in fairly high oxidation states, many metals bond strongly with oxygen, forming molecular compounds. These bonds must be regarded as largely covalent, on account of the evident instability of true M^{5+} and M^{6+} ions. An O^{2-} ion is capable of donation of electron density to the metal ion via one σ-bond and two p-d π-bonds, using empty t_{2g} metal d orbitals. There will thus be three fairly strong bonds and considerable neutralisation of the energetically unfavourable, high positive charge on the metal. It is necessary, however, to write the bond as M=O rather than as M≡O because of valency considerations; perhaps M$\leqq$O expresses the true situation best. Such a bond is characterised by an infrared stretching frequency usually within the range 800–1000 cm^{-1}. Molybdenum and tungsten form stable oxohalides and oxohalide complexes, many of which contain this linkage. Some examples are MOF_4, $MOCl_4$, MO_2Cl_2, $MOCl_5^{2-}$ and $MOBr_5^{2-}$, where M = Mo and W. These compounds sometimes arise as impurities in the preparation of binary halides if traces of oxygen or water are present. Intentional preparation may be as follows

$$Mo \xrightarrow{Cl_2, O_2} MoO_2Cl_2$$

$$MoO_3 \xrightarrow{SOCl_2} MoOCl_4$$

$$MoCl_5 \xrightarrow{conc. aq. HCl} MoOCl_5^{2-}$$

There are also the tungsten oxofluoro-complex ions WOF_5^-, $WO_2F_4^{2-}$ and $WO_3F_3^{3-}$, which have been shown to involve octahedral co-ordination. $MoOF_4$ is a chain polymer with cis-F bridges while WOF_4 is a tetramer (NbF_5 structure). $WOCl_4$ acquires its six co-ordination by having linear . . . O—W—O—W—O . . . chains, while one form of $MoOCl_3$ has polymeric chains involving four bridging chlorines. Another form of this compound is isomorphous with $MoOBr_3$, $WOCl_3$ and $WOBr_3$; they have the $NbOCl_3$ structure in which planar Nb_2Cl_6 units are joined by M—O—M bridges into infinite chains. MO_2Cl_2 contain metal ions with asymmetrical oxygen bridges completing a distorted octahedral array.

Solutions of Mo(V) in HF, HCl and HBr have been studied in some detail by e.s.r., for which Mo^{5+} ($4d^1$, $I = 5/2$ for ^{95}Mo and ^{97}Mo) is well suited. In HCl $[MoOCl_5]^{2-}$ and a paramagnetic dimer predominate at [HCl] ⟩ 10M. Below 2M HCl only diamagnetic species are present. In HBr there is evidence for $[MoOBr_5]^{2-}$ as well as for a paramagnetic dimer. The e.s.r. spectra of $MoOX_5^{2-}$ show strong in-plane π-bonding with four halogens; no hyperfine splitting from the other is seen on $g_{\parallel}$ because the very strong Mo=O π-bond repels unpaired spin from the bond to the halogen trans to it.

The free acids H_2MoOX_5 and $HMoOX_4$ (X = Cl, Br) have been reported to result from reaction of MoO_3 with HX; MoO_3 forms cis-$[MoO_2F_3(OH_2)]^-$ in 5M

HF (tungsten behaves similarly). The crystal structure of $(NH_4)[MoO_2F_3]$ shows it to be fluorine bridged and have *cis*-MoO_2 groups (better for π-bonding in d^0).

3.5 Complexes of Molybdenum and Tungsten

The main difference between the co-ordination chemistry of chromium and that of molybdenum and tungsten lies in the poor stability of the heavier metals' lower oxidation states relative to their higher oxidation states. Thus some of the co-ordination chemistry of Mo(IV), Mo(V), Mo(VI), W(V) and W(VI) is absent for chromium, but there is an almost complete absence (except for a few Mo(III) compounds) of anything corresponding to the massive chemistry of Cr(III). We shall now consider each oxidation state in turn.

3.5.1 Oxidation State VI

This is represented mainly by tungsten (note the relative stabilities: WCl_6 is the commonest tungsten halide; $MoCl_5$ has a similar position for Mo). The complex fluorides MF_7^- and MF_8^{2-} (M = Mo, W) are obtained by reaction of MF_6 with alkali-metal fluoride in iodine pentafluoride. The same reaction in liquid SO_2 gives reduction to MoF_8^{3-}.

WF_6 forms a number of adducts, for example: WF_6L (L = Me_3N, py, Me_3P) and WF_6L_2 (L = py, Et_2S). With ether, however, oxygen is abstracted and $WOF_4(OEt_2)$ forms. Similarly

$$WF_6 + Et_2NSiMe_3 \rightarrow WF_5.NEt_2 + Me_3SiF$$

WCl_6 is less amenable to adduct formation; reduction normally occurs to form, for example, $[WCl_4(bipy)]^+Cl^-$. $MoNCl_3$ is known (as is the tungsten analogue); crystallographic study shows it to have a tetrameric structure involving both nitrogen and chlorine links. The Mo—N distance is about 165 pm, and may reasonably be regarded as a triple bond ($\nu(Mo-N) = 1039$ cm^{-1}). With Et_4NCl, $MNCl_3$ form $(Et_4N)_2.MNCl_5$, ($\nu(Mo-N) = 1023$ cm^{-1}). These can be compared with K_2OsNCl_5, but their structure is not yet definitively known.

Thiohalides may be prepared using antimony trisulphide

$$WCl_6 \xrightarrow{Sb_2S_3} WSCl_4$$

$$MCl_5 \xrightarrow{Sb_2S_3} MSCl_3 \ (M = Mo, W).$$

The tetrahalide is dimeric with *cis*-halogen bridges.

A number of oxohalide complexes with oxygen ligands are known, which are probably octahedral. They are of the type $MO_2Cl_2L_2$ and some examples are: $MoO_2Cl_2(Ph_3PO)_2$ and the corresponding tungsten compound; $WO_2Cl_2(Me_2SO)_2$, and $MoO_2Cl_2(PhCOCH_3)_2$. The interesting complex $[MoO_3NH(CH_2CH_2NH_2)_2]$ has been the subject of an X-ray structural determination (figure 3.5), as has $[MoO_3(oxalate)]^{2-}$, where one oxide ion bridges across two Mo atoms, thus achieving a molybdenum co-ordination of six. The acetylacetonate $MoO_2(acac)_2$ is *cis* octahedral. The extraordinarily stable $W(OPh)_6$, whose benzene rings may be nitrated *in situ*, may best be mentioned here. It is made by phenolysis of WCl_6. Another novel compound is $W(NMe_2)_6$; reaction of WCl_6 with $LiNR_2$ or amines nor-

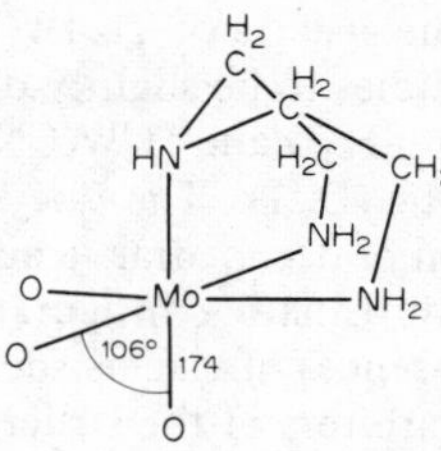

Figure 3.5 The *fac*-MoO$_3$ linkage in [MoO$_3$\{NH(CH$_2$CH$_2$NH$_2$)$_2$\}]

mally produces mainly W(III) or W(IV) compounds, but use of lithium dimethyl-amide gives significant amounts of W(NMe$_2$)$_6$. X-ray diffraction has shown it to be octahedral. The point group is the rare T$_h$, expected for undistorted M(AB$_2$)$_6$ units (for example hexanitrites), owing to the planarity of the WNC$_2$ entities.

Compounds such as Mo(S$_2$C$_2$Ph$_2$)$_3$ and Mo\{Se$_2$C$_2$(CF$_3$)$_2$\}$_3$, may formally in-volve Mo(VI) or Mo(0) according as the ligand is written

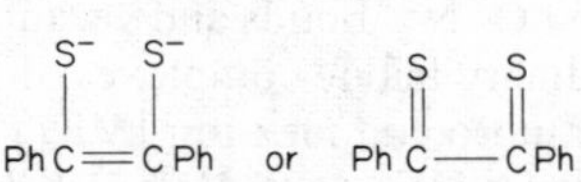

$$
\begin{array}{cc}
\underset{\displaystyle |}{\text{S}^-} \; \underset{\displaystyle \|}{\text{S}^-} & \underset{\displaystyle \|}{\text{S}} \; \underset{\displaystyle \|}{\text{S}} \\
\text{Ph C} = \text{C Ph} \quad \text{or} \quad \text{Ph C} - \text{C Ph}
\end{array}
$$

The formal oxidation state of these (and other) complexes of dithiolene-type ligands is rather uncertain, since extensive delocalisation over metal-chelate rings doubt-less occurs. More importantly, they exhibit the rare trigonal prismatic geometry (D$_{3h}$), found in MoS$_2$; their structures involve some rather short intermolecular contacts, suggesting that nonbonded interactions influence the co-ordination geo-metry.

3.5.2 Oxidation State V

Apart from the pentahalide adducts MoCl$_5$L (L = POCl$_3$, NMe$_3$), the compounds have mainly halide, nitrogen-bonded thiocyanate, cyanide or oxide ligands. The most investigated complexes are undoubtedly [Mo(CN)$_8$]$^{3-}$ and [W(CN)$_8$]$^{3-}$ and their reduction products, [Mo^{IV}(CN)$_8$]$^{4-}$ and [W^{IV}(CN)$_8$]$^{4-}$. The Mo compounds have been especially closely examined, but the W compounds appear to be very similar in properties

$$\text{K}_3[\text{MoCl}_6] \xrightarrow[\text{O}_2]{\text{KCN}} \text{K}_4[\text{Mo(CN)}_8] \xrightarrow{\text{Ce}^{4+}} \text{K}_3[\text{Mo(CN)}_8]$$

The compounds are prepared as indicated in the equation above; the [Mo(CN)$_8$]$^{3-}$ ion may be reduced to [Mo(CN)$_8$]$^{4-}$ by ferrocyanide ion. The structure of K$_4$Mo(CN)$_8$.2H$_2$O, as indicated by X-ray diffraction, shows dodecahedral co-ordination. This geometry is expected to be rather similar in energy to the square antiprism and there is some vibrational spectroscopic evidence which tends to support the latter configuration in solution, but this type of question cannot be finally decided by these means. The Mo(IV) complex is diamagnetic, which is in accord with the expected increased stability of the d$_{x^2-y^2}$ orbital (compared with the other four d orbitals) in a dodecahedral arrangement. Most recent Raman spectral work on the potassium salt suggests that the dodecahedral structure is in fact retained in solution, as do the spectra of $Tl_4[W(CN)_8]$. $(Bu^n{}_4N)_3[Mo(CN)_8]$ has also been shown by X-ray studies to be slightly distorted from dodecahedral (D_{2d}) symmetry but $Na_3W(CN)_8 \cdot 4H_2O$ and $H_4W(CN)_8 \cdot 6H_2O$ (whose Mo analogues are isomorphous) are antiprismatic (D_{4d}). The isocyanide complexes $M(MeNC)_4(CN)_4$ adopt distorted dodecahedral structures. Suffice it to say that the dodecahedral and square antiprismatic configurations have nearly the same energies, and the presence (or absence) of effects such as hydrogen-bonding and lattice-packing may be the final arbiters of the structure adopted.

Besides being thermodynamically stable, the $[Mo(CN)_8]^{4-}$ ion is kinetically inert, refusing exchange with $^{14}CN^-$ in solution. The $[Mo(CN)_8]^{3-}$ ion is quite photosensitive, being reduced to $[Mo(CN)_8]^{4-}$. The electron-transfer reaction

$$[Mo(CN)_8]^{4-} + [Mo(CN)_8]^{3-} \longrightarrow [Mo(CN)_8]^{3-} + [Mo(CN)_8]^{4-}$$

is rapid. The $[Mo(CN)_8]^{3-}$ ion has a magnetic moment close to the spin-only one-electron value; e.s.r. measurements indicate delocalisation of the d electron onto the ligands. The unusual stability of these complexes possibly reflects considerable $d\pi$ overlap between the filled C–N π bonds and vacant Mo d orbitals.

We now revert to the ordinary halide complexes of Mo(V) and W(V). The two elements each form an octafluoro complex ion $[MF_8]^{3-}$ of unknown structure, in addition to hexahalo complexes MF_6^- and MCl_6^-. KWF_6 is known to have octahedral co-ordination of the tungsten (X-ray). These complexes are usually made directly; for example

$$MoCl_5 + KCl \xrightarrow{fuse} KMoCl_6$$

or

$$MX_5 \xrightarrow[CH_2X_2]{R_4NX} R_4N^+(MX_6)^- \qquad (M = Mo, X = Cl; M = W, X = Cl, Br)$$

Disproportionation can occur; for example

$$2KWCl_6 \xrightarrow{250°} K_2WCl_6 + WCl_6$$

Pentachloride complexes are rare, but examples such as $WCl_5 \cdot Et_2O$ do exist; $MoCl_5 \cdot Ph_3AsO$ is, however, found to have the constitution $[Ph_3AsCl][MoOCl_4]$. There are a number of thiocyanato complexes of Mo(V) and W(V), which are believed to be bonded as M–NCS rather than M–SCN on account of infrared spectral evidence. These complexes usually also contain oxide as ligand. The reduction of tungstate by thiocyanate gives species such as $WO_2(NCS)_3{}^{2-}$ and $WO(NCS)_5{}^{2-}$; direct substitution of halide is also possible; for example

$$MoOCl_5{}^{2-} \xrightarrow{NCS^-} MoO(NCS)_5{}^{2-}$$

Molybdenum(V) and (VI) form multiple links with nitrogen. However, this behaviour seems less marked than is the case with Re and Os.

Amino-acid complexes are known, often of the type $MoO_2(acid)_2$; the histidine complex is a dimer with oxygen bridges. Sulphur ligands such as dithiocarbamate or xanthate form complexes that are also dimeric but with only a single oxygen atom in the bridging position (see figure 3.6).

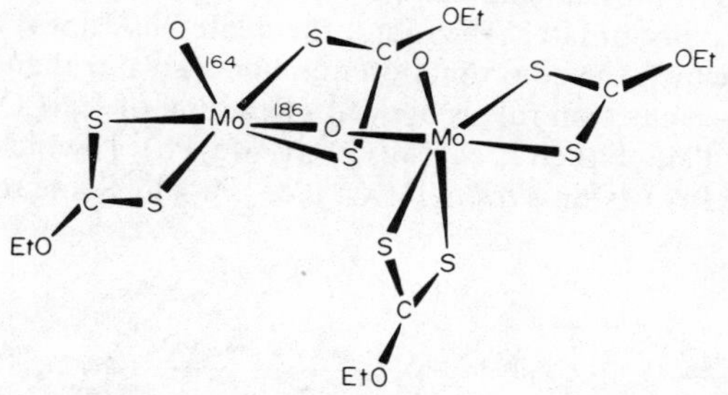

Figure 3.6 The dimeric xanthato complex, $Mo_2O_3(S_2OEt)_4$ (after A.B. Blake, F.A. Cotton and J.S. Wood, *J. Am. chem. Soc.*, **86** (1964), 3025)

3.5.3 Oxidation State IV

This is predominantly represented by molybdenum, although both elements form the hexahalogen complexes MoX_6^{2-} (X = F, Cl, Br, NCS) and WX_6^{2-} (X = Cl, Br, I). Some of these have been investigated by X-ray diffraction and are octahedral. The magnetic moments for MCl_6^{2-} are 2.2 B.M. (M = Mo) and 1.5 B.M. (M = W), respectively, the larger spin-orbit coupling constant of tungsten leading to a greater deviation from the spin-only value of 2.8 B.M. Pyridine converts these complex halides into pyridine adducts

$$WCl_6^{2-} + 2py \rightarrow WCl_4py_2 + 2Cl^-$$

These may also be obtained by pyridine reduction of the pentahalide

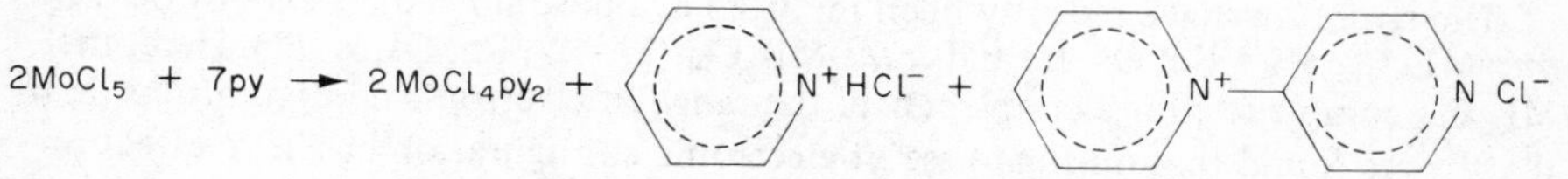

or by direct addition

$$MoCl_4 + 2py \rightarrow MoCl_4py_2$$

Other amines, Ph_3PO and acetylacetone also react, the last substituting for Cl^-

$$MoF_4 + 2NMe_3 \rightarrow MoF_4(NMe_3)_2$$

$$MoCl_4 + 2CH_3COCHCOCH_3^- \rightarrow MoCl_2(acac)_2 + 2Cl^-$$

The diarsine complex $MoBr_4$ diars$_2$ (μ = 1.96 B.M.) is obtained by the action of bromine on $Mo(CO)_4$ diars. Mo(IV) and W(IV) do not form ammines, ammonolysis to MNH_2 being predominant.

Phosphine complexes of molybdenum and tungsten   These compounds have been the subject of much study of late. They (and the related hydrides) include a number of complexes in which the co-ordination number is greater than six, with the result that the eighteen-electron rule is obeyed. Reaction of $[M(CO)_6]$ with phosphines (for example, PMe_2Ph) gives *cis*-$[M(CO)_4(PMe_2Ph)_2]$, which on treatment with excess X_2 (Cl_2, Br_2) affords *trans*-$[MX_4(PMe_2Ph)_2]$. Some reactions are as follows

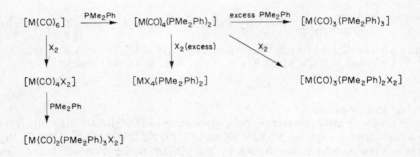

The complexes $[MX_4(PMe_2Ph)_2]$ add one further mole of phosphine; the room-temperature magnetic moment of $[WCl_4(PMe_2Ph)_2]$ is 2.05 B.M., and that of $[WCl_4(PMe_2Ph)_3]$ is 2.68 B.M., despite it being for a $5d^2$ ion, with concomitant large spin–orbit coupling. The high moment is probably a consequence of the low symmetry of the seven-co-ordinate complex.

$[WCl_4(PMe_2Ph)_2]$ reacts with borohydrides to form $[WH_6(PMe_2Ph)_3]$; this hydride also results from reaction with sodium amalgam under hydrogen. Under the appropriate atmospheres, *fac*-$[W(CO)_3(PMe_2Ph)_3]$ and $[W(N_2)_2(PMe_2Ph)_4]$ are formed.

The structures have recently been reported of phosphine complexes of the type *mer*-ML_3Cl_3 (M = Re, Os, Ir) and *trans*-ML_2Cl_4 (M = W, Re, Os, Ir, Pt). Here, the ML_2Cl_4 complexes, for example, cover the range of electronic configurations d^2 to d^6. It was found that unit increase of electronic configuration has little effect on M–Cl bond lengths but rather more on M–P bond lengths (a contraction of 5 pm). Also, the M^{IV}–Cl and M^{III}–Cl bond distances for any one d^n configuration differ by about 3 pm while the M^{IV}–P and M^{III}–P distances are nearly identical; this is rationalised in terms of the more 'ionic' M–Cl bonds being more sensitive to the formal oxidation state of the metal. These interesting results form one of the few cases in which a number of structurally related complexes have been subject to a detailed study of the effects of oxidation state and electronic configuration.

Phosphine complexes containing the Mo=O linkage are readily obtained.

$$\textit{trans-}Mo(PR_3)_2Cl_4 \xrightarrow{\text{EtOH}} \textit{mer-}[MoOCl_2(PR_3)_3]$$

The preparation of the Mo(IV) diphosphine complexes proceeds via an isolable Mo(V) complex

$$MoOCl_3 \xrightarrow{\text{diphos}} [MoOCl_3(\text{diphos})] \xrightarrow[\text{boil}]{\text{MeCN}} MoOCl_2(\text{diphos})$$
$$(\mu = 1.73 \text{ B.M.}) \qquad (\text{diamag})$$

with vertical arrows:

$MoOCl_3 \downarrow$ MeOH, R_3P → $[MoOCl_2(PR_3)_3]$

$MoOCl_2(\text{diphos}) \downarrow$ diphos/MeOH →
$[MoOCl_3(\text{diphos})]$
$[MoOCl(\text{diphos})_2]\,Cl$
(with a $+$ sign before the second line)

Hydrido complexes Reaction of *cis*-[Mo(PMe$_2$Ph)$_2$Cl$_4$] with excess ligand and borohydride gives [Mo(PMe$_2$Ph)$_4$H$_4$]; the n.m.r. spectrum shows the expected high-field line at τ 12.20 as a quintet (1:4:6:4:1) split by coupling with four equivalent phosphorus nuclei. The complex has a nonrigid structure at ambient temperatures with all phosphines and hydrides equivalent. The tungsten analogue, however, has a complex n.m.r. spectrum in the region τ 10-13, demonstrating that rapid intramolecular rearrangement does not occur. On warming to 70°, however, a 'fluxional' quintet spectrum is obtained with $\tau = 11.66$.

Dinitrogen complexes A number of molybdenum complexes has attracted attention recently; the reason for this is that the enzyme nitrogenase, which is involved in nitrogen fixation, contains iron and molybdenum. The iron probably forms the initial bond to the nitrogen, while the role of molybdenum is as yet unknown. One potential model system is provided by [ReCl(N$_2$)(PMe$_2$Ph)$_4$] (playing the part of the iron function) reacting with MoCl$_4$.2Et$_2$O to form Cl(PMe$_2$Ph)$_4$Re.N$_2$.MoCl$_4$.Et$_2$O, which has ν(N—N) at 1795 cm^{-1} (compared to 2331 in dinitrogen gas), an evident weakening of the N—N bond. This has been ascribed to Re(I) pushing t_{2g} density into π^* orbitals of nitrogen, while simultaneously Mo(IV) withdraws π-electron density.

Reduction of MoOCl$_2$(diphos)$_2$ by zinc in a nitrogen atmosphere gives *trans*-Mo(N$_2$)$_2$(diphos)$_2$. In an important group of reactions, this last complex gives *trans*-Mo(N$_2$)$_2$(diphos)$_2$. In an important group of reactions, this last complex and its tungsten analogue react with HCl to give MCl$_2$(N$_2$H$_2$)(diphos)$_2$ and with H$_2$SO$_4$ to give what is thought to be M(SO$_4$H)(N$_2$H$_2$)(diphos)$_2$. Furthermore, M(N$_2$)$_2$(PMe$_2$Ph)$_4$ and similar complexes react with H$_2$SO$_4$ in methanol to give ammonia in up to 90 per cent yield

$$M(N_2)_2(PR_3)_4 \xrightarrow[\text{MeOH}]{H_2SO_4} 2NH_3 + N_2 + \text{other products}$$

These reactions may well be relevant to naturally occurring processes of nitrogen fixation.

Molybdenum alkylamides The alkylamides of molybdenum are prepared from MoCl$_5$ and LiNR$_2$ (for example, R = Me or Et). The reaction involves reduction to an initial product approximating to Mo(NR$_2$)$_3$ (notice the similarity and differences of the tungsten system, page 34). On heating, disproportionation occurs, and the volatile $Mo(NR_2)_4$ distils off. This probably has a distorted tetrahedral structure; reaction with CS_2 yields $Mo(S_2CNR_2)_4$ (probably eight-co-ordinate).

3.5.4 Oxidation State III

Molybdenum provides nearly all the complexes in this oxidation state, although its performance in so doing is only a pale shadow of that of its lighter neighbour chromium. Thus there is no air-stable hydrated ion $[Mo(H_2O)_6]^{3+}$. However, a number of standard types of six-co-ordinate complex are known, notably $[MoX_6]^{3-}$ (X = Cl, Br), $[Mo\ acac_3]$, $[Mo\ phen_3]^{3+}$, $[Mo\ bipy_3]^{3+}$ and $[Mo(NCS)_6]^{3-}$. A few adducts, for example, $MoCl_3(THF)_3$, $WCl_3(diphos)_2$ and the dimeric WCl_3py_2, also exist.

The hexahalides are prepared by electrolytic reduction of Mo(VI) in HCl or HBr, which also yields the monoaquo derivatives $[MoX_5(H_2O)]^{2-}$. Magnetic moments are 3.7 – 3.8 B.M.; this is close to the spin-only value, since no orbital contribution is expected in the case of an octahedral d^3 complex. The electronic spectra have been interpreted, Δ values lying in the region 17 000 – 21 000 cm^{-1}. The complexes are oxidised readily by air to Mo(V). This behaviour contrasts with that of tungsten; reduction of tungstate in hydrochloric acid gives salts of the dimeric anion $[W_2Cl_9]^{3-}$, which forms the dimeric pyridine complex $W_2Cl_6py_4$ (see page 32).

There are very few ammines, $[MoCl_5NH_3]^{2-}$ being an example

$$(NH_4)_2[MoCl_5(H_2O)] \xrightarrow{\text{liq. } NH_3} (NH_4)_2[MoCl_5(NH_3)]$$

It is at once hydrolysed in water, again contrasting with the inert chromium complexes.

The acetylacetonate may be prepared by heating a salt of $[MoCl_6]^{3-}$ with acetylacetone. This product, $[Mo\ acac_3]$, has $\mu = 3.8$ B.M. and, like the other Mo(III) complexes, it is oxidised by air to Mo(V).

Treatment of $(NH_4)_3[MoCl_6]$ with 1,10-phenanthroline or 2,2'-bipyridyl gives the complexes $[Mo\ phen_3]Cl_3$ or $[Mo\ bipy_3]Cl_3$. These complexes are stable to oxidation. Complexes $K_4M(CN)_6$ (M = Mo, W) are formed by hydrogen reduction of the corresponding M(IV) compounds $K_4M(CN)_8$. They are rapidly oxidised in methanol to $K_3M(CN)_6$. The structures of these compounds are unknown.

3.5.5 Oxidation States II and 0

These rank as low oxidation states, needing π-acceptor ligands for stabilisation, and hence most of their chemistry is dealt with in chapter 9. However, in addition to the cyanides just mentioned, two or three compounds may be noted here.

Reaction of $[MoCl_6]^{3-}$ with the diarsine gives a low-spin, six-co-ordinate complex $[MoCl_2(diars)_2]$, which doubtless owes its stability to d–d π bonding. The corresponding iodide is isomorphous with a tungsten analogue $[WI_2(diars)_2]$. Lithium/THF reduction of $[Mo\ bipy_3]Cl_3$ gives $Mo\ bipy_3$, an air-sensitive compound formally containing Mo(0). However, this compound, together with many others of the types M phen$_x$ and M bipy$_x$, is perhaps best formulated Mo^{3+} (bipy$^-$)$_3$, the conjugated aromatic system acting as a co-ordinated anion. Other examples of this type of behaviour may be provided by the dithiolato-complexes $M\{PhC(S)C(S)Ph\}_3$

(M = Mo, W) (page 35), which can be reduced to $M\{PhC(S)-C(S)Ph\}_3^{1- \text{ or } 2-}$, the additional electrons of the anions residing in orbitals that are mainly ligand in character. Molybdenum acetate formally contains molybdenum(II) and is a dimer (compare chromous acetate) with Mo—Mo = 211 pm; this and other carboxylates $Mo_2(O_2CR)_4$ may be prepared from $Mo(CO)_6$ and the carboxylic acid in diglyme. $W(OAc)_2$ may be obtained similarly.

3.6 Organo-Compounds

It is reported that tungsten forms a phenyl, which is simple in the sense that no π-bonding ligand is present to confer stability. Thus treatment of WCl_6 with phenyl lithium gives a black, air-sensitive substance, which is either $Li_3WPh_6 \cdot 3Et_2O$ or $Li_2WPh_6 \cdot 3Et_2O$; the observed diamagnetism suggests a distinct preference for the latter formulation. However, the structure of this interesting compound is uncertain. The C_6F_5 group often forms stable metal–carbon σ-bonds, and reaction of C_6F_5Li with WCl_6 yields bright green crystals of $LiW(C_6F_5)_5 \cdot 2Et_2O$. This has a magnetic moment of 2.70 B.M. at room temperature, reacts with water to give C_6F_5H and with $HgCl_2$ to form $Hg(C_6F_5)_2$. On heating to 100° under vacuum, $W(C_6F_5)_5$ (and possibly also $W(C_6F_5)_4$) forms.

Reaction of $MoCl_5$ or WCl_6 with Me_3SiCH_2MgCl in ether yields $\{M(CH_2SiMe_3)_3\}_2$. These alkyls adopt a dimeric structure with the terminal alkyl groups staggered, presumably due to steric effects. The Mo—Mo distance in the molybdenum compound is 216.7 pm, and may be said to correspond to a triple bond, with overlap of d_{z^2}, d_{xz} and d_{yz} orbitals. The use of unorthodox aryl or alkyl groups is, however, not essential for a W—C σ-bond to be obtained. Thus WMe_6 can be made by the simple reaction

$$WCl_6 \xrightarrow{\text{MeLi}} WMe_6 + LiCl$$

It is a red solid, m.p. 30°, subliming at room temperature under reduced pressure, and reacting with nitric oxide to give $WMe_4\{ON(Me)NO\}_2$ (see figure 3.7).

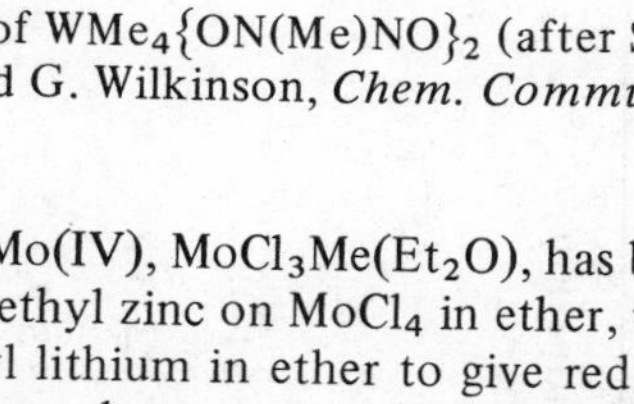

Figure 3.7 The structure of $WMe_4\{ON(Me)NO\}_2$ (after S.R. Fletcher, A. Shortland, A.C. Skapski and G. Wilkinson, *Chem. Commun.* (1972), 922)

An alkyl derivative of Mo(IV), $MoCl_3Me(Et_2O)$, has been obtained as a brown solid by the action of dimethyl zinc on $MoCl_4$ in ether, while molybdenum(II) acetate reacts with methyl lithium in ether to give red crystals of $Li_4[Mo_2Me_8] \cdot 4Et_2O$. The analogous THF adduct has an eclipsed structure like that of $Re_2Cl_8^2$ (page 54) with a Mo—Mo distance of 214.8 pm.

MANGANESE

Metal: b.c.c.; m.p. 1247°; I_1: 7.43 eV; I_2: 15.64 eV; I_3: 33.69 eV

Oxides: MnO, Mn$_2$O$_3$, MnO$_2$

Halides: MnX$_2$ (X = F, Cl, Br, I), MnX$_3$ (F, Cl), MnF$_4$

			Oxidation State and Representative Compounds					
	0	1	2	3	4	5	6	7
Typical donor atom/group	CO	CN	O,N,Hal	O,N,Hal	O,Hal	O^{2-}	O^{2-}	O^{2-},F
Co-ordination number								
3			Mn{N(SiMe$_3$)$_2$}$_2$·THF $\checkmark$					
4 tet			[MnCl$_4$]$^{2-}$ $\checkmark\checkmark\checkmark$					
4 planar			Mn(II)phthalocyanine $\checkmark$					
5 T.B.P.			[MnBrN(C$_2$H$_4$NMe$_2$)$_3$]$^+$ $\checkmark\checkmark$					
5 S.P.			[Mn(Ph$_2$MeAsO)$_4$ClO$_4$]$^+$ $\checkmark\checkmark\checkmark$	[MnCl$_5$]$^{2-}$ $\checkmark$				
6	Mn$_2$(CO)$_{10}$ $\checkmark$	[Mn(CN)$_6$]$^{5-}$ $\checkmark\checkmark$	[Mn(OH$_2$)$_6$]$^{2+}$† $\checkmark\checkmark\checkmark$	[MnCl$_6$]$^{3-}$ $\checkmark\checkmark\checkmark$	[MnF$_6$]$^{2-}$ $\checkmark\checkmark$	[MnO$_4$]$^{3-}$ $\checkmark$	[MnO$_4$]$^{2-}$ $\checkmark$	[MnO$_4$]$^-$ $\checkmark\checkmark$
7			[Mn(EDTA)(OH$_2$)]$^{2-}$ $\checkmark$	Mn(NO$_3$)$_3$(bipy)				
8			[Mn(NO$_3$)$_4$]$^{2-}$					

† More than one spin-state

$\checkmark$ Known; $\checkmark\checkmark$ Several examples; $\checkmark\checkmark\checkmark$ Very common

4 Technetium and Rhenium

These two metals are less well known than many other of the second- and third-row transition metals, although interest in rhenium chemistry has recently increased. Technetium, as its name suggests, is an artifically produced element, the only such among all the d transition metals. It is thus not an element of which the majority of chemists have any direct experimental knowledge. Rhenium is a relatively rare element with few uses but with several interesting features, for example its unusual complex hydrides. These metals mark the half-way stage in the progression along the transition-metal series. The stability of the d^5 Mn^{2+} ion is not reflected in Tc^{2+} and Re^{2+}, however; the heavier metals, as usual, are more stable in higher oxidation states.

Technetium was first obtained by Perrier and Segré in 1937 by the bombardment of molybdenum with deutrons, the first previously unknown element to be produced artificially. The most useful isotope is ^{99}Tc, although it is not the longest-lived (^{97}Tc: $t_{1/2} = 2.6 \times 10^6$ years)

$$^{99}Tc \xrightarrow[21\ 000\ \text{years}]{\beta} {}^{99}Ru$$

^{99}Tc is stable enough for all practical purposes and has the advantage of being relatively easily obtained from uranium reactor fission products; annual production is on a multi-kilogram scale.

Rhenium was discovered in gadolinite by Noddack, Tacke and Berg in 1925; a similar discovery was made by Loring and Druce. It is a rare metal, the abundance being about 10^{-8}. As there are no rhenium minerals, it is extracted from flue dusts from roasted molybdenum or copper ores.

4.1 The Metals and their Aqueous Chemistry

Both metals may be obtained by the reduction of ammonium perrhenate and pertechnetate in hydrogen at high temperature (600°). They have hexagonal close-packed lattices. The metals dissolve relatively easily in oxidising acids, being soluble in dilute nitric acid. They also dissolve in hydrogen peroxide and are attacked by hot chlorine, sulphur or oxygen but retain their silver grey lustre in dry air. The melting points are high, 2250° (Tc) and 3180° (Re), respectively. Rhenium has the very high density of 20.5 g cm^{-3}.

With the likely exception of Tc^{3+}aq., the probable species obtained as a green solution by electrolytic reduction of pertechnetate, there are no conventional hydrated cations.

4.2 Halides

The known halides of the elements are shown in table 4.1 They are formed in oxi-

Table 4.1

	F	Cl	Br	I
Tc	6,5	4		
Re	7,6,5,4	6,5,4,3	5,4,3	4,3,2,1

dation states ranging from the group valency of +7 down to +3. Some of the lower halides continue the tendency of Mo and W to form oligomers. It is clear that further work with technetium should reveal further halides which are unknown at present (1974).

Few heptafluorides exist, but rhenium, as the heaviest metal in d-Group 7, duly forms one when heated to 400° with fluorine under moderate pressure (3 atmospheres). The volatile orange solid (b.p. 74°) is thermally stable but is decomposed into perrhenate by water. The molecular symmetry is unknown. The hexafluoride is octahedral in the vapour phase and is formed when the elements react under milder conditions than for ReF_7; the latter is formed at the same time and is removed by treatment with Re metal. Reduction of the hexafluoride with $W(CO)_6$ gives ReF_5 and ReF_4. The former disproportionates on heating

$$2ReF_5 \xrightarrow{50°} ReF_6 + ReF_4$$

The hexa-, penta-, and tetrafluorides are all hydrolysed by water, for example

$$3ReF_6 + 10H_2O \longrightarrow 2HReO_4 + 18HF + ReO_2$$

This reaction emphasises the stability of the $+7(d^0)$ and $+4(d^3)$ oxidation states of rhenium.

With the possible exceptions of neptunium and astatine, rhenium is the only element which might have formed a heptachloride; it apparently does not do so but forms $ReCl_6$ by the reaction

$$ReF_6 + 2BCl_3 \xrightarrow{0°} ReCl_6 + 2BF_3$$

Rhenium hexachloride is a dark green solid, m.p. 29°. It is rapidly hydrolysed in air and has a magnetic moment of 1.88 B.M. at room temperature. On heating, it forms $ReCl_5$, which is a dimer, $Cl_4ReCl_2ReCl_4$, with two bridges establishing octahedral co-ordination; the Re—Re distance is 374 pm so that metal—metal bonding is minimal. β-$ReCl_4$, prepared from equimolar quantities of $ReCl_3$ and $ReCl_5$ in a sealed tube at 300°, has an interesting chain structure in which $ReCl_6$ octahedra are linked with alternate triple and single chlorine bridges. α-$ReCl_4$ is not now believed to exist. A different polymorph, γ-$ReCl_4$, of unknown structure is obtained, however, on dechlorinating $ReCl_5$ with unsaturated hydrocarbons.

Rhenium trichloride has a structure that contains clusters of three directly bonded Re atoms (see figure 4.1). It forms red crystals, which volatilise to give a green vapour, the colour change probably corresponding with depolymerisation. It may be made by heating the pentachloride in nitrogen. Rhenium tribromide and tri-iodide have similar although not isomorphous structures, also based on a Re_3X_9 unit.

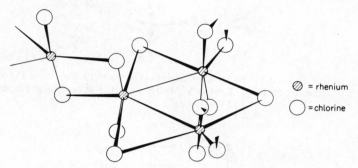

= rhenium

= chlorine

Figure 4.1 The cluster structure of Re_3Cl_9 (after F.A. Cotton and J.T. Mague, *Inorg. Chem.*, **3** (1964), 1402)

Recently, $ReCl_3$ has been shown to react with BBr_3 in a sealed tube at 280° to form $Re_3Cl_3Br_6$; repeated sublimation affords, *inter alia*, a compound believed to be Re_3Br_8Cl. ReF_6 and BBr_3 react to form $ReBr_5$, which decomposes above 100° giving $ReBr_3$.

The deep yellow technetium hexafluoride is made by direct combination at 400°. The pentafluoride is formed as a byproduct and is more stable than ReF_5, which easily disproportionates as mentioned above.

Technetium tetrachloride is formed by the action of chlorine on the metal at 400°C. The tetrachloride is stable to heat, instantly hydrolysed by water, and gives $TcCl_6^{2-}$ in hydrochloric acid. The crystal structure of $TcCl_4$ has chains of octahedra sharing two edges.

4.3 Oxides

Both metals form the heptoxide fairly readily by burning in oxygen. Both heptoxides are volatile yellow solids, which dissolve in water to give the corresponding acid HMO_4. They are, however, not isostructural; Re_2O_7 has a structure, unusual in showing the same ion with two different co-ordination numbers, consisting of distorted ReO_6 octahedra and ReO_4 tetrahedra linked in layers (see figure 4.2). On hydrolysis, $Re_2O_7(OH_2)_2$ is formed; this may have a structure closely related to the Re_2O_7 unit. In the vapour phase, the structure of Re_2O_7 consists of two ReO_4 tetrahedra sharing an apex, the structure established for Tc_2O_7 in the solid state.

Rhenium trioxide ReO_3 may be prepared by CO reduction of Re_2O_7. It has the very simple structure depicted (see figure 4.3). ReO_3 will not form an oxyanion, disproportionation into ReO_4^- and ReO_2 occurring in alkali. Technetium trioxide is made by heating TcO_3Br.

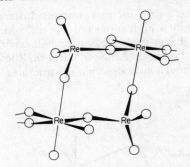

Figure 4.2 A dimeric unit in the structure of Re_2O_7 showing the 'mixed' co-ordination of rhenium (after B. Krebs, A. Muller and H. Beyer, *Inorg. Chem.*, **8** (1969), 436)

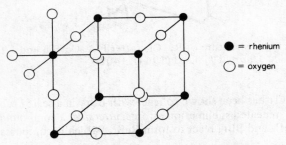

Figure 4.3 The cubic structure of ReO_3

The dioxides of both metals are stable, and may be made by reduction of the MO_4^- anions by zinc and hydrochloric acid; this gives the hydrated oxides, which may then be dehydrated.

4.4 Oxohalides

A number of investigations have recently been made into the oxohalides; MOF_4 (M = Re, Tc) adopt dimeric structures consisting of distorted MOF_5 groups linked by *cis*-fluorine bridges (in the case of Tc, a trimeric polymorph, again *cis*-F bridged, is known). $ReOCl_4$ prepared from Re and SO_2Cl_2 at 350° is square pyramidal (Re–O, 163 pm; Re–Cl, 226 pm). It reacts with MeLi to give the red–purple $ReOMe_4$, m.p. 44°, which is a member of the growing class of transition-metal alkyls that have no stabilising π-bonding ligand present. $ReOMe_4$ is stable to 150° and, notably, to water but reacts with oxygen. $ReOCl_3$ exists in α- and β-forms; it may be prepared from $ReCl_5$ and ReO_2 in a sealed tube at 300° or by irradiation (350 nm) of $ReOCl_4$. $TcOX_3$ (X = F, Cl, Br) result from halogenation of TcO_2.

The higher oxofluorides MO_3F may be prepared as follows

$$TcO_2 \xrightarrow[150°]{F_2} TcO_3F \qquad \text{and} \qquad KReO_4 \xrightarrow{IF_5} ReO_3F$$

The oxohalides TcO_3Cl and $TcOCl_4$ are obtained on chlorinating the metal in a Pyrex tube.

4.5 Perrhenates and Pertechnetates

These colourless salts contain the d^0 oxyanion and are expected to be stable by inference from the corresponding d^0 anions of Nb, Ta and Mo, W and from RuO_4, OsO_4. The strong oxidising properties of MnO_4^-, not surprisingly, are absent in TcO_4^- and ReO_4^-, but perrhenate, for example, will oxidise Ti^{3+}. Pertechnetic acid $HTcO_4$ has been isolated; the product of dissolving the metals in 100 volume H_2O_2 is in fact HMO_4. Both are strong acids and may be titrated using methyl red. Many salts are known; they may be made from M_2O_7 by the action of aqueous alkali. The lithium salts are formed by reaction of the metals with molten $LiClO_4$ at 250° The anions are tetrahedral with Tc–O = 175 pm and Re–O = 177 pm. The bonding is expected to be similar to that in OsO_4, in which there are four σ bonds using Os sp^3 hybrids with an Os–O π-system using d(Os)–p(O) overlap. The overall bond order is about two, judging from the ReO_4^- stretching frequencies of 971 and 918 cm^{-1}. ReO_6^{5-} and ReO_5^{3-} are also known, being prepared from ReO_4^- in very basic solution or by fusion of MO_2 (M = K, Rb, Cs) with ReO_2 at high temperatures. Reduction of TcO_4^- affords TcO_4^{2-} isolated as the Ba salt. The ion tends to disproportionate into TcO_4^- and TcO_4^{3-}.

Potassium nitridorhenate, $K_2[ReO_3N]$, may be made by the action of potassamide on Re_2O_7 in liquid ammonia. It may be compared with potassium osmiamate, $K[OsO_3N]$, with which it is isoelectronic.

4.6 Complexes of Technetium

It is difficult to compare the co-ordination chemistries of technetium and rhenium, since owing to the activity of technetium insufficient work has been done to give a clear overall picture. However, the complexes of technetium, where known, resemble those of rhenium and may be discussed fairly briefly as follows.

Although $KTcF_6$ is known, and TcF_6 forms $(NO)_2TcF_8$ and $(NO_2)TcF_7$ on reaction with NOF and NO_2F respectively, the most important complex halides are those of Tc(IV). The stable, octahedral $TcCl_6^{2-}$ ion may be prepared by reduction of TcO_4^- with concentrated HCl, and the corresponding complexes of the other halogens have also been made. The bromo- and iodotechnetates(VI) are virtually indistinguishable from the rhenium analogues; also the unit-cell dimensions of all the halo-complexes are virtually identical to the rhenium analogues.

Like rhenium, technetium forms a remarkable complex hydride K_2TcH_9. Both these compounds are discussed below.

The best-known technetium complexes are probably those of the ubiquitous diarsine o-$C_6H_4(AsMe_2)_2$. Thus $TcCl_6^{2-}$ is reduced by excess diarsine in aqueous alcoholic HCl to give $[Tc^{III}Cl_2diars_2]$ Cl as orange crystals. This ion may be oxidised or reduced

$$TcCl_2 \text{ diars}_2 \xleftarrow{H_3PO_2} [TcCl_2 \text{ diars}_2]^+ \xrightarrow{Cl_2} [TcCl_4 \text{ diars}_2]^+$$

All the products (bromides and iodides were also made) are isostructural with the corresponding rhenium compounds. The Tc(V) complexes appear as a rare example of an eight-co-ordinated transition metal in a 5+ oxidation state; a cation of such high charge is rather small to accommodate eight ligand atoms, although a few other examples, such as TaF_8^{3-} and $Mo(CN)_8^{3-}$ are known.

Triphenylphosphine and bipyridyl react with $TcCl_4$ to give, respectively, $TcCl_4(PPh_3)_2$ and $TcCl_4(bipy)$. With bis-(diphenylphosphino)ethane, reduction to Tc(III) occurs and $[TcCl_2(Ph_2PC_2H_4PPh_2)_2]$ Cl is isolated, isomorphous with the Re analogue. This Tc(III) compound is reduced by ethanolic sodium borohydride to $TcCl_2(diphos)_2$, which is rather unstable, especially in solution, towards oxidation to the Tc(III) complex. Magnetic results for these complexes are in accord with their containing d^3, low-spin d^4 or low-spin d^5 ions for the Tc(IV), (III) and (II) complexes respectively.

A dinuclear species $Tc_2Cl_8^{3-}$, similar to the corresponding Re compound mentioned below, has been shown by X-ray diffraction to have the eclipsed D_{4h} structure with Tc—Tc = 213 pm. This ion has the fractional mean oxidation state of 2.5. Magnetically it obeys the Curie law ($\mu = 2.0$ B.M.; $\theta = 0$). Polarographic results for $ReCl_8^{2-}$ show the existence of $ReCl_8^{n-}$ ($n = 3,4$) so it appears that the lighter metal adopts a slightly lower oxidation state in the most stable complex, a fairly general rule. $Tc_2Cl_8^{2-}$ does not appear to exist.

We may finally note the low oxidation-state cyanide $K_5[Tc(CN)_6]$, isostructural with the corresponding Mn and Re compounds, and made by reduction of TcO_4^- by potassium in the presence of KCN.

4.7 Complexes of Rhenium

The co-ordination chemistry of rhenium of the nonorganometallic type has not been investigated so extensively as that of many other elements. The general picture is fairly orthodox, perhaps complicated by the oligomeric tendencies of Re(III) and by the relative stability of the Re=O and Re≡N groupings and of the Re—H linkage, though most of the hydride chemistry is of the type stabilised by π-bonds. Complexes in oxidation states I, II, **III**, **IV**, **V**, VI and VII are known, but those in bold type are the most usual.

4.7.1 Rhenium(VII)
There are only a few complexes (other than the exceptional $[ReH_9]^{2-}$) with this oxidation state, for example

$$4BrF_3 + 3KReO_4 \rightarrow 3KReO_2F_4 + 2Br_2 + 3O_2$$

This complex, as would be expected, is readily hydrolysed. The yellow $Cs_2[ReO_3Cl_3]$, probably the *cis* isomer, is formed thus

$$Re \xrightarrow{\text{100 vol. } H_2O_2} HReO_4 \xrightarrow[\text{CsCl}]{\text{HCl (saturate)}} Cs_2ReO_3Cl_3$$

The oxofluoride complexes $[ReO_2F_4]^-$ and $[ReO_3F_3]^{2-}$ also exist, as do some substituted oxochlorides, for example

$$HReO_4 \text{ or } Re_2O_7 \xrightarrow[\text{Me}_2SO]{\text{HCl}} ReO_3Cl(Me_2SO)_2$$

The most spectacular heptavalent compounds are the hydrides, K_2MH_9 (M = Tc, Re), first prepared (1954) by potassium reduction of ReO_4^- in aqueous ethylenediamine. The product was said to be $KRe.4H_2O$; a report in 1959 agreed with this, but the following year n.m.r. evidence appeared, which showed the product to be a hydride, whereupon the formulation was altered to $KReH_4.2H_2O$. Infrared data and further analytical data indicated K_2ReH_8 (1961); this should be d^1 and hence paramagnetic, which the compound is not. Finally, neutron diffraction and careful analysis established K_2ReH_9 (1963).

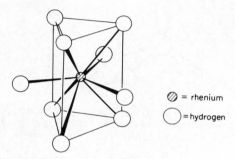

= rhenium

= hydrogen

Figure 4.4 The structure of $[ReH_9]^{2-}$ (after S.C. Abrahams, A.P. Ginsberg and K. Knox, *Inorg. Chem.*, **3** (1964), 558) Re–H = 168 pm

The structure is based on a trigonal prism with three hydrogens lying outside the rectangular faces (see figure 4.4). Only one n.m.r. line is seen in solution, due to rapid interchange of equatorial and prismatic hydrogens. The best method of preparation, for the sodium salt, consists of treating sodium perrhenate with sodium in ethanol. On heating to 245°, hydrogen is liberated, followed by sodium. Decomposition of the tetraethylammonium salt (above about 115°) yields ethane and hydrogen.

4.7.2 Rhenium(VI)
This oxidation state is still in the region of oxide or fluoride stabilisation. Thus we have

$$ReF_6 + 2KF \rightarrow K_2ReF_8$$

where the complex anion is shown by X-ray diffraction to be a square antiprism. Controlled hydrolysis gives $K[ReOF_5]$.

The chemistry of dithiolene complexes has been the subject of much interest in recent years; some difficulty attends the assignment of oxidation state of the metal. Thus may be viewed as either tris-(*cis*-stilbenedithiolato)rhenium(VI) or tris-(dithiobenzil)-rhenium(0). This may be prepared from the analogous thio-ester $(Ph_2C_2S_2PS_2)^-$ and $ReCl_5$; it has one unpaired electron (μ_{eff} = 1.79 B.M.) and adopts the trigonal prismatic configuration (see figure 4.5).

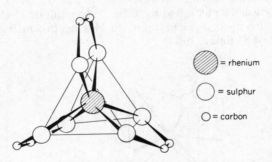

= rhenium

= sulphur

= carbon

Figure 4.5 The co-ordination geometry of $Re(S_2C_2Ph_2)_3$; one phenyl group (omitted) is attached to each carbon (after R. Eisenberg and J.A. Ibers, *Inorg. Chem.*, 5 (1966), 411)

The e.s.r. signal seen in solution is centred at g = 2.015, near the free-electron value. Thus it seems the unpaired electron is predominantly localised on the ligands but this does not enable anything to be said about the oxidation state of the metal. $[Re\{S_2C_2(CN)_2\}_3]^{2-}$, formally Re(IV) ($d^3$), is readily prepared from $ReCl_6^{2-}$ and $Na_2S_2C_2(CN)_2$ in methanol. It has a magnetic moment at room temperature of only 1.65 B.M., against the spin-only value of 3.8 B.M.

4.7.3 Rhenium(V)

With rhenium(V) we enter the region of stability of halides other than fluoride and of other types of ligand. This is particularly so if the Re=O or Re≡N groups are present; probably the function of these is to reduce the charge on what is formally Re^{5+} by π-donation.

$K[ReF_6]$ is the product of the reaction of ReF_6 with KI in liquid sulphur dioxide. On hydrolysis it disproportionates into Re(IV) and Re(VII). The reduction of Re_2O_7 with KI in hydrochloric acid gives the oxo-complex $K_2[ReOCl_5]$; this ion may also be prepared by the reaction

$$ReCl_5 \xrightarrow[M^+]{aq.\ HCl} M_2ReOCl_5 \qquad (M = Rb,\ Cs)$$

Reaction of $KReO_4$ in sulphuric acid–methanol and zinc in the presence of the appropriate hydrohalic acid leads to the formation of salts of the anion $[ReOX_4]^-$; these are square pyramidal with the oxygen at the apex (the short Re–O distance, approx. 170 pm, confirms that there is Re–O multiple bonding).

The ReO^{3+} group forms a number of complexes, especially with phosphines. It has a characteristic infrared stretching frequency at 920–985 cm^{-1}. The group *trans* to Re=O is quite labile, this behaviour recalling the weak *trans* bonding in vanadyl VO^{2+} complexes, which are typically square pyramidal. Reduction of Re_2O_7 with tertiary phosphines in ethanolic HCl gives $ReOCl_3(PR_3)_2$, obtainable in isomeric forms with *cis* or *trans* RR_3 groups. These are readily differentiated by their dipole moments. Substitution reactions are possible, such as

$$[ReOCl_3(PR_3)_2] \xrightarrow{\text{HCl}} PHR_3[ReOCl_4(PR_3)]$$

$$[ReOI_3(PR_3)_2] \xrightarrow{\text{EtOH}} [ReO(OEt)I_2(PR_3)_2] + HI$$

The action of wet pyridine is to substitute the phosphine ligands by pyridine, while Re—Cl bonds are hydrolysed and deprotonated.

$$[ReOCl_3(PR_3)_2] \xrightarrow[\text{H}_2\text{O}]{\text{C}_5\text{H}_5\text{N}} [ReO_2py_4]Cl.2H_2O + Re_2O_3Cl_4py_4$$

The presence of hydrazine as reducing agent in the original reaction mixture can lead to the formation of nitrido-complexes; for example

$$Re_2O_7 + PPh_3 + N_2H_4 \xrightarrow[\text{HCl}]{\text{EtOH}} ReNCl_2(PPh_3)_2$$

These have a characteristic Re—N stretch at 1010–1070 cm^{-1} and the Re—N bond is thus very slightly stronger than Re—O in the ReO^{3+} group (or more precisely, the force constant is rather greater, after allowing for the slightly smaller reduced mass). Reaction with other phosphines (for example PEt_2Ph) affords $ReNCl_2(PR_3)_3$; the structures of these two systems are shown in figure 4.6.

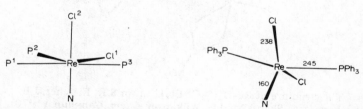

Figure 4.6 The co-ordination spheres of $ReNCl_2(PPhEt_2)_3$ and $ReNCl_2(PPh_3)_2$. The octahedron is moderately distorted. Interbond angles: Cl^1—Re—P^2 = 165.2°: Cl^2—Re—P^2 = 80.9°; P^1—Re—P^3 = 170.8°; Cl^1—Re—N = 99.2°; other distortions $\lesssim$5° (after P.W.R. Corfield, R.J. Doedens and J.A. Ibers, *Inorg. Chem.*, **6** (1967) 197, 204

ReNBr$_2$(PPh$_3$)$_2$ reacts with KCN in methanol to form $[ReN(CN)_4]_n^{2-}$ and $[ReN(CN)_5]^{3-}$; the cyanide groups in the former compound appear from X-ray evidence to be bonding via nitrogen (Re—$\widehat{N}$—C = 136°), an extremely unusual situation. There is a —Re—N—Re—N— chain with alternating Re—N distances of 224 and 153 pm; the latter is unusually short.

A range of other compounds containing rhenium-nitrogen bonds may be prepared.

$$cis\text{-}ReOCl_3(PR_3)_2 \xrightarrow{\quad PhNH_2 \quad} ReCl_3(NPh)(PR_3)_2$$

$$\xrightarrow[\text{HCl}]{\text{MeNHNHMe}} ReCl_3(NMe)(PR_3)_2$$

$$\xrightarrow[\text{HCl}]{\text{PhCONHNH}_2}$$

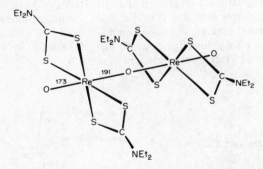

The crystal structure of $ReCl_3(NMe)(PEt_2Ph)_2$ shows $Re-N = 168.5$ pm; with the phosphines *trans*.

Reaction of $ReNCl_2(PPh_3)_2$ with NaS_2CNEt_2 affords $ReN(S_2CNEt_2)_2$. This compound has a square-pyramidal structure with the Re atom raised 73 pm above the S_4 basal plane; this arrangement is well known in other square-pyramidal complexes, for example $VO(acac)_2$.

Reaction of $ReOCl_3(PPh_3)_2$ with NaS_2CNEt_2 in acetone affords $Re_2O_3(S_2CNEt_2)_4$. This has a structure involving a staggered configuration of the

Figure 4.7 The structure of $Re_2O_3(S_2CNEt_2)_4$ (after S.R. Fletcher, J.F. Rowbotham, A.C. Skapski and G. Wilkinson, *Chem. Commun.* (1970), 1572)

dithiocarbamate ligands, which is probably dictated by interligand repulsions (see figure 4.7). Terminal Re—O distances are commensurate with some degree of triple bonding, while the bridging distances correspond to a situation intermediate between a double and a single Re—O bond (in $Re_2OCl_{10}^{4-}$, Re—O is 186 pm).

The eight-co-ordinate cyano-complex $K_3[Re^V(CN)_8]$ deserves mention. It is the product of the interaction of CN^- and $[Re^{IV}Cl_6]^{2-}$. Spontaneous oxidation from Re(IV) to Re(V) takes place, doubtless because eight-co-ordinated Re(IV) has 19 electrons, one of which must lie in an unstable orbital. Of course, it should next be asked why an eight-co-ordinated Re(IV) species is formed as an intermediate rather than $[Re(CN)_6]^{2-}$. The answer to *this* probably lies in the high nucleophilicity and small size of the cyanide ion. Further oxidation of $[Re^V(CN)_8]^{3-}$ to $[Re^{VI}(CN)_8]^{2-}$ can be achieved.

4.7.4 Rhenium(IV)

The complexes of Re(IV) are not numerous but include the important octahedral halogen species $[ReX_6]^{2-}$ (X = F, Cl, Br, I), which form stable and convenient starting materials for subsequent reactions. Preparation is as follows

$$K[ReO_4] \xrightarrow{\text{aqueous KI/HX}} K_2[ReX_6] \qquad (X = Cl, Br, I)$$

$$K_2[ReCl_6] \xrightarrow[300°]{KHF_2} K_2[ReF_6]$$

These anions may be converted into phosphine complexes; for example

$$(PHR_3)_2[ReBr_6] \xrightarrow{\text{heat}} [ReBr_4(PR_3)_2] + 2HBr$$

Another useful route to phosphine complexes is

$$ReO_4^- \xrightarrow[\text{HCl/EtOH}]{Ph_3P} [ReOCl_3(PPh_3)_2] \xrightarrow[\text{EtOH}]{R_3P} \textit{mer}\text{-}ReCl_3(PR_3)_3 \xrightarrow{CCl_4} \textit{trans}\text{-}ReCl_4(PR_3)_2$$

(PR_3 = alkyldiarylphosphine or aryldialkylphosphine).
Pyridine is also an effective ligand here

$$K_2[ReCl_6] \xrightarrow[200°]{C_5H_5N} [ReCl_4py_2]$$

$[ReX_6]^{2-}$ reacts with HX and formic acid to give a number of carbonyl halides, for example, $Re(CO)_5X$ and $Re(CO)_4X_2^-$.

4.7.5 Rhenium(III)

The chemistry of rhenium(III) contains a number of interesting polynuclear halide complexes. Thus a compound of empirical formula $CsReCl_4$ is obtained by the interaction of rhenium trichloride with HCl and CsCl in water; X-ray analysis shows it to have a trimeric structure (see figure 4.8). This structure should be compared

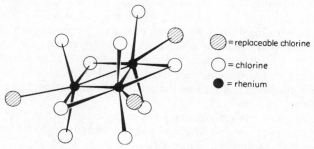

= replaceable chlorine

◯ = chlorine

● = rhenium

Figure 4.8 The $[Re_3Cl_{12}]^{3-}$ unit in $CsReCl_4$ (after J.A. Bertrand, F.A. Cotton and W.A. Dollase, *Inorg. Chem.*, **2** (1963), 1166). The three chlorines shaded are those that may be replaced by phosphines to give $Re_3Cl_9(PR_3)_3$

with that of Re_3Cl_9 which is derived from it by superposing terminal chlorines to form

bridges. The structures clearly involve metal–metal bonds. Re_3Cl_9 reacts with acetonitrile to yield $Re_3Cl_9(MeCN)_3$. Oxygen-donors such as Me_2SO and Ph_3PO likewise form complexes $Re_3Cl_9L_3$. Reduction of perrhenate in aqueous HCl with hypophosphite gives the dimeric ion $[Re_2Cl_8]^{2-}$ (see figure 4.9). This is a clear

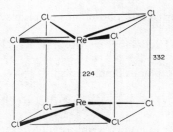

Figure 4.9 The $[Re_2Cl_8]^{2-}$ ion

example of metal atoms closely bound together by metal–metal bonding only. However, there seems to be more than one such bond, as shown by the curious eclipsed configuration. A simple molecular-orbital treatment is as follows: the Re–Re bond defines the z axis, and we take $d_{x^2-y^2}$; p_x, p_y and s orbitals of each rhenium atom, which are of the right symmetry to form σ-bonds to chlorine. The d_{z^2} and p_z orbitals will form a hybrid of σ-symmetry relative to the z axis, d_{xz} and d_{yz} have π-symmetry and d_{xy} has δ-symmetry. Overlap of the σ hybrid on each Re gives us a σ-bond, the π orbitals give two π-bonds, while d_{xy}–$d_{xy'}$ overlap gives a δ bond. We now have 12 bonding orbitals (8 Re–Cl and 4 Re–Re). Counting the electrons to be placed in them: each Re contributes 7, each Cl 1, and with two negative charges the total is 24. Of these, 16 are assigned to 8 Re–Cl bonds, giving us 8 to fill the remaining 4 Re–Re bonding orbitals. This results in a quadruple bond (Re–Re 224 pm), while the δ bond will prefer an eclipsed configuration, as observed – if the configuration were staggered, d_{xy}–d_{xy} overlap would be nil. More recent M.O. data, from extended overlap calculations, suggest that the π contribution to the bond is five times that of the δ-contribution, and nearly three times that of the σ-contribution.

The bromide ion forms similar complexes to the chlorides, but there are no corresponding fluorides or iodides. $Re_2Cl_8^{2-}$ and the bromine analogue undergo a number of reactions as shown in figure 4.10.

The structure of $\{ReCl_3(PEt_3)\}_2$ is shown in figure 4.11. This also possesses the eclipsed configuration. The benzoate $Re_2(OCOPh)_2I_4$ has a bridging structure, derived from $Re_2X_8^{2-}$ by replacement of four equatorial halogens by two carboxy-

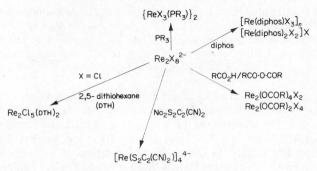

Figure 4.10 Some reactions of $[Re_2X_8]^{2-}$ (X = Cl, Br)

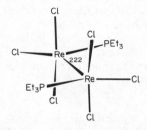

Figure 4.11 The co-ordination sphere of $Re_2Cl_6(PEt_3)_2$ (after F.A. Cotton and B.M. Foxman, *Inorg. Chem.*, **7** (1968), 2135)

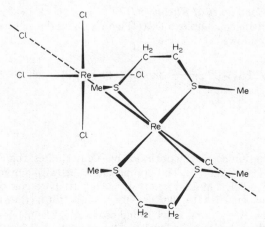

Figure 4.12 The novel dimeric 2,5-dithiahexane complex $Re_2Cl_5\{MeS(CH_2)_2SMe\}_2$ (after M.J. Bennett, F.A. Cotton and R.A. Walton, *J. Am. chem. Soc.*, **88** (1966), 3866)

lates (Re—Re = 220 pm). Raman data show that for a large number of these systems, ν(Re—Re) occurs at about 285 cm^{-1}. Now in Re$_2$(CO)$_{10}$ (Re—Re bond order 1.0) the analogous vibration occurs at 128 cm^{-1} and for a system with a quadruple bond, the analogous vibration might be expected at $\sqrt{4}$ times this value, as is observed for Re$_2$X$_8{}^{2-}$. The good agreement with simple theory may well be due to chance.

Re$_2$Cl$_5$(DTH)$_2$, the 2,5-dithiahexane complex, has a staggered structure (see figure 4.12), with Re—Re 229 pm. The concomitant loss of δ-bonding on attaining this configuration is responsible for the lengthening of the metal–metal bond.

Reaction of NCS$^-$ with Re$_2$Cl$_8{}^{2-}$ in acidified methanol gives Re$_2$(NCS)$_8{}^{2-}$ while in acetone, some Re(NCS)$_6{}^{2-}$ is formed due to oxidation.

As a final example of the versatility of these systems, Re$_2$X$_8{}^{2-}$ react with X$_2$ to form Re$_2$X$_9{}^-$, which also appear to contain Re—Re bonds.

There are a number of phosphine and arsine complexes of Re(III). Some are obtained by mild replacement by monodentate ligand molecules of the three chlorine ions in [Re$_3$Cl$_{12}$]$^{3-}$ shown shaded in figure 4.8, giving [Re$_3$Cl$_9$(PR$_3$)$_3$]. However, monomeric complexes of an ordinary octahedral type can be obtained by reaction of o-C$_6$H$_4$(AsMe$_2$)$_2$ (diarsine) with perrhenic acid and HCl, using hypophosphorous acid as reducing agent. In this way golden-yellow [ReCl$_2$(diars)$_2$]ClO$_4$ has been isolated. The reduction of perrhenate with excess monodentate phosphine also gives a monomeric complex [ReCl$_3$(PR$_3$)$_3$].

4.7.6 Rhenium (II)

There are few complexes of rhenium(II) but the diarsine complexes mentioned above may be reduced to the +2 state by Sn(II), giving [Re^{II}Cl$_2$(diars)$_2$] (or for that matter they may be oxidised to [Re^{V}Cl$_4$(diars)$_2$]$^+$). Complexes of the polydentate arsines PhAs(o-C$_6$H$_4$AsPh$_2$)$_2$ and As(o-C$_6$H$_4$AsPh)$_3$ are also known.

4.7.7 Dinitrogen Complexes of Rhenium(I)

An octahedral dinitrogen complex of Re(I) which obeys the eighteen-electron rule has been made as follows

$$[\text{ReCl}_2(\text{N}_2\text{COPh})(\text{PPh}_3)_2] \xrightarrow[\text{MeOH}]{\text{PMe}_2\text{Ph}} trans\text{-}[\text{ReCl}(\text{N}_2)(\text{PMe}_2\text{Ph})_4] + \text{PhCOOMe}$$

$$+ \text{HCl} + 2\text{PPh}_3$$

The *trans* configuration has been confirmed by X-ray studies; the Re—N distance is 196.6 pm and N—N 105.5 pm. This compound reacts with a number of molecules such as ZrCl$_4$, TaF$_5$, MoCl$_4$(THF)$_2$ or CrCl$_3$(THF)$_3$ to give adducts involving dinitrogen bridging, such as (Me$_2$PhP)$_4$ ClRe—N—N—MoCl$_4$(OMe). This last compound has a comparatively long N—N bond distance (121 pm) possibly caused by delocalisation of the bonding N$_2$ π electrons into empty Mo(V) 4d orbitals. The system Re—N—N—Mo is linear and the molecular conformation (two eclipsed octahedra) indicates a delocalised π system.

4.7.8 Phosphine Hydride Complexes

Finally, we may mention the interesting rhenium phosphine hydride complexes, which are formed by reduction of the rhenium chloride phosphine complexes.

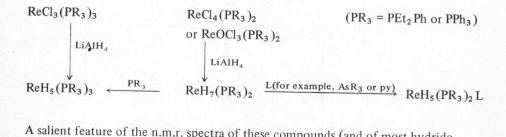

A salient feature of the n.m.r. spectra of these compounds (and of most hydrido phosphine complexes) is that there seems to be intramolecular fluxional behaviour, that is some kind of rapid interchange so that each hydride is equivalent with respect to each phosphine — thus for $ReH_5(PR_3)_3$ the high-field ^1H n.m.r. line (due to the hydrides) is a quartet (splitting from three equivalent phosphines; ^{31}P has $I = \frac{1}{2}$). Hydrido carbonyls are also known.

$$ReH_3(PPh_3)_4 \xrightarrow[C_6H_6]{CO} \begin{cases} ReH(CO)_2(PPh_3)_3 \begin{cases} \xrightarrow{HX} Re(CO)_2(PPh_3)_2 X \ (X = Cl, Br) \\ \xrightarrow{I_2} Re(CO)_2(PPh_3)_2 I \end{cases} \\ \text{and} \\ ReH(CO)_3(PPh_3)_2 \xrightarrow{HCl} Re(CO)_3(PPh_3)_2 Cl \end{cases}$$

Analytical data are not a reliable guide to the number of hydridic hydrogens in such complexes and the method frequently adopted is careful integration of the n.m.r. signal and comparison with the integrated part due to, say, the phosphine ligands.

A dinitrogen complex hydride of Re(I) that obeys the eighteen-electron rule may be obtained by treatment of the nonahydridorhenate anion $ReH_9{}^{2-}$ with $Ph_2PCH_2CH_2PPh_2$ in 2-propanol under nitrogen. The product is *trans*-$ReHN_2(diphos)_2$. In accord with its structure, this complex has a hydrido n.m.r. resonance at τ 19.94, a dinitrogen stretching frequency at 2006 cm^{-1} and a Re—H stretching frequency at 1818 cm^{-1} in the infrared. The N_2 group is displaced by CO or C_2H_4.

IRON

Metal: b.c.c.; m.p. 1528°; I_1: 7.90 eV; I_2: 16.18 eV; I_3: 30.64 eV
Oxides: FeO, Fe_2O_3, Fe_3O_4 (all often nonstoichiometric)
Halides: FeX_2 (X = F, Cl, Br, I); FeX_3 (X = F, Cl, Br)

Oxidation State and Representative Compounds

	0	1	2	3	4	5	6
Typical donor atom/group	CO,PR_3	NO	O,N,Hal	O,S,Hal	As	O^{2-}	O^{2-}
Co-ordination number							
3				$Fe\{N(SiMe_3)_2\}_3$ √			
4 planar			Fe(II)phthalocyanine √	$[Fe(Ph_3PO)_4]^{3+}$† √√√			
4 tet	√		$[FeCl_4]^{2-}$ √√√	$[FeCl_4]^-$ √√√			
5 T.B.P.	$Fe(CO)_5$		$[FeBrN(C_2H_4NMe_2)_3]^+$ √√	$[Fe(N_3)_5]^{2-}$ √√			
5 S.P.	√	$Fe(NO)(S_2CNR_2)_2$ √√	$[Fe(Ph_2MeAsO)_4(ClO_4)]^+$	$Fe(acac)_2Cl$‡ √√√			
6	$[Fe(CO)_5H]^+$ √	$[Fe(NO)(H_2O)_5]^{2+}$ √√	$Fe(OH_2)_4Cl_2$‡	$Fe(acac)_3$‡ √√√	$[Fe(diars)_2Cl_2]^{2+}$ √√	$[FeO_4]^{3-}$ √	$[FeO_4]^{2-}$ √
7				$[Fe(EDTA)(OH_2)]^-$			
8			$[Fe(naphthyridine)_4]^{2+}$ √	$Fe(NO_3)_4^-$			

† Not definitely characterised; ‡ More than one spin-state known
√ Known; √√ Several examples; √√√ Very common

5 Ruthenium and Osmium

Ruthenium and osmium are the first pair of the *platinum metals,* the others being rhodium, iridium, palladium and platinum itself. The metals in this group occur native, often as alloys with each other, and besides resembling each other chemically in a general way, had to be placed together (as Group VIII) in the earlier forms of the Periodic Table.

Ruthenium was finally isolated by K.K. Klaus in 1844; the name had been chosen by G.W. Osann 18 years earlier for one of the three new metals he had isolated from Russian platinum residues (Ruthenia — a district in Russia). It may also be obtained from laurite, $(Ru, Os)S_2$, or the native alloy osmiridium, both of which are also sources of osmium.

Both ruthenium and osmium, besides being unattacked by the atmosphere, also resist all acids including aqua regia. They are, however, attacked by oxygen, fluorine or chlorine, but only at elevated temperatures. They will also dissolve in molten sodium peroxide to give oxy-salts. The metals are hexagonally close-packed, and have melting points of $2310°$ (Ru) and $3050°$ (Os).

These two metals show one of the larger ranges of oxidation states, from +8 in their tetroxides to zero in carbonyls and related compounds. As is to be expected, higher oxidation states are rather more stable for osmium than for ruthenium. Like iron, ruthenium has +2 and +3 as its principal states with their relative stability depending on the particular ligand. Osmium(III), however, has a certain lack of stability, since high-field ligands allow easy reduction to d^6 osmium(II) while low-field ligands cannot resist easy oxidation to osmium(IV), a stable state for the heaviest metal of the triad.

Rather curiously, the zero oxidation state (d^8) is not well represented, the carbonyls being the chief examples. This situation contrasts with the d^8 electron configurations of Rh(I) and Ir(I) which are very versatile, forming a variety of complexes with different ligands.

The higher ranges of oxidation are peopled by rather unusual chemical creatures — the volatile osmium tetroxide and its ruthenium analogue, osmium heptafluoride, osmyl(VI) complexes formally resembling those of the better-known entity $UO_2{}^{2+}$, and nitrido-complexes with the $Os\equiv N$ grouping. In addition there are the usual hexafluorides.

A special peculiarity of ruthenium is the very stable Ru—NO grouping. Very many nitrosyls are formed by this metal.

5.1 Aqueous Cationic Chemistry

This is limited to ruthenium, the heavier element forming no aquo-ions. This distinction between second- and third-row elements is fairly general for the right-hand side of the transition-metal series. Ruthenium(IV) perchlorate is believed to exist in aqueous solution as $[RuO(H_2O)_5](ClO_4)_2$. There is evidence for the existence of $[Ru(H_2O)_6]^{3+}$ though hydrolysis of the stable chloro-complexes $[RuCl_n(H_2O)_{6-n}]^{(3-n)+}$ does not proceed beyond $[RuCl(H_2O)_5]^{2+}$.
$Cs_2RuCl_5(H_2O)$ and the red $(Ph_4As)cis$-$[RuCl_4(H_2O)_2]$ are isolated from $RuCl_3$ in HCl on treatment with the appropriate cation: both contain octahedrally co-ordinated ruthenium. A green form of $(Ph_4As)[RuCl_4(H_2O)_2]$ is isolated from treatment of reduced methanolic ruthenium trichloride with Ph_4AsCl. It may be the *trans* isomer.

$[Ru(H_2O)_6]^{2+}$ definitely exists in aqueous solution, and may be obtained by electrolytic reduction of aqueous Ru(III) chlorides followed by ion exchange.

5.2 Halides

The following undoubted halides exist (one or two of the lower ones are not yet fully characterised).

Table 5.1

	+7	+6	+5	+4	+3	+2
Ru	–	F	F	F	F Cl Br I	Cl Br I
Os	F	F	F	F Cl Br	Cl Br I	I

Osmium heptafluoride is noteworthy for the stabilisation of the very high +7 state by fluorine. This yellow solid may be made by direct reaction at 600° and 400 atmospheres. The infrared spectrum suggests that the molecule may be a pentagonal bipyramid, but a perfectly regular polyhedron would not be expected. The Os(VII) oxofluoride $OsOF_5$ is obtained from oxyfluorination of the metal; it is reduced to grey-green $OsOF_4$ by a hot tungsten filament.

Ruthenium and osmium hexafluorides are low-melting (54° and 32°) solids, made by direct fluorination. Their molecules are octahedral. The two pentafluorides are polymeric in the solid state†, M_4F_{20}, achieving six-co-ordination by formation of a cyclic tetramer. They dissociate in the vapour phase. Preparation is from Ru and BrF_3 or by $W(CO)_6$ reduction of OsF_6; the latter is an interesting method. Other methods are u.v. irradiation of OsF_6; and reaction of OsF_6 with iodine, giving OsF_5 and IF_5. Salts $K[RuF_6]$ and $K[OsF_6]$ are obtained straightforwardly by the action of bromine trifluoride on mixtures of potassium and ruthenium or osmium bromides.

The only stable halide of Ru(IV) is RuF_4, a yellow solid of unknown structure made by reduction of RuF_5 with I_2. A similar fluoride OsF_4 is obtained with

† There exist three groups of isomorphous pentafluorides: they are MF_5 (Tc, Re); $M'F_5$ (Nb, Ta, Mo, W); and $M''F_5$ (Ru, Os, Rh, Ir).

OsF_5 when the latter is prepared as first described above, but the stability of Os(IV) shows itself in the isolation of red osmium tetrachloride from high-pressure chlorination of the metal at 600°. The tetrachloride is hydrolysed to the dioxide in water; on heating in vacuo, it forms Os_2Cl_7, which is reportedly isomorphous with α-$RuCl_3$. A tetrabromide, stable to 350°, is obtained by analogous methods.

5.2.1 Halide Complexes

Hexahalide complexes of metals in fairly high oxidation states are often more stable than the corresponding simple halides. Thus the Ru(IV) and Os(IV) complexes $[MX_6]^{2-}$ are known for both metals for X = F, Cl and Br; $[OsI_6]^{2-}$ also exists. They are made in various ways; $K_2[RuCl_6]$ and $K_2[RuBr_6]$ can be made by halogen oxidation of Ru(III) halide solutions; they may fairly easily be reduced again to Ru(III). The corresponding Os(IV) complexes are very stable; thus $[OsBr_6]^{2-}$ is the product of the reduction of osmium tetroxide by hydrobromic acid. All these ions are octahedral. Salts of $[OsCl_6]^{2-}$ are particularly useful as starting materials for the preparation of many other osmium compounds.

A related compound of special interest is $K_4[Ru_2OCl_{10}]$. It contains a linear Ru—O—Ru linkage and, although a Ru(IV) compound, is diamagnetic. These facts are accounted for if there are two three-centre bonding systems as shown in figure 5.1. Each of these will have four electrons — one from each Ru and two from oxygen —

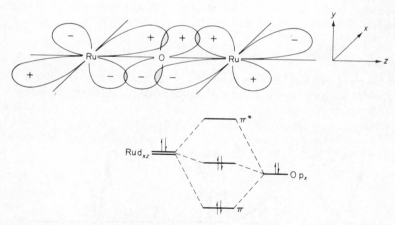

Figure 5.1 Bonding in $[Ru_2OCl_{10}]^{4-}$ (there is another system formed by d_{yz}-p_y overlap). For clarity in this and following diagrams, not all of the nodal surfaces of the 4d or 5d orbitals are shown

which will occupy the bonding and associated nonbonding M.O.s, and the remaining two d electrons on each Ru will pair up, presumably in d_{xy}, thus explaining the diamagnetism. An exactly analogous osmium anion is present in $Cs_4Os_2OCl_{10}$, which has linear Os—O—Os with Os—O = 178 pm.

5.2.2 Other Polynuclear Systems

Some μ-nitrido complexes, which are rather analogous to the above M—O—M systems, are known

$$K_2RuO_4 \xrightarrow[OH^-]{NH_4OH} Ru_2N(OH)_5 \cdot nH_2O \xrightarrow{MBr} [Ru_2NBr_8(H_2O)_2]^{3-}$$

or

$$[Ru(NO)Cl_5]^{2-} \xrightarrow{SnCl_2} [Ru_2NCl_8(H_2O)_2]^{3-}$$

The latter compound has been shown by X-ray studies to have a linear structure with the eclipsed configuration (a consequence of the M.O. scheme outlined in figure 5.1). There are osmium analogues; for example

$$OsCl_6{}^{2-} \xrightarrow[100°]{NH_4OH} [Os_2N(NH_3)_8Cl_2]^{3+}$$

Related to these systems are the compounds known as 'ruthenium reds', produced by reaction of ruthenium chlorides in air with ammonia to give a red solution, which affords a diamagnetic product of the formula $Ru_3O_2(NH_3)_{14}Cl_6$. This is probably $[(NH_3)_5 Ru^{III}-O-Ru^{IV}(NH_3)_4-O-Ru(NH_3)_5]^{6+}$; on oxidation $[(NH_3)_5 Ru-O-Ru(NH_3)_4-O-Ru(NH_3)_5]^{7+}$ is formed. This oxidation in acid solution of the intense red 6+ ion to the yellow 7+ ion forms a very delicate test for even mild reducing agents. Recent ^{99}Ru Mössbauer studies support these formulations, as does the following reaction

$$\text{'Ruthenium red'} \xrightarrow[45°]{en} [Ru_3O_2(NH_3)_{10}en_2]Cl_6$$

The dark red-green product has been shown by X-ray analysis to have a linear Ru—O—Ru—O—Ru system with the overall structure of the cation as shown in figure 5.2.

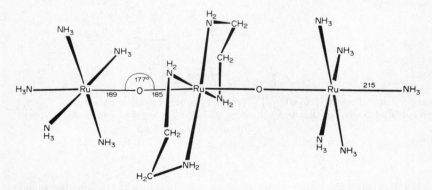

Figure 5.2 The structure of the $[(NH_3)_5RuORu(en)_2ORu(NH_3)_5]^{6+}$ ion (after P.M. Smith, T. Fealey, J.E. Earley and J.V. Silverton. *Inorg. Chem.*, **10** (1971), 1943)

5.2.3 Other Halide Complexes

Halogeno carbonyl complexes are conveniently prepared by refluxing either RuX_3 or $OsX_6{}^{2-}$ with HX and formic acid (a suitable source of CO). A typical sequence is

$$RuCl_3 \cdot xH_2O \xrightarrow{\ HCO_2H\ } [Ru(CO)Cl_4(OH_2)]^{2-} \text{ (green)}$$

$$\downarrow HCl/HCO_2H$$

$$[Ru(CO)Cl_5]^{2-} \quad \text{(red)}$$

$$\downarrow$$

$$cis\text{-}[Ru(CO)_2Cl_4]^{2-} \quad \text{(yellow)} \xrightarrow{\ PPh_3\ } [Ru(CO)_2(PPh_3)_2Cl_2]$$

$$\downarrow \qquad \qquad \nearrow_{PPh_3}$$

$$fac\text{-}[Ru(CO)_3Cl_3]^- \quad \text{(yellow)}$$

Similar sequences afford the bromo- and iodo-complexes. The carbonyl groups almost certainly take up *cis* positions in these systems; this comment applies fairly generally to carbonyl and nitrosyl complexes, and is explicable on the basis of the fact that both these ligands are very good π acceptors and that a *cis* configuration minimises competition for the available d electron density. Nitrosyl halides are also known (see section 5.5.2); comparison of the position of the $\nu_2(Ru-N)$ and $\nu_2(Ru-C)$ stretches in analogous compounds shows that, despite the greater mass of NO, $\nu(M-N)$ is consistently some 80 cm^{-1} higher.

5.2.4 Lower Halides

Ruthenium trifluoride is obtained by iodine reduction of the pentafluoride at 150°. The trichloride $RuCl_3$ is an important ruthenium compound, made by direct combination. It exists in two forms, one of them isostructural with $CrCl_3$ (layer lattice, octahedral co-ordination) and like $CrCl_3$, insoluble in water. A water-soluble form, possibly $RuCl_3(H_2O)_3 \cdot xH_2O$, is obtainable by reaction of RuO_4 with hydrochloric acid; the tri-iodide may be obtained by addition of iodide ion to this solution. The tribromide is obtained in a manner similar to the hydrated chloride. The nature of hydrated ruthenium chloride, as commonly obtained, may be noted here. The use of material from different commercial preparations in carrying out, for example, the reactions detailed in section 5.2.3 results in considerable variation in reaction times. The reason for this is that '$RuCl_3 \cdot xH_2O$' can be a mixture of species, some of which may be polymers, and some of which contain ruthenium (IV). This material can be converted into a genuine Ru(III) chloride by several times evaporating to dryness with concentrated HCl.

Except for the unknown OsF_3, there is a complete range of trihalides as we should expect for this principal oxidation state. The osmium trihalides are dark, insoluble substances made by heating the tetrahalide ($OsCl_3$ and $OsBr_3$) or by the action of iodine on the di-iodide (OsI_3). $OsCl_3$ has a magnetic moment of 1.77 B.M. at room temperature, falling to 1.65 B.M. at 77 K, which is fairly typical behaviour for a $5d^5$ ion.

Ruthenium(III) and osmium(III) hexahalide complexes may be obtained; those of osmium are easily oxidised. Such series (including isomers) as $[OsCl_xX_{6-x}]^{2-}$ ($x = 0 - 6$) (X = Br, I) have been isolated.

$$K_2OsO_4 \xrightarrow{\text{HCl}} K_2OsCl_6$$

$$(NH_4)_2OsCl_6 \xrightarrow[\text{NH}_4\text{I}]{\text{HI}} (NH_4)_2OsI_6$$

$$OsCl_6{}^{2-} \xrightarrow{\text{HI}} OsCl_5I^{2-} \longrightarrow trans\text{-}OsCl_4I_2{}^{2-} \longrightarrow mer\text{-}OsCl_3I_3{}^{2-}$$

$$\longrightarrow trans\text{-}OsCl_2I_4{}^{2-} \longrightarrow OsClI_5{}^{2-} \longrightarrow OsI_6{}^{2-}$$

$$OsI_6{}^{2-} \xrightarrow{\text{HCl}} OsClI_5{}^{2-} \longrightarrow cis\text{-}OsCl_2I_4{}^{2-} \longrightarrow fac\text{-}OsCl_3I_3{}^{2-}$$

$$\longrightarrow cis\text{-}OsCl_4I_2{}^{2-} \longrightarrow OsClI_5{}^{2-} \longrightarrow OsI_6{}^{2-}$$

The individual complexes can be separated from the mixtures by high-voltage ionophoresis and isolated as their Cs^+ salts; careful choice of conditions can result in obtaining just one isomer, thus heating $OsI_6{}^{2-}$ with concentrated HCl for one hour at 40° results in 100 per cent fac-$[OsCl_3I_3]^{2-}$.

Alcoholic solutions of ruthenium trichloride, bromide or iodide may be reduced to the dihalides, which are dark blue or green in colour, by hypophosphite or hydrogen. Little seems to be known about them. The only dihalide of osmium is OsI_2, made by heating $(H_3O)_2[OsI_6]$. Salts of $Ru_5Cl_{12}{}^{2-}$ have recently been isolated from the blue solutions formed by reduction of Ru(III) or (IV) chloride solutions; it appears to be a cluster complex ion with one unpaired electron. It can be used as a convenient source of ruthenium complexes; with pyridine it gives $RuCl_2py_4$ and with cyclopentadiene (π-C_5H_5)$_2$Ru.

5.3 Oxides and Oxyanions

Undoubtedly the outstanding compounds are the tetroxides RuO_4 and OsO_4, the only such in the chemistry of any metal. Ruthenium tetroxide may be made by periodate oxidation of the dioxide. The low-melting (25°), volatile (b.p. 40°) yellow solid is less stable than OsO_4. It dissolves in carbon tetrachloride to give a solution that will cleave C=C linkages and perform other organic oxidations. The molecule is presumably tetrahedral. Osmium tetroxide (m.p. 41°, b.p. 131°) is also yellow and soluble in organic solvents. It is obtained by nitric acid oxidation of the dioxide. The tetrahedral molecule has an Os—O distance of 177 pm. It will oxidise C=C linkages, even aromatic ones, to the *cis* diol. Its vapour is *extremely* toxic, affecting vision.

The only other well-defined oxides are the dioxides RuO_2 and OsO_2, though the sesquioxides M_2O_3 have been reported. Ruthenium dioxide, made by direct combination at 1000°, is a blue solid with the rutile structure. A brown oxide OsO_2 of similar structure is made from the metal and NO at 650°.

These two metals form a number of fairly stable oxyanions. As in the tetroxides, the metal–oxygen bonding is to be considered as covalent, with a σ-bond usually regarded as arising from metal s and p orbitals and a π-bond formed by overlap of metal d orbitals with oxygen p orbitals.

Ruthenium, fused with KNO_3 and KOH, gives black $K[Ru^{VII}O_4]$, potassium perruthenate, which decomposes in water into the ruthenate $K_2[Ru^{VI}O_4]$ and oxygen.

The perosmate ion $[Os^{VIII}O_4(OH)_2]^{2-}$, however, is formed from OsO_4 and aqueous KOH. Complex oxides of Os(VII) are also known; thus sodium hexaoxo-osmate(VII), $Na_5[OsO_6]$, is made by the action of oxygen at 550° on the metal and sodium monoxide. The purple osmates(VI), unlike the ruthenates(VI), are really octahedral *osmyl* complexes, and are readily made by the reduction of perosmates. The osmyl group $[OsO_2]^{2+}$, has, as has been mentioned, a formal resemblance to the uranyl group $[UO_2]^{2+}$ and is similarly linear. The four hydroxyl groups in $K_2[OsO_2(OH)_4]$ complete the octahedron; the larger uranyl group is seldom octa-hedral. The three orbitals used in one of the two O–Os–O π bonds are shown in figure 5.3. The OsO_2^{2+} ion has one remaining t_{2g} orbital, which is occupied by the two electrons of the d^2 Os(VI). The two three-centre π bonds, their corresponding non-bonding orbitals, and the two Os–O σ bonds will be filled by the 12 electrons

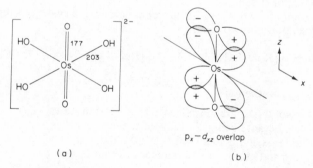

$p_x - d_{xz}$ overlap

(a)   (b)

Figure 5.3 The osmate ion (a) and its π bonding (b)

notionally from the two O^{2-} ions. This particular type of ion is therefore most likely to be stable in the case of a d^n ion where $n = 0$ to 2. The ruthenyl group RuO_2^{2+} is less known than the osmyl group, but has been definitely characterised.

$$RuO_4 \xrightarrow{HCl/CsCl/CCl_4} trans\text{-}Cs_2[RuO_2Cl_4] \quad (\nu(RuO_2) = 830 \text{ cm}^{-1})$$

5.4 Nitrido Complexes of Osmium

Osmium forms these more readily than does any other metal. As was mentioned in section 5.3 osmium readily forms multiple bonds with oxygen, so that similar bond-ing with nitrogen is not unreasonable. Thus potassium osmiamate $K[OsO_3N]$, prepared by treating $K_2[OsO_4(OH)_2]$ with ammonia, is isoelectronic with OsO_4. A comparison of the infrared stretching frequencies is instructive.

$[OsO_3N]^-$: $Os-N = 156$ pm $\nu_1 = 1021$ cm^{-1} (largely Os–N)

$\nu_2, \nu_4 = 897, 871$ cm^{-1} (largely Os–O)

OsO_4: $Os-O = 174$ pm $\nu_1 = 965$ cm^{-1}; $\nu_3 = 954$ cm^{-1}

The nitrogen character of the 1021 cm^{-1} band was demonstrated by ^{15}N substitution (shift to 993 cm^{-1}).

Potassium osmiamate is reduced by HCl to $K_2[Os^{VI}NCl_5]$, the pentachloro-nitrido-osmate(VI) anion, which has $Os-N = 161$ pm and $\nu(Os-N) = 1080$ cm^{-1}. Considered formally as OsN^{3+} co-ordinated octahedrally by five chloride ions, six electrons are required for the σ and the two π Os–N bonds and this leaves just two electrons in the stable nonbonding d_{xy} orbital (the Os–N axis being taken as z). It may be noted that the *trans* (to N) Os–Cl bond is 260 pm, while *cis* Os–Cl = 236 pm; in weakly acid solution, the ion $[OsNCl_4(OH_2)]^-$ forms, so that the *trans* chloride is probably labile, in accord with the structure of $OsNCl_5{}^{2-}$. Furthermore, the five-co-ordinated C_{4v} ion $[OsNCl_4]^-$ is also known (Os–N = 160 pm, Os–Cl = 231 pm, $\nu(Os-N) = 1123$ cm^{-1}), confirming the weakness of the binding of the fifth chloride ion in $[OsNCl_5]^{2-}$.

Although osmium nitrides are the more numerous, the analogous terminal Ru≡N group is not unknown, as in

$$\text{\textit{trans}-Cs}_2[RuO_2X_4]^{2-} \xrightarrow{\text{HX/N}_3^-} Cs_2[RuNX_5] + N_2 \qquad (X = Cl, Br)$$

5.5 Complexes of Ruthenium

Most of these compounds are in the +2 or +3 oxidation states. The +2 state is particularly versatile because being d^6, it obeys the eighteen-electron rule when octahedrally co-ordinated and is stabilised by π-bonding ligands, while also it is a high enough oxidation state to bond with amines or other ligands that rely on σ electron donation only.

5.5.1 Ammine Complexes

There exists a considerable variety of amine complexes of Ru(II) and Ru(III), both ammines and complexes of heterocyclic amines. The hexammines $[Ru(NH_3)_6]^{3+}$ (colourless) and $[Ru(NH_3)_6]^{2+}$ (orange) are both easily made

$$RuCl_3 \xrightarrow{\text{NH}_3 + \text{Zn} + \text{NH}_4\text{Cl}} [Ru(NH_3)_6]^{2+} \xrightarrow{\text{Cl}_2} [Ru(NH_3)_6]^{3+}$$

The Ru(II) compound is readily oxidised

$$[Ru(NH_3)_6]^{2+} \longrightarrow [Ru(NH_3)_6]^{3+} + e; E_o = 0.21 \text{ V}$$

Crystallographic study on $Ru(NH_3)_6I_2$ and $Ru(NH_3)_6(BF_4)_3$ has recently shown them to contain octahedral $[Ru(NH_3)_6]^{n+}$ ions. The Ru–N bond is slightly longer in the former (t_{2g}^6, Ru–N = 214.4 pm; t_{2g}^5, Ru–N = 210.4 pm); this is ascribed to the electrostatic effect (Ru^{3+} versus Ru^{2+})rather than to the change in electronic configuration. The tris-*o*-phenanthroline and tris-bipyridyl complex ions [Ru phen$_3$]$^{2+}$ and [Ru bipy$_3$]$^{2+}$ are formed by the action of the diamine on RuCl$_3$, an additional reducing agent being unnecessary in these cases; the stability of the Ru(II) complexes doubtless is owed to the higher-field ligands in the octahedral d^6 situation. These complexes may be oxidised to Ru(III) by ceric ion (compare with corresponding Fe(II) and Os(II) complexes).

$$[\text{Ru bipy}_3]^{2+} + Ce^{4+} \longrightarrow [\text{Ru bipy}_3]^{3+} + Ce^{3+}$$

Mixed complexes such as [RuCl(NH$_3$)$_5$]$^{2+}$, *cis* and *trans*-[RuCl$_2$(NH$_3$)$_4$]$^+$ and [RuCl$_2$ bipy$_2$] are also known, but the chemistry is by no means so extensive as that of, say, Co(III). Some of these complexes are fairly inert (as is to be expected from the crystal-field stabilised d^6 configuration) and have been used for kinetic studies. Electron-transfer reactions have also been investigated; for example

$$[\text{Ru bipy}_3]^{3+} + [\text{Fe(CN)}_6]^{4-} \longrightarrow [\text{Ru bipy}_3]^{2+} + [\text{Fe(CN)}_6]^{3-}$$

At 18°, $k = 10^5$ mol^{-1} s^{-1}; it is an outer-sphere reaction involving only t_{2g} electrons and consequently fast.

5.5.2 Nitrosyl and Nitrogen Complexes

It is a characteristic of Ru(II) that it will complex with two groups, SO$_2$ and N$_2$, which have a rather limited co-ordination chemistry. Thus, aqueous hydrazine reacts with RuCl$_3$ to give [Ru(NH$_3$)$_5$N$_2$]$^{2+}$, isolated as the iodide (see figure 5.4). [Ru(NH$_3$)$_5$N$_2$Ru(NH$_3$)$_5$]$^{4+}$ ions result from reduction of [Ru(NH$_3$)$_6$]Cl$_3$ and reaction with N$_2$; the ion contains a linear Ru–N–N–Ru unit. The N$_2$ stretching

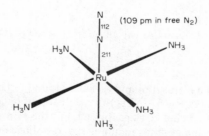

Figure 5.4 The [Ru(NH$_3$)$_5$N$_2$]$^{2+}$ cation

frequency is about 2150 cm^{-1} (compare 2331 cm^{-1} in free N$_2$). The stability of this complex may be related to the exceptional ability of Ru(II) to bond with the NO group. Thus the N$_2$ complex is isoelectronic with [Ru^{II}(NH$_3$)$_5$NO]Cl$_3$, which is obtained by the action of nitrite and acid on [Ru(NH$_3$)$_6$]$^{2+}$. Other examples of

Ru(II) complex nitrosyls are $[Ru^{II}Cl_5NO]^{2-}$ and $[Ru^{II}(H_2O)_5NO]^{3+}$. The nitrosyl group is formally considered to bond as NO^+, this idea being supported by the diamagnetism.

The following data for $Na_2[Ru(NO_2)_4(OH)NO].2H_2O$ (see figure 5.5) suggest considerable back-bonding, qualitatively similar to that in complexes of the iso-electronic CO group, but greater in extent because of the formal positive charge on NO^+.

Table 5.2

	NO	NO^+	$Na_2[Ru(NO_2)_4(OH)NO].2H_2O$
$NO(cm^{-1})$	1878	2200	1907
N–O distance (pm)	114	106	113

The i.r. and bond-length data indicate that the two electron pairs concerned, in two orbitals formed by overlap of the filled d_{xz} and d_{yz} with the two empty π^* NO^+ orbitals, weaken the N–O link considerably compared with NO^+. The Ru–N link will, however, be strengthened.

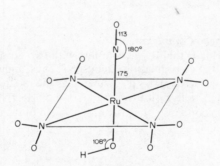

Figure 5.5 The structure of the anion in $Na_2(NO_2)_4(OH)NO].2H_2O$

Another example of a nitrosyl bound as NO^+ is found in $Ru(NO)(S_2CNEt_2)_3$. The related Ru^{3+} complex $Ru(S_2CNEt_2)_3$ has a distorted octahedral RuS_6 core, due to the four-membered chelate rings formed, but the nitrosyl complex has one *mono*dentate dithiocarbamate, thus retaining six-co-ordination. The NO group is bound in the manner associated with $M–NO^+$ groupings (Ru–$\widehat{N}$–O, about 170°; Ru–N = 172 pm). Here N–O is 117 pm and ν(N–O) occurs at 1803 cm^{-1} in the i.r.

Recent evidence indicates that Ru^{2+} is a very strong π donor – thus NO and (to a lesser extent) N_2, which are good π-acceptors, will form strong bonds to Ru^{2+}. It is notable that $[Ru(NH_3)_5N_2]^{3+}$ cannot be made; attempted oxidation of $[Ru(NH_3)_5N_2]^{2+}$ leads to the evolution of the bound dinitrogen; this accords with the weaker π-donor properties of Ru^{3+}.

A number of ruthenium nitrosyl complexes may be obtained starting from the tetroxide or the trichloride as can be seen from the following reaction scheme

$$RuO_4 \xrightarrow[HX]{HNO_3} (Ru(NO)X_3)_n \xrightarrow{X^-} [Ru(NO)X_5]^{2-}$$

$$X = Cl, Br, I$$

$$X = Cl \;\Big|\; R_3P/EtOH$$

$$RuCl_3(NO)(PR_3)_2 \text{ and } [Ru(NO)(PR_3)_3Cl_2]^+$$

$$RuCl_3.3H_2O \xrightarrow[EtOH]{NO/PPh_3} RuCl_3(NO)(PPh_3)_2$$

Another reaction involves protonation of a nitro-complex

$$[Ru(NO_2)_2(bipy)_2] \xrightleftharpoons[NaOH]{HPF_6/MeOH} [Ru(bipy)_2(NO)(NO_2)] (PF_6)_2$$

[99]Ru exhibits the Mössbauer effect. Few data are available yet, largely because the recoilless transition has rather high energy so that work needs to be conducted at 4.2 K, and because the precursor, [99]Rh, has a half-life of 16 days. The series $[RuL_5(NO)]^{-\text{ or }4+}$ (L = CN, NH$_3$, NCS, Cl, Br) has been examined and the isomer shift was found to decrease as the ligand-field strength of L increased. The cyanide showed an appreciable quadrupole splitting (0.40 mm s^{-1}) concomitant with marked d_{xz}, $d_{yz} \to \pi^*$ NO back-donation. Quadrupole splittings are much smaller for [99]Ru than for [57]Fe. It seems certain, from X-ray results on K$_2$[Ru(NO)Cl$_5$] and other data, that the Ru$-$N$-$O group is linear in all these compounds.

Ru(NO)$_2$(PPh$_3$)$_3$ may be prepared from alcoholic RuCl$_3$ on treatment with N-methyl-N-nitroso-*p*-sulphonamide (source of NO) and the tertiary phosphine, while [RuCl(NO)$_2$(PPh$_3$)$_2$]$^+$PF$_6^-$ is prepared by the following method

$$RuCl_3(NO)(PPh_3)_2 \xrightarrow{Zn} RuCl(NO)(PPh_3)_2 \xrightarrow{NO^+} [RuCl(NO)_2(PPh_3)_2]^+$$

These complexes, some of which seem to be formally Ru(0), undergo some interesting reactions. Thus RuCl(NO)(PPh$_3$)$_2$ reacts with oxygen or SO$_2$ to form adducts, which react with SO$_2$ or O$_2$, respectively, to form RuCl(NO)(SO$_4$)(PPh$_3$)$_2$. Reaction with CO yields RuCl(CO)(NO)(PPh$_3$)$_2$, which may also be prepared from the trichloride

$$RuCl_3 \xrightarrow[\text{2-methoxy ethanol}]{PPh_3/reflux} RuClH(CO)(PPh_3)_2 \xrightarrow{NO} RuCl(CO)(NO)(PPh_3)_2$$

$$\Big\downarrow X_2 \,(X = halogen)$$

$$RuClX_2(NO)(PPh_3)_2$$

The chloride ligand in these complexes is very labile; with H$_2$O, RuCl(CO)(NO)(PPh$_3$)$_2$ forms Ru(OH)(CO)(NO)(PPh$_3$)$_2$. One complex of this type, RuI(CO)(NO)(PPh$_3$)$_2$, has a trigonal bipyramidal structure with a linear Ru$-$N$-$O group. In contrast $[RuCl(NO)_2(PPh_3)_2]^+$ exhibits 'mixed' nitrosyl co-ordination. The Ru–N distances and Ru–N̂–O angles (see figure 5.6) typify the behaviour of NO^+ and NO^- co-ordination (this topic is more fully discussed in section 6.10.5).

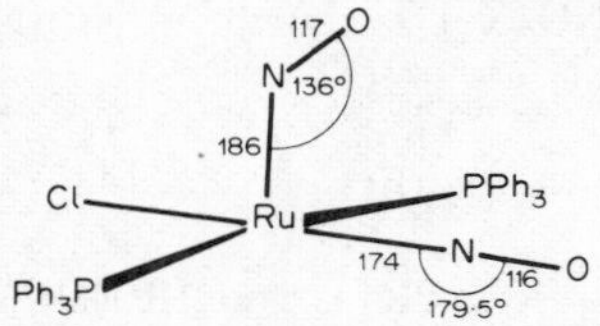

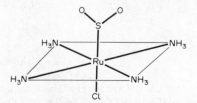

Figure 5.6 The structure of $[RuCl(NO)_2(PPh_3)_2]^+$ ions, which exhibit two types of nitrosyl co-ordination (after C.G. Pierpoint, D.G. Van Derveer, W. Durland and R. Eisenberg, *J. Am. chem. Soc.*, **92** (1970), 4761)

Reaction of $RuCl_3(NO)(PR_3)_2$ ($PR_3 = PPh_3$, PPh_2Me) with ethanolic KOH and excess phosphine yields complexes $RuH(NO)(PR_3)_3$. X-ray diffraction shows the triphenylphosphine complex to have a trigonal bipyramidal structure (equatorial phosphines, linear Ru–N–O). Its n.m.r. behaviour is normal; the hydride proton gives a resonance at τ 16.6 (quartet due to splitting from the equivalent P nuclei with $I = \frac{1}{2}$; $J[P–H] = 30$ Hz) but the methyldiphenylphosphine analogue exhibits stereochemically nonrigid ('fluxional') behaviour. At room temperature, the hydride resonance is a singlet (τ15.0) but on cooling to 160 K it becomes a $1:2:2:2:1$ multiplet, due to two phosphines *cis* to H ($J = 32$ Hz) and one *trans* ($J = 64$ Hz). Thus the structure is probably as shown in figure 5.7.

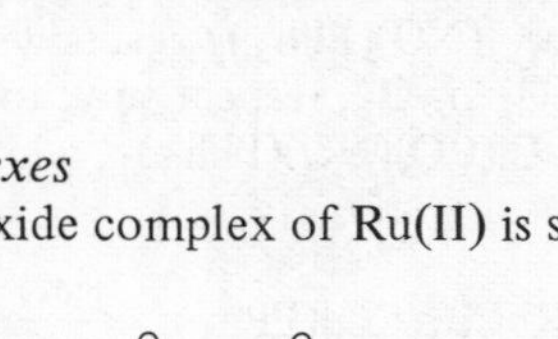

Figure 5.7 Possible structure of the 'nonrigid' complex $RuH(NO)(PPh_2Me)_3$

5.5.3 Sulphur Dioxide Complexes
The structure of a sulphur dioxide complex of Ru(II) is shown in figure 5.8.

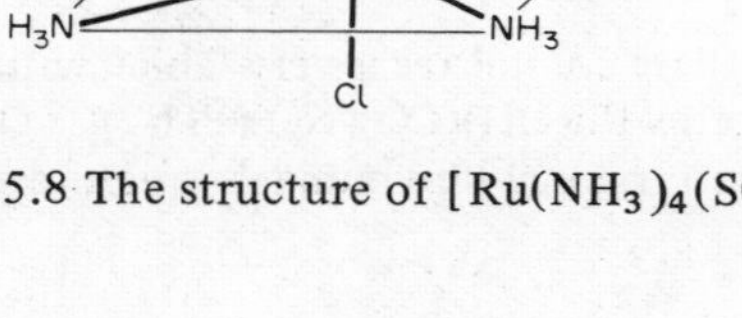

Figure 5.8 The structure of $[Ru(NH_3)_4(SO_2)Cl]^+$

The structure of the co-ordinated SO_2 molecule is similar to that of the free molecule. This complex, which is typical of several such formed by Ru(II), may be made as follows

$$Ru(HSO_3)_2(NH_3)_4 + HCl \longrightarrow [RuCl(NH_3)_4 SO_2]^+ + H_2O + HSO_3^-$$

5.5.4 Carboxylates

Reaction of ruthenium trichloride with acetic acid and sodium acetate gives the cation $[Ru_3O(CO_2Me)_6(OH_2)_3]^+$ based on a Ru_3O core with bridging acetates (see figure 5.9), and having a magnetic moment of 1.77 B.M. This is analogous to carboxylates formed by Cr(III), Fe(III) and some other metals; it forms a diamagnetic phosphine adduct $[Ru_3O(CO_2Me)_6(PPh_3)_3]$, which formally has one Ru(II) and two Ru(III) atoms.

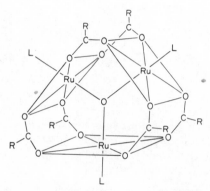

Figure 5.9 A representation of the structure of the ruthenium carboxylate complexes based on an Ru_3O core

The bonding in the phosphine adduct may be explained as follows: first the Ru_3O core is built up by the coupling of each Ru σ orbital with a sp^2 hybrid oxygen orbital; this gives the basic σ-bonding framework. Five orbitals on each metal ion are employed in σ-bonding to four bridging carboxylates and one phosphine (or other donor). There are now three thus far unused d orbitals on each metal ion; one of these from each ruthenium forms linear combinations with the p_z orbital of the central oxygen, giving a total of one bonding, eight nonbonding and one antibonding molecular orbitals. The available electrons that are unused in σ-bonding (2 x 5 + 6 from Ru and 2 from O) thus fill the bonding and nonbonding M.O.s.

A blue pyridine complex $[Ru_3O(O_2CMe)_6py_3]^+X^-$ (X = ClO_4^-, BF_4^-) exists; an extensive redox sequence occurs in some of these complexes, doubtless facilitated by the fact that there are up to 16 nonbonding electrons in these systems, whose partial removal should have little effect on the stability of the system.

With carboxylic acid-acid anhydride mixtures, ruthenium trichloride forms dimeric complexes $[Ru_2(O_2CR)_4Cl]$ (R = Me, Et, Pr^n, Bu^n) containing one Ru(II)

and one Ru(III) ion; the magnetic moment per dimer unit is about 4 B.M., implying *three* unpaired electrons. The structure (see figure 5.10), determined by X-ray analysis, has chlorines bridging adjacent units. It is necessary to explain the para-magnetism by postulating that the three unpaired electrons are in three almost degenerate molecular orbitals, thus preventing the spin-pairing expected for a second-row metal.

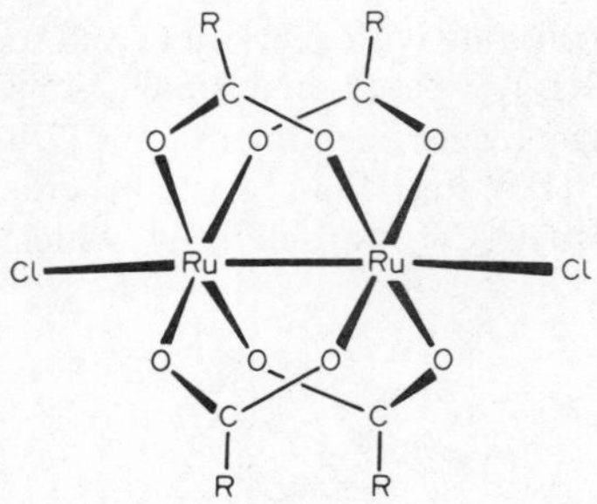

Figure 5.10 The structure of $Ru_2(O_2CR)_4Cl$ (after M.J. Bennett, F.A. Cotton and K.G. Çaulton, *Inorg. Chem.,* **8** (1969), 1)

5.5.5 Complexes with oxygen- and sulphur-containing ligands

Ruthenium(III) complexes of dialkyl sulphides are well known, and are prepared by refluxing ruthenium trichloride with the disulphide; from e.s.r. and n.m.r. data, they appear to possess the *mer* configuration

$$RuCl_3 \xrightarrow[EtOH]{Et_2S} mer\text{-}[RuCl_3(Et_2S)_3] \xrightarrow{LiBr} mer\text{-}[RuBr_3(Et_2S)_3]$$

When hydrogen is bubbled through a solution of hydrated $RuCl_3$ in Me_2SO, reduction occurs and $RuCl_2(Me_2SO)_4$ is isolated. This appears to be *trans.* As expected for a low-spin d^6 complex, it is substitution-inert, and the bromide analogue has to be prepared by hydrogen reduction of $RuBr_3$ in Me_2SO.

5.5.6 Complexes with π-Bonding Ligands

As may be expected, these complexes are mainly formed by the d^6 ion Ru(II) and are mostly octahedral. However, there are some well-defined examples formed by Ru(III), and these will be noted first.

Complex ions $[RuCl_4(diars)]^-$, $[RuCl_2(diars)_2]^+$ and $[Ru(diars)_3]^{3+}$ are known, together with the uncharged complex $RuCl_3(AsPh_2Me)_3$ and a number of others of similar constitution. They are made either from $RuCl_3$ and arsine by direct com-bination, or by oxidation of the corresponding Ru(II) compound. Carbonyls are also known in this oxidation state; for example, $RuBr_3(CO)(PPh_3)_2$, made from $(NH_4)_2RuBr_5$ treated with the phosphine in methanol (no CO is present in this reaction; it presumably comes from the methanol).

Reaction of 'RuCl$_3$.xH$_2$O' with triphenylphosphine (see figure 5.11) leads to a number of complexes, dependent on the conditions, excess phosphine causing reduction to the divalent state.

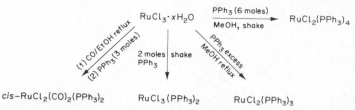

Figure 5.11 Reactions of ruthenium trichloride with triphenylphosphine

Metathetical treatment of RuCl$_3$ with LiBr, followed by reaction with the appropriate ligand yields among others RuBr$_3$(PPh$_3$)$_2$, RuBr$_3$(AsPh$_3$)$_2$, RuBr$_2$(PPh$_3$)$_3$ and RuBr$_2$(PPh$_3$)$_4$. The structure of RuCl$_2$(PPh$_3$)$_3$ is interesting, in that the sixth co-ordination position is 'blocked' by a hydrogen atom of a phenyl group, which may be why six co-ordination does not occur.

Many phosphine complexes of Ru(II) are of the type RuX$_2$(diphos)$_2$, where X = halogen, H or alkyl. Geometrical isomers have been isolated; some examples are

$$[Ru_2Cl_3(PR_3)_6]Cl + diphos \xrightarrow{150°} cis\text{-}RuCl_2\ diphos_2$$

$$\Big\downarrow LiAlH_4 \Big| THF$$

$$trans\text{-}RuH_2diphos_2 \xleftarrow[benzene]{LiAlH_4} trans\text{-}RuClH\ diphos_2$$

$$RuCl_3.xH_2O \xrightarrow[EtOH]{diphos} trans\text{-}RuCl_2(diphos)_2 \xrightarrow{liq.\ K} cis\text{-}RuH_2(diphos)_2$$

The dimeric [Ru$_2$Cl$_3$(PPhEt$_2$)$_6$]Cl is formed by boiling RuCl$_3$.xH$_2$O and the phosphine in ethanol (the osmium analogue is conveniently prepared from OsCl$_6^{2-}$) and is thought to contain a triply bridged unit (see figure 5.12).

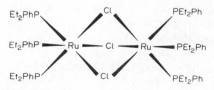

Figure 5.12 The probable structure of [Ru$_2$Cl$_3$(PPhĖt$_2$)$_6$]$^+$

A most interesting isomerisation has been found to take place on reduction of *trans*-[RuCl$_2$(diphos)$_2$] with sodium naphthalenide (diphos = Me$_2$PCH$_2$CH$_2$PMe$_2$).

The $Ru^0(diphos)_2$, which might be the expected product of this reaction, isomerises into the hydride of Ru(II) depicted in figure 5.13. This is inferred from n.m.r. and i.r. studies, which indicate the presence of a Ru—H group, and preliminary deuteration of the ligand shows that to be the origin of the resulting deuteride. The structure has been confirmed by X-ray diffraction.

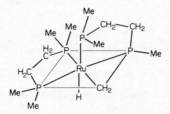

Figure 5.13 The structure of $Ru(diphos)H \{CH_2P(Me)(CH_2)_2PMe_2\}$

$RuCl_2(PPh_3)_3$, previously mentioned, is one of the most important ruthenium (II) compounds; it is a versatile catalyst for the hydrogenation of alk-1-enes. It is readily converted into $RuHCl(PPh_3)_3$ (probably the 'active' catalyst). Some reactions are given in figure 5.14. X-ray diffraction shows $RuHCl(PPh_3)_3$ to have a tri-

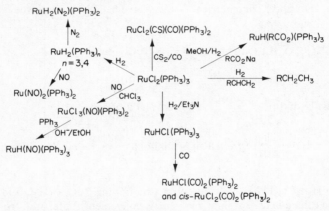

Figure 5.14 Reactions of $RuCl_2(PPh_3)_3$

gonal bipyramidal structure (Ru—H = 170 pm) while $RuH(O_2CR)(PPh_3)_3$, selective catalysts for the hydrogenation of alk-1-enes, contain a bidentate carboxylate and *mer*-disposed phosphines. $[RuCl_3(PEt_2Ph)_3]$ reacts on heating with esters, ketones or aldehydes to give either $[Ru_2Cl_3(PEt_2Ph)_6]^+ [RuCl_3(PEt_2Ph)_3]^-$ or $[Ru_2Cl_4(PEt_2Ph)_5]$. This latter compound has been shown to be a dimer with three bridging and one terminal chlorines; the Ru—Ru separation is 336 pm (little interaction). The complex salt has also been examined by X-rays. Here, the cation contains two octahedrally co-ordinated Ru atoms with three bridging chlorines, while the anion adopts the *mer* configuration, generally favoured for $M(PR_3)_3X_3$ units. In the above reaction with aldehydes, the mechanism is fairly complex since both α and β hydrogens are labilised and CO is abstracted.

Another novel phosphine complex is obtained from reaction of ruthenium tri-chloride with tri-*n*-butylphosphine in ethanol; $[RuCl_3(PBu^n_3)_2]_2$ is precipitated first and the mixed Ru(II)–Ru(III) dimer $Ru_2Cl_5(PBu^n_3)_4$ is obtained from the mother liquor. This has the structure shown in figure 5.15, with a Ru—Ru distance

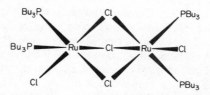

Figure 5.15 The trinuclear complex $Ru_2Cl_5(PBu^n_3)_4$

of 311.5 pm; Ru(II) and Ru(III) ions are, of course, not distinguishable. The mag-netic moment is 1.50 B.M.

The dimeric complexes can often be converted into monomeric hydrides

$$[Ru_2Cl_3(PMe_2Ph_2)_6]Cl \xrightarrow[N_2H_4/H_2]{PMe_2Ph} RuH_2(PPh_2Me)_4$$

The cabonyls of Ru(II) are quite numerous but are mostly of the type $RuX_2(CO)_nL_{(4-n)}$, where X = Cl, Br or I; L = a phosphine or arsine and n = 1 to 4. Preparation is often from $RuCl_3$, excess phosphine or arsine to achieve reduction, and CO. However, the specific addition of CO is not always necessary, an alcohol often sufficing, as indicated above in the case of $[RuBr_3(CO)(PPh_3)_2]$. Thus $[RuCl_2(CO)_2(PEt_3)_2]$ may be obtained by treatment of $RuCl_3$ in allyl alcohol with triethylphosphine and KOH.

When $RuCl_2(PPh_3)_3$ reacts with CO in N,N'-dimethylacetamide at room temperature, $RuCl_2(CO)(PPh_3)_2(dma)$ is obtained; the molecule of solvent of crystallisation is lost on recrystallisation from dichloromethane–methanol. The resulting five-co-ordinate complex is an efficient isomerisation catalyst; for example, oct-1-ene is isomerised into oct-2-ene, oct-3-ene and oct-4-ene.

A final example of novel ruthenium chemistry is shown in the following sequence of reactions involving a diarsine complex

$[Ru(NO)Cl_3]_n \xrightarrow{diars} trans-[Ru(NO)Cl(diars)_2]^{2+} \xrightarrow{N_2H_4} trans-[RuN_3Cl(diars)_2]$

$trans-[RuN_3Cl(diars)_2] \downarrow NO^+PF_6^-$

$\xrightarrow{O_2, hv} trans-[RuN_2Cl(diars)_2]^+$

$trans-[Ru(NO_2)Cl(diars)_2]$

and

$trans-[Ru(NO)Cl(diars)_2]^+$

5.6 Complexes of Osmium

The favoured oxidation states are again +3 and, especially, +2. Although the +4 state is common and stable for halides and halide complexes, and OsO_2 is the stable oxide, it is largely avoided by osmium in its co-ordination chemistry with the exception of a few phosphine and arsine complexes. The general outline of the co-ordination chemistry of osmium is closely similar to that of ruthenium.

Complex ammines are, except for osmyl complexes, represented by the colourless $[Os(NH_3)_6]^{3+}$ (obtained by the action of ammonia on $[OsBr_6]^{2-}$ at 290°) and by $[OsCl(NH_3)_5]^{2+}$. A corresponding complex ion incorporating ethylenediamine, $[Os\ en_3]^{3+}$, is known. This ion is, of course, of a conventional nature, but both Os(III) and Os(IV) tend to form complexes with deprotonated ethylenediamine, as exemplified by the ion $[Os^{IV}(HNCH_2CH_2NH_2)_2\ en]^{2+}$ (this process is rather analogous to the better known loss of a proton by an aquo-complex to give a hydroxy complex). Other ethylenediamine complexes have recently been reported. Thus

$$K_2OsO_4(OH)_2 \xrightarrow{\text{enHCl}} [OsO_2\ en_2]Cl_2 \xrightarrow{\text{Zn/Hg}} cis\text{-}[OsH_2\ en_2]ZnCl_4$$

Both Os(III) and Os(II) form many complexes with bipyridyl, 1,10-phenanthroline and $2,2'6',2''$-terpyridyl; the chemistry is similar to that of Ru. Examples are $[Os\ phen_3]^{3+}$, $[Os\ bipy_3]^{3+}$, $[OsCl_2\ phen_2]^+$ and $[OsCl\ (bipy)(terpy)]^{2+}$ for Os(III); Os(II) also forms these same complex ions, with the positive charge reduced by one unit. The d^6 Os(II) compounds are the more stable; thus $K_2[Os^{IV}Cl_6]$ gives $[Os^{II}bipy_3]^{2+}$ on treatment with bipyridyl; the Os(III) compound is obtained by oxidation with chlorine

$$[Os\ bipy_3]^{2+} \longrightarrow [Os\ bipy_3]^{3+} + e \qquad E_0 = -0.88\ V$$

The complexes $[OsL_3]^{n+}$ (L = phen or bipy; n = 3 or 2) are substitution-inert and may be obtained as optical enantiomers.

5.6.1 Complexes with π-bonding ligands

These are rather similar to the corresponding compounds of ruthenium. The main difference is that the Os(IV) complexes $[OsX_2\ diars_2](ClO_4)_2$ may be obtained by oxidation of a corresponding Os(III) complex; no Ru(IV) analogues are known (yet). Osmium(III) forms uncharged complexes such as $[OsCl_3(AsPh_3)_3]$ as well as the unipositive ion $[OsCl_2(diars)_2]^+$.

Osmium(IV) and (III) phosphine complexes may conveniently be prepared from OsO_4 by refluxing with HCl and the phosphine in ethanol: the sequence is as follows

$$OsO_4 \longrightarrow (R_3PH)_2[OsCl_6] \longrightarrow trans\text{-}[OsCl_4(R_3P)_2] \longrightarrow mer\text{-}[OsCl_3(R_3P)_3]$$

$$\downarrow$$

$$[(R_3P)_3OsCl_3Os(R_3P)_3]Cl$$

It is not always possible, especially with the strongly reducing trialkyl phosphines, to isolate the osmium(IV) compound, but the latter may be obtained from the osmium(III) complex on boiling with CCl_4. The *mer* → *fac* conversion can be accomplished by initial reaction with borohydride or hydrazine (probably forming $Os(PR_3)_3H_4$) and then adding HCl. Osmium(III), like ruthenium(III), is low-spin d^5 in its complexes, and *fac*- and *mer*-$OsCl_3(PR_3)_3$ can be distinguished on the basis of their e.s.r. spectra: the former (C_{3v}) will have an axially symmetric g tensor, with only two g values while the latter (C_{2v}) will have three g values. Thus, for $OsCl_3(PBu^n{}_2Ph)_3$, the *fac* isomer has $g_\perp = 1.83$, $g_\parallel = 1.28$ and the *mer* isomer has g values of 3.30, 1.65 and 0.36.

Alkyl sulphide complexes also exist and are readily prepared.

$$OsCl_4 \xrightarrow[\text{EtOH}]{Et_2S} mer\text{-}[OsCl_3(Et_2S)_3]$$

5.6.2 Hydrido and Related Complexes

Reduction of *mer*-$[OsCl_3(PR_3)_3]$ with zinc amalgam leads to the formation of some unusual compounds, the nature of the product depending on the gas used to provide the atmosphere in the reaction vessel! Thus, under argon, the chloride-bridged dimer $[Os_2Cl_3(PR_3)_6]Cl$ is formed; with a nitrogen atmosphere, the di-nitrogen complex $[OsCl_2(N_2)(PR_3)_3]$ obtains; spectroscopic evidence suggests the chlorides are *cis* and the phosphines *mer,* as in the starting material. Hydrogen and carbon monoxide afford $[OsH_2Cl_2(PR_3)_3]$ and $[OsCl_2(CO)(PR_3)_3]$ respectively, while $[OsCl_2(N_2)(PR_3)_3]$ reacts with borohydride to form $[OsHCl(N_2)(PR_3)_3]$. The stability of the osmium–dinitrogen bond is well demonstrated by the reaction of the latter compound with HCl

$$[OsHCl(N_2)(PEt_2Ph)_3] + HCl \longrightarrow OsCl_2(N_2)(PEt_2Ph)_3 + H_2$$

There are some indications here that N_2 is a rather *weak* σ donor, so that its bonding ability appears to depend on its π-acceptor properties (see also section 5.5.2).

Complexes of Os(II) of the type $[OsCl_2 \text{ diphos}_2]$, made from $[OsCl_6]^{2-}$ and the diphosphine, may be converted into dihydrides, hydrido-chlorides, hydrido-alkyls, chloro-alkyls or dialkyls by appropriate use of a reducing agent, such as $LiAlH_4$, or an alkylating agent, such as methyl lithium or trimethyl aluminium. In most cases, geometrical isomers can be isolated, and their configuration is conveniently assigned from the ^1H or ^{31}P n.m.r. or from electric dipole moment measurement. The hydrido-monophosphine complexes are notable for their ability to increase their co-ordination number while retaining the eighteen-electron configuration.

For example (the hydride resonances are given parenthetically)

$$OsCl_3 \xrightarrow[\text{EtOH, reflux}]{PPh_3/BH_4^-} OsH_4(PPh_3)_3 \quad (\tau\ 17.85;\ 1{:}3{:}3{:}1 \text{ quartet})$$

$$\textit{trans-}OsCl_4(PMe_2Ph)_2 \xrightarrow[\text{EtOH}]{BH_4^-} OsH_6(PMe_2Ph)_2 \quad (\tau\ 18.6;\ 1{:}2{:}1 \text{ triplet})$$

$MH_2(PR_3)_4$ (M = Ru, Os) also exist; the number of hydrides in these complexes is most conveniently estimated from integration of the proton n.m.r. resonances (or from ^{31}P n.m.r. data). $OsH_6(PR_3)_2$ fits nicely into the $ReH_7(PR_3)_2 \rightarrow IrH_5(PR_3)_2$ isoelectronic series.

5.6.3 Carbonyl Complexes

Apart from the binary carbonyls, the carbonyls of osmium are virtually confined to the +2 oxidation state. They are of the type $[OsX_2(CO)_nL_{(4-n)}]$, where X = Cl, Br, I or H; L = a phosphine or arsine and n =1–4. $[OsCl_2(CO)_4]$ is made from $OsCl_3$ and CO at 200 atmospheres and 160°, while treatment of $OsCl_3$ with CO and PPh_3 gives $[OsCl_2(CO)_2(PPh_3)_2]$. A remarkable example of the simultaneous appearance of a carbonyl and a hydride group which were not introduced as such is the following

$$OsCl_3(PEt_2Ph)_3 \xrightarrow[\text{boil}]{\text{KOH/EtOH}} [OsClH(CO)(PEt_2Ph)_3]$$

As a consequence of the discovery of this and related reactions, the possibility of the unconventional introduction of groups such as M≡N, M–H or M–CO, often from the solvent, is now always routinely considered in investigations in this area of chemistry.

Osmium forms a number of nitric oxide complexes (although not as many as ruthenium): $[Os(NO)_2(OH)(PPh_3)_2]PF_6$ has been examined crystallographically, and is another example of 'mixed' NO^-–NO^+ behaviour (see section 5.5.2) with one 'straight' and one 'bent' Os–N–O group. It (and other nitrosyls) may be prepared according to the reaction

$$MCl_2(CO)_2(PPh_3)_2 \xrightarrow{NO_2^-} M(ONO)_2(CO)_2(PPh_3)_2 \xrightarrow[-CO_2]{} M(NO)_2(PPh_3)_2$$

(M = Ru Os)

$\Big\downarrow O_2/HPF_6$

$$[M(NO)_2(OH)(PPh_3)_2]PF_6$$

$OsCl(NO)(CO)(PPh_3)_2$ undergoes an interesting reaction with mercuric chloride; the initial reaction is oxidative addition, to form $OsCl_2(NO)(HgCl)(PPh_3)_2$, which is then photolysed to form $OsCl_3(NO)(PPh_3)_2$ and mercury.

COBALT

Metal: h.c.p.; m.p. 1490°; I_1: 7.86 eV; I_2: 17.05 eV; I_3: 33.49 eV
Oxides: CoO
Halides: CoX_2 (X = F, Cl, Br, I); CoF_3

Oxidation State and Representative Compounds

	0	1	2	3	4
Typical donor atom/group		P,C,S	O,N,Hal	O,N,Hal	F
Co-ordination number					
3			√ $Co\{N(SiMe_3)_2\}_2.PPh_3$		
4 tet	? $Co(PMe_3)_4$		√√√ $[CoX_4]^{2-}$		
4 planar					
5 T.B.P.		√√ $[Co(CNMe)_5]^+$ √	√√ $[Co(pyO)_5]^{2+†}$		
5 S.P.		$Co(NO)(S_2CNR_2)_2$	√√ $[Co(Ph_2MeAsO)_4(ClO_4)]^{+†}$		
6		?	√√√ $[Co(OH_2)_6]^{2+†}$ √ $[Co(NO_3)_4]^{2-}$	√√√ $Co(CN)_6^{3-†}$	√ $[CoF_6]^{2-}$
8	√ $Co_2(CO)_8$				

† More than one spin-state; ? Suspected; √ Known; √√ Several examples; √√√ Very common

6 Rhodium and Iridium

These metals form the second pair of the platinum metals. They have these metals' characteristic properties of nobility, rarity of conventional hydrated ions, and ready formation of a wide variety of stable complexes. The complexes may have conventional ligands such as Cl^- or NH_3, but metal–carbon σ bonds are readily formed and there is a wide variety of π-bonded low oxidation-state complexes. The co-ordination chemistry of these two metals, therefore, forms rather an extensive area of study, which will be surveyed in some detail in the following pages. Much of the work to be described is quite recent.

Rhodium was discovered by W.H. Wollaston in 1804 (Greek, *rhodon*, a rose – it has rose-coloured compounds), and iridium by Smithson Tennant also in 1804 (Greek *iris*, a rainbow – it has compounds of many different colours).

The principal oxidation states are +3 for both metals, with +1 for complexes of π-bonded ligands and +4 for iridium. Other oxidation states are known (Rh, 0 to +6; Ir, −2 to +6), but are of infrequent occurrence. These are the first heavy transition metals that do not form the oxyanions of the type $[MO_4]^{n-}$ which are characteristic of the elements V, Nb, Ta; Cr, Mo, W; Mn, Tc, Re; Fe, Ru, Os; Co.

The virtual absence of the +2 oxidation state, so common for cobalt, is to be specially noted, and may be attributed to the increased stability of the spin-paired d^6 +3 state, whose crystal-field stabilisation energy is even greater than for cobalt(III) as a result of the higher value of Δ for the heavier metals.

6.1 The Metals and their Aqueous Chemistry

Rhodium (m.p. 1960°) and iridium (m.p. 2443°) are silvery-white metals, untarnished in air but slowly attacked by the halogens and oxygen at red heat. The metals are remarkable in being unattacked by all acids, including aqua regia. The structure of both is cubic close packed.

Rhodium alone forms a hydrated ion. Thus treatment of $RhCl_3$ with excess perchloric acid gives $[Rh(H_2O)_6](ClO_4)_3$ whose high stability contrasts with the highly oxidising nature of the corresponding cation of cobalt. There is also some evidence for the formation of an aquated rhodium(II) species on reduction of $[RhCl(H_2O)_5]^{2+}$ with chromous ion. IrO_2, the most stable oxide of iridium, and Ir_2O_3 will both dissolve in acids but no simple hydrated cations have been authenticated.

6.2 Halides

It is difficult to give a definitive list of authentic binary halides, but the following are well defined.

Table 6.1

	+3	+4	+5	+6
Rh	F Cl Br I	F	F	F
Ir	F Cl Br I	F	F	F

In addition, IrX, IrX$_2$ (X = Cl, Br, I), RhCl, RhCl$_2$ and IrX$_4$ (X = Cl, Br, I) have been reported but perhaps not fully characterised; many of them are probably authentic. The salient feature is the concentration on the +3 state, fluorine alone having the ability to give +5 and +6 states. The comparative lack of well-defined IrX$_4$ halides is surprising, since Ir(IV) is a very stable state of that metal as shown by its stable dioxide and the wealth of Ir(IV) complexes, including halogen complexes. The stability of the $[\text{IrCl}_6]^{2-}$ species combined with a slight oxidising tendency of Ir(IV) towards halide ions appears to render the IrX$_4$ species difficult to prepare; once prepared they should probably be stable.

RhCl$_3$, the most common compound of rhodium, is prepared by direct combination at 300°. The red product, volatile at 800°, has the six-co-ordinate chlorine-bridged AlCl$_3$ layer structure. It dissociates in the gas phase at about 1000° into RhCl$_2$(g) + Cl$_2$. It is insoluble in water, but a soluble form results from

$$\text{Rh}_2\text{O}_3 + \text{HCl(aq)} \rightarrow \text{RhCl}_3\text{(aq)} \xrightarrow[180°]{\text{HCl}} \text{RhCl}_3$$

This form of the trichloride gives the yellow hydrated ion on boiling in much water.

RhBr$_3$ is prepared by direct combination at 300°, while RhI$_3$ is made in the wet way using potassium iodide.

IrCl$_3$, a red, water-insoluble compound, which exists in two crystal structures both having six-co-ordinated metal atoms, is made by direct combination at 500°. IrBr$_3$ is made by the action of bromine on IrBr$_2$, the probable product of heating IrBr$_3$(aq) in HBr gas (the IrBr$_3$(aq) is obtained from IrO$_2$ and aqueous HBr). IrI$_3$ is obtained by dehydration of its hydrate, obtained from IrO$_2$ (note this second example of the instability of Ir(IV) in this context) and aqueous HI.

The fluorides form two series of compounds, which are fairly orthodox in their preparations and properties. Thus direct combination yields rhodium hexa-, penta-, and trifluorides, while the tetrafluoride is conveniently obtained by the treatment of RhBr$_3$ with bromine trifluoride — if the use of the latter can ever be termed convenient! The hexafluoride is a black low-melting (70°) solid, while the pentafluoride is tetrameric, Rh$_4$F$_{20}$, having the same structure as M$_4$F$_{20}$ (M = Ru, Os, Ir). The red RhF$_3$ adopts an AB$_3$ 6:2 co-ordination structure similar to that of ReO$_3$. IrF$_6$ and IrF$_5$ are both obtained by direct combination, while IrF$_4$ is obtained by reduction of IrF$_6$ vapour by a red-hot tungsten wire.

6.3 Oxides

Anhydrous rhodium dioxide is reported to be formed by heating Rh_2O_3 under an atmosphere of oxygen and has been assigned the (expected) rutile structure. The more stable product of heating the metal in oxygen is the sesquioxide Rh_2O_3, which has the 6:4 co-ordinate corundum (α-Al_2O_3) structure. In the case of iridium, the black dioxide is obtained in this manner, or alternatively by hydrolysis of Ir(IV) solutions, followed by dehydration of the product. It has the rutile structure (as do RuO_2 and OsO_2) with 6:3 co-ordination. The sesquioxide Ir_2O_3 is obtainable in the dry way by heating $K_2[IrCl_6]$ with sodium carbonate.

Heat decomposes the oxides as follows

$$2Ir_2O_3 \xrightarrow{500°} Ir + 3IrO_2$$

$$IrO_2 \xrightarrow{1200°} Ir + O_2$$

$$2Rh_2O_3 \xrightarrow{1200°} 4Rh + 3O_2$$

6.4 Some Other Binary Compounds

Rhodium and iridium both combine with the heavier elements of Groups V and VI on being heated with them. A considerable number of different phases have been established. Thus Rh_2P and Ir_2P have the anti-fluorite structure; this type of compound retains considerable metallic character and no definite oxidation state can be assigned. The sulphides Rh_9S_8, Rh_3S_4, Rh_2S_3, Rh_2S_5 and RhS_3, and a somewhat similar Ir series, are known. Many of these compounds, as might be expected, are semiconductors.

6.5 Complexes of Rhodium

Most of these have d^6 rhodium in the +3 oxidation state, but the d^8 rhodium(I), usually square planar in configuration, is also well represented by a variety of interesting compounds. Rhodium is noteworthy for recent developments in its hydride chemistry in both these oxidation states.

6.5.1 Rhodium(IV)

This is an unstable state represented only by halide complexes. The reason is probably that low-field ligands that are not too easily oxidised will be required in this low-spin d^5 situation where the reduction to the stable d^6 state is the immediate consequence of any attempt to introduce a medium- or high-field ligand. Thus chlorine oxidation of $Cs_3[RhCl_6]$ gives $Cs_2[RhCl_6]$, which is isomorphous with $(NH_4)_2[PtCl_6]$. The corresponding fluoride $Cs_2[RhF_6]$ is also known, but no bromo- or iodo-complexes.

6.5.2 Rhodium(III)

Rhodium (III) forms the yellow aquo-ion $[Rh(H_2O)_6]^{3+}$, which is quite stable; for the exchange with solvent water

$$[Rh(H_2O)_6]^{3+} + H_2^{18}O \rightarrow [Rh(H_2O)_5 H_2^{18}O]^{3+} + H_2O$$

the rate is given by $k_1[Rh(H_2O)_6]^{3+} + k_2[RhOH(H_2O)_5]^{2+}$. The ion is acidic; thus for the equilibrium

$$[Rh(H_2O)_6]^{3+} \rightleftharpoons [RhOH(H_2O)_5]^{2+} + H^+$$

$pK = 3.3$ at $25°$. The aquo-ion is subject to substitution by aqueous chloride ion

$$[Rh(H_2O)_6]^{3+} \xrightarrow{S_N 1} [RhCl(H_2O)_5]^{2+} \xrightarrow[\text{fast}]{S_N 1} \textit{trans-}[RhCl_2(H_2O)_4]^+$$

The *trans*-labilising effect of the first-entering chloride ion is noteworthy.

A number of other chloroaquo complexes are known. Thus dissolution of $[Rh(H_2O)_6](ClO_4)_3$ in hydrochloric acid gives a mixture of all species $[RhCl_n(H_2O)_{6-n}]^{(3-n)+}$, some of which have been isolated. Thus *cis*- and *trans*-$[RhCl_4(H_2O)_2]^-$ are obtained by ion exchange; while $K_3[RhCl_6]$, itself obtained by the action of concentrated aqueous potassium chloride on rhodium sesquioxide, is converted into $K_2[RhCl_5(H_2O)]$ on crystallisation from water.

The inertness of these octahedral d^6 complexes, which can be related to the considerable loss of crystal-field stabilisation energy on formation of a five-co-ordinate reaction intermediate, is considerable. Thus the orange-yellow trisacetylacetonate $[Rh\ acac_3]$, which can be resolved into its optical isomers chromatographically by passage through a long column of d-lactose, shows optical stability even on nitration and bromination.

Rhodium(III) iodocarbonyl complexes are known, being prepared by reaction with HI and formic acid (a convenient source of CO); with chloride and bromide, analogous species are formed

$$RhCl_3 \xrightarrow[\text{reflux}]{HX/HCO_2H} [Rh(CO)X_5]^{2-} \qquad (X = Cl, Br, I)$$

Ammine complexes There are a number of ammines $[RhX_n(NH_3)_{6-n}]^{(3-n)+}$ where X is an acid group such as Cl or CH_3COO. It would be tedious to attempt an enumeration of them, but a preparative entry into the series is possible by boiling $RhCl_3$ with ammonium chloride and ammonium carbonate, which gives a separable mixture of $[RhCl(NH_3)_5]Cl_2$ and *trans*-$[RhCl_2(NH_3)_4]Cl$. The former may be converted into the colourless $[Rh(NH_3)_6]Cl_3$ by hot aqueous ammonia, but more interestingly the reaction

$$[RhCl(NH_3)_5]Cl_2 \xrightarrow[(NH_4)_2SO_4]{Zn\ dust} [RhH(NH_3)_5]SO_4$$

affords a rather uncommon example of a complex hydride where the metal is in a moderately high oxidation state and there is no stabilisation by π-bonding ligands. The complex hydride shows the usual characteristic physical properties, namely an infrared stretching frequency at 2079 cm^{-1} and a n.m.r. absorption at τ 27.1; $J[Rh-H] = 14$ Hz. In dilute aqueous ammonia, the species *trans*-$[RhH(H_2O)(NH_3)_4]^{2+}$ is formed ($\tau 32.0$; $J[Rh-H] = 25$ Hz), demonstrating the *trans*-labilising effect of the hydride ligand. *Trans* influence appears in the bond distances in $[RhH(NH_3)_5]^{2+}$; Rh$-$N = 205.8 pm (*cis*), 224 pm (*trans*). Both complexes add C_2H_4 and C_2F_4 reversibly to give species of the type

$[Rh(C_2H_5)(NH_3)_5]^{2+}$; the structure of this latter compound is known — the Rh–N bond *trans* to ethyl is 18 pm longer than the others, which illustrates the *trans* influence of alkyl groups (see section 7.6.1 for a discussion of *trans* influence). Another example of a complex containing a metal–carbon σ bond and a water molecule is $[Rh(OH_2)Cl(AsMe_3)_2\{C_4(CF_3)_4\}]$, which contains the system

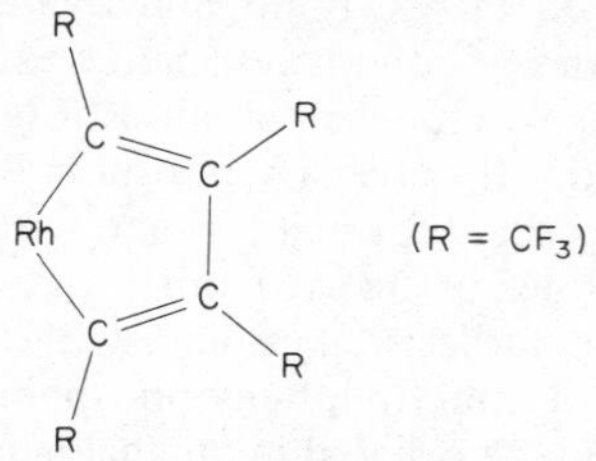

(R = CF₃)

The stability of these complexes containing what are normally thought of as 'mutually exclusive' ligands is attributable to the inertness of the low-spin d^6 configuration. The visible spectrum of $[RhH(NH_3)_5]^{2+}$ indicates that the position of H in the spectrochemical series is between NH_3 and H_2O. There is evidence, however, that in complexes such as $[PtHCl(PR_3)_2]$, H takes a much higher position near to CN^-.

In addition to the ammines, there are a number of complexes with pyridine as ligand, but no more than four pyridine groups may be attached. Thus the complexes $[Rh(C_5H_5N)_{4-n}Cl_{n+2}]^{(1-n)+}$ are known in some of their isomeric forms. Pyridine appears here to be a weaker ligand than ammonia because of the lack of hexapyridine and pentapyridine complexes and because of its ready replacement

$$trans\text{-}[Rh\,py_4Cl_2]^+ \xrightarrow{NH_3} trans\text{-}[Rh(NH_3)_4Cl_2]^+$$

This weakness has been attributed to steric hindrance.

The reaction of $[RhCl_6]^{3-}$ with pyridine to give $[Rh\,py_4Cl_2]^+$ is a very slow process; it is, however, catalysed by ethanol. This more rapid reaction appears to involve a rhodium(I)-hydride intermediate — an example of the ready formation of hydrido-complexes of rhodium (and iridium) in alcohols.

Complexes of sulphur ligands   Rhodium(III) and iridium(III) form a number of complexes with sulphur donors; thus dialkyl sulphides react with ethanolic $RhCl_3$ or $IrCl_6^{3-}$ to yield $MCl_3(SR_2)_3$. From n.m.r. data it appears that the *mer*-isomer is formed. Reaction of $RhCl_3$ in acid ethanol with acetylacetone and H_2S at 0° yields $[Rh(MeCS.CH.CS.Me)_3]$, while rhodium trichloride in aqueous NaCl reacts with ammonium pentasulphide to form $(NH_4)_3[Rh(S_5)_3]$. This last compound is thought to contain

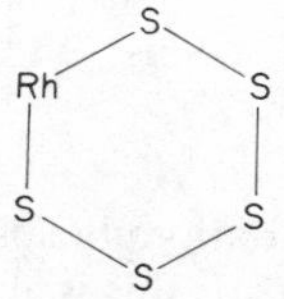

rings (the actual interbond angles would, of course, not be 120°).

[RhCl(PPh$_3$)$_3$] and NaS$_2$CNMe$_2$ react to form a number of complexes; when the dithiocarbamate is in excess, [Rh(S$_2$CNMe$_2$)$_3$] is obtained.

Phosphine complexes Many phosphine and arsine complexes of Rh(III) have been isolated. They are of particular interest because the presence of the AsR$_3$ or PR$_3$ ligands stabilises a metal–H or metal–CO bond, and several rhodium carbonyls and hydrides are known in this series of complexes. Some examples follow.

On treatment with triethylphosphine, alcoholic RhCl$_3$ gives *mer*-[RhCl$_3$(PEt$_3$)$_3$]. Analogous reactions occur with other tertiary phosphines or arsines. With Me$_2$PhP, both the *fac*- and *mer*-complexes are formed (in fact, with many phosphines, a little of the *fac*-complex is formed along with a majority of the *mer*-isomer). On reaction with LiBr in acetone, metathesis takes place with the formation of *mer*-[RhBr$_3$(PR$_3$)$_3$]. In the case of dimethylphenylphosphine, *mer*-RhCl$_2$Y(PMe$_2$Ph)$_3$ (Y = Br, I, N$_3$, SCN, NCO) may be isolated in an analogous way after a few minutes reaction; refluxing for some hours is required to form *mer*-[RhY$_3$(PMe$_2$Ph)$_3$]. *Mer*-[RhCl$_2$(NCS)(PMe$_2$Ph)$_3$] is probably N-bonded, but on heating at 150° it appears to transform into the S-bonded isomer. Figure 6.1 shows the methyl resonances in the ^{1}H n.m.r. spectrum of *mer*-[RhCl$_3$(PMe$_2$Ph)$_3$]; this topic is more fully discussed in section 6.9.1. This type of octahedral chloro-complex can be reduced to a complex hydride, thus

$$RhCl_3L_3 \xrightarrow{H_3PO_2} RhHCl_2L_3 \text{ or } RhH_2ClL_3$$

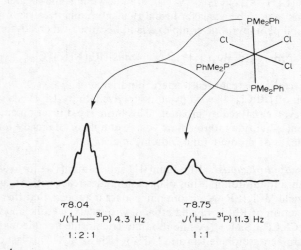

$\tau 8.04$ $\qquad$ $\tau 8.75$

$J(^1H{-\!\!-}^{31}P)$ 4.3 Hz $\qquad$ $J(^1H{-\!\!-}^{31}P)$ 11.3 Hz

1 : 2 : 1 $\qquad$ 1 : 1

Figure 6.1 The ^{1}H n.m.r. spectrum of *mer*-[RhCl$_3$(PMe$_2$Ph)$_3$] showing the methyl resonances only, to demonstrate the 'virtual' coupling of the *trans* phosphines (after G.M. Intille, *Inorg. Chem.*, **11** (1972), 695)

Reaction of phosphines such as triethylphosphine with RhCl$_3$ in 2:1 ratio yields the presumably dimeric Rh$_2$Cl$_6$(PEt$_3$)$_4$; this is also formed from appropriate ratios of RhCl$_3$ and *mer*-RhCl$_3$(PEt$_3$)$_3$.

As might be expected, $o\text{-}C_6H_4(AsMe_2)_2$ gives the tetra-substituted $[RhCl_2(diars)_2]Cl$. Complexes of the terdentate ligand $(o\text{-}Me_2AsC_6H_4)_2AsMe$ (TTAS) are readily formed with $RhCl_3.3H_2O$ and $HX(X = Cl, Br, I)$ in ethanol; *fac* and *mer* isomers are obtained for $[RhX_3(TTAS)]$ $(X = Cl, Br)$, but only the *fac* for $X = I$. A Rh—CO linkage is easily obtained

$$RhCl_3 + 2PPhEt_2 + CO \xrightarrow{\text{EtOH}} RhCl_3(CO)(PPhEt_2)_2$$

These complexes may often be alternatively prepared by oxidative-addition reactions of Rh(I) complexes (compare the rather analogous reactions of the d^8 Pt(II) complexes)

$$RhCl(CO)(PPh_3)_2 + Cl_2 \rightarrow RhCl_3(CO)(PPh_3)_2$$

$$RhCl(CO)(PBu_3)_2 + CH_3Cl \rightarrow Rh(CH_3)Cl_2CO(PBu_3)_2$$

$$RhCl(PPh_3)_3 + H_2 \rightarrow RhH_2Cl(PPh_3)_3$$

The last example, formally an oxidation by molecular hydrogen, is particularly striking. Reaction of hydrated rhodium trichloride with monoalkyl ditertiarybutyl-phosphines when heated for prolonged periods under reflux gives $[RhHCl_2(PBu^t_2R)_2]$ (for example, R = Me, Et, Pr^n). These complexes are thought to have a five-co-ordinate square pyramidal structure with mutually *trans* phosphines and chlorides, and an apical hydride.

They react with base in ethanol or methanol to form *trans*-$[RhCl(CO)(PBu^t_2R)_2]$, but in isopropanol containing base, a species which may be $[RhH_2Cl(PBu^t_2R)_2]$ is formed, which is a very active hydrogenation catalyst.

6.5.3 Rhodium(II)

These complexes may be dealt with very briefly. Those reported should be scrutinised rather carefully, since there have been some instances of incorrect assignment of what were in fact Rh(III) complexes to the unstable +2 state. However, *trans*-$[RhCl_2\{P(o\text{-}CH_3C_6H_4)_3\}_2]$ appears to be a genuine Rh(II) complex and is square planar; it is formed from direct reaction between tris(o-tolyl)phosphine and rhodium trichloride in ethanol at room temperature. The blue-green compound is isomorphous with the Pd(II) and Pt(II) analogues, has a magnetic moment of slightly over 2 B.M., and gives rise to an e.s.r. signal. Similar reaction of $RhCl_3$ with monoalkyl-ditertiarybutylphosphines in alcohol at room temperature yields $[RhCl_2(PBu^t_2R)_2]$ (R = Me, Et, Pr^n); evidence suggests that these are genuine rhodium(II) complexes, too. It is thought that these complexes may owe their stability to steric effects, in that the bulky ligands block the 'vacant' fifth and sixth co-ordination positions from attack. The diamagnetic rhodium acetate appears to be a rhodium(II) complex; it may be prepared from Rh_2O_3 with acetic acid or from rhodium trichloride and sodium acetate in ethanol and has a dimeric structure (see figure 6.2); donor molecules such as H_2O readily fill the vacant co-ordination positions. The rhodium-rhodium distance is 238.6 pm, which accounts for the observed diamagnetism since it is calculated that a Rh—Rh single-bond distance would be about 278 pm. In the dimeric dimethylglyoximate (DMG) complex, $Rh_2(DMG)_4(PPh_3)_2$, there are no

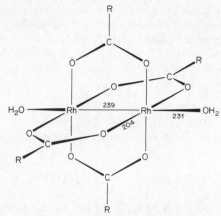

Figure 6.2 The dimeric structure adopted by $Rh_2(O_2CMe)_4 \cdot 2H_2O$ and other carboxylates (distances shown for rhodium acetate)

bridging ligands, the observed Rh—Rh distance being 291 pm (probably lengthened by a few pm owing to inter-ligand repulsions). In the 'intermediate' compound $Rh_2(OAc)_2(DMG)_2(PPh_3)_2$, the Rh—Rh distance is 261.6 pm, an 'intermediate' distance.

On reaction with HBF_4, rhodium acetate is protonated to give green cationic species, which may be described as Rh_2^{4+}; these react with PPh_3 to form $Rh^I(PPh_3)_3BF_4$, which in turn reacts with LiCl forming $[Rh^I(PPh_3)_3Cl]$; CO reacts to form $[Rh(CO)(PPh_3)_3]BF_4$. The red solutions containing 'Rh_2^{4+}' and PPh_3 are active catalysts; the optimum activity is reached with $Rh:PPh_3$ ratios of 1:2. Typical compounds hydrogenated include hex-l-ene, hex-l-yne, and hexa-1,5-diene.

The scarcity of rhodium(II) complexes must be caused by the great stability of Rh(I) and Rh(III); thus Rh(II) complexes will be subject to easy oxidation, reduction or disproportionation.

6.5.4 Rhodium(I)

This is an important oxidation state. Since rhodium(I) is a d^8 species, simple crystal-field considerations lead to the expectation of a square planar geometry, which expectation is realised in many instances. However, the +1 state ranks as a low oxidation state, requiring π bonding for stability in addition to covalent σ bonds. This would lead to the expectation that the eighteen-electron rule ought to be obeyed, with the consequences of five-co-ordination; sometimes this is observed also.

Phosphine complexes   Complexes with phosphines are readily obtained by refluxing $RhCl_3$ with excess of the phosphine in ethanolic solution. The best known example is the square planar $[RhCl(PPh_3)_3]$, one of the most interesting complexes known. The bromide and iodide complexes $[Rh(PPh_3)_3X]$ (X = Br, I) are prepared by refluxing $RhCl_3$ with triphenyl phosphine in ethanol for 5 minutes, and adding the lithium halide. They resemble the chloride in their properties, for example in being homogeneous hydrogenation catalysts. $[RhCl(EPh_3)_3]$ (E = Sb, As) are also known.

The main types of reaction of [RhCl(PPh₃)₃] are (a) the ready replacement of one PPh₃ ligand and (b) addition reactions giving a Rh(III) complex; both (a) and (b) may or may not be reversible, depending on the added ligands. Typical reactions are shown in figure 6.3.

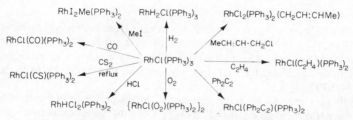

Figure 6.3 Some reactions of RhCl(PPh₃)₃

The reversible reaction with hydrogen is of some importance, since it leads to the catalytic *cis* hydrogenation of olefins. The mechanism is believed to involve the intermediate shown in figure 6.4, which explains the stereospecific nature of the addition.

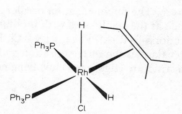

Figure 6.4 Suggested structure of the intermediate in olefin hydrogenation

The reaction with CO is not reversible. Even the CS molecule is stabilised by this versatile complex.

The mechanism of the hydrogenation process has been the subject of some controversy. It has been clearly shown from ^{31}P and ^{1}H n.m.r. studies that dissociation of [RhCl(PPh₃)₃] under the experimental conditions used is negligible (as proved for the iridium analogue); however, the adduct *cis*-[RhClH₂(PPh₃)₃], formed by oxidative addition of dihydrogen, does readily dissociate, so that the mechanism may be

$$RhCl(PPh_3)_3 \xrightarrow{H_2} RhH_2Cl(PPh_3)_3 \xrightarrow{-PPh_3} RhH_2Cl(PPh_3)_2$$

$$\updownarrow R_2CCR_2$$

$$\left\{ \begin{array}{l} RhCl(PPh_3)_3 \\ R_2CHCHR_2 \end{array} \right. \longleftarrow H_2RhCl(PPh_3)_2(R_2CCR_2)$$

[RhCl(PPh$_3$)$_3$], whose distorted planar structure minimises nonbonded interactions, is an extremely versatile catalyst. It decarbonylates aldehydes as follows

$$RhCl(PPh_3)_3 + RCH_2.CH_2CHO \rightarrow RhCl(CO)(PPh_3)_2 + PPh_3 + \begin{cases} RCH_2CH_3 \\ or\ RCH{=}CH_2 + H_2 \end{cases}$$

It also reacts with methylmagnesium halides to give [RhMe(PPh$_3$)$_3$], which on heating forms [Rh(PPh$_3$)$_2$(Ph$_2$P.C$_6$H$_4$)] in which there is a σ bond to an *ortho* carbon. With excess iodomethane, it forms [RhI$_2$Me(PPh$_3$)$_2$], which has a square pyramidal structure with an apical methyl and *trans* phosphines; this reacts with CS$_2$ to form [RhI$_2$(CS$_2$Me)(PPh$_3$)$_2$].

[RhCl(PPh$_3$)$_3$] reacts with hydrazine at 80° in the presence of triphenylphosphine and dihydrogen to give [RhH(PPh$_3$)$_4$]; X-ray investigation shows that the RhP$_4$ core is tetrahedral, so the hydrogen presumably lies on a threefold axis.

RhCl(CS)(PPh$_3$)$_2$ involves square planar co-ordination of rhodium, the phosphines being *trans,* as steric considerations would suggest. Rh—C—S back-bonding contributes to a short Rh—C distance (178.7 pm) compared with 186 pm in the CO analogue. The dioxygen adduct of [RhCl(PPh$_3$)$_3$] is in fact dimeric, as shown by crystallographic analysis; in [RhCl(O$_2$)(PPh$_3$)$_2$]$_2$ the Rh—Rh distance is 334 pm implying very limited interaction. The dioxygen molecules are asymmetrically bonded, in a manner that implies they bond π to one rhodium and σ to the other.

Allene and alkene complexes. The polymerisation of allenes in the presence of metal–allene complexes, such as the stable ones with rhodium(I), is of some industrial importance. Thus reaction of [RhX(PPh$_3$)$_3$] (X = Cl, Br, I) with allene gives [RhX(PPh$_3$)$_2$C$_3$H$_4$], which catalyses the tetramerisation and polymerisation of allene. This complex has a structure (figure 6.5) involving nearly square planar co-ordination.

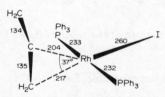

Figure 6.5 The co-ordination geometry in the 'square planar' allene complex RhI(PPh$_3$)$_2$(C$_3$H$_4$), showing the bonding of one double bond to the metal (after C.A. Reilly and H. Thyret, *J. Am. chem. Soc.,* **89** (1967), 5144)

Alkenes also bind strongly to rhodium; an interesting comparison has been made of [Rh(acac)(C$_2$H$_4$)$_2$] and [Rh(acac)(C$_2$H$_4$)(C$_2$F$_4$)] both being square planar with the alkenes perpendicular to the RhO$_2$ plane. It is notable that the tetrafluoroethylene is bound much more closely (Rh—C 201 ± 1 pm) than the ethylene (Rh—C 219 pm), implying that the bonding is dominated by π-acceptor rather than σ-donor effects. A more complicated example of rhodium–alkene co-ordination is found in bromo(tri-*o*-vinylphenyl)phosphine rhodium(I), where the rhodium is at the centre of a trigonal bipyramid involving co-ordination to phosphorus, bromine

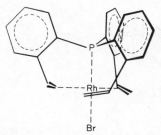

Figure 6.6 Bromo (tri-*o*-vinylphenyl) phosphine rhodium (I) (redrawn from C. Nave and M.R. Truter, *Chem. Commun.* (1971), 1253)

and three olefinic double bonds (see figure 6.6). Reaction of $[Rh(PPh_3)_3Cl]$ with allyl magnesium chloride affords a π-allyl complex

$$[RhCl(PPh_3)_3] + C_3H_5MgCl \rightarrow [Rh(PPh_3)_2(\pi\text{-}C_3H_5)] + PPh_3 + MgCl_2$$

The n.m.r. spectra of this and related compounds provide confirmation of their structures; the rhodium couples only to one hydrogen, on the central carbon of the π-allyl group, implying a strong σ interaction with that carbon. The structure of this compound is thought to be that shown in figure 6.7, where the allylic plane is inclined at an angle rather greater than 90° to the RhP_2 plane.

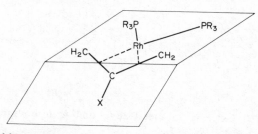

Figure 6.7 Probable geometry of the π-allyl complexes $Rh(PR_3)_2(H_2C.CH.CH_2)$

Carbonyl halides and other complexes An important rhodium(I) compound is rhodium carbonyl chloride, $Rh_2Cl_2(CO)_4$. This can be made by the action of CO on $RhCl_3$ at 105°. The structure (see figure 6.8) suggests that Rh—Rh bonding may occur, but the extent of any such bonding is difficult to assess. It is a reactive compound and some principal conversions are shown in figure 6.9. One reaction of this compound that is of especial interest occurs with allyl chloride, when a dimeric π-allyl complex is formed

$$Rh_2Cl_2(CO)_4 \xrightarrow{C_3H_5Cl} Rh_2(\pi\text{-}C_3H_5)_4Cl_2$$

X-ray examination shows it to be a chlorine-bridged dimer, the salient feature being that the two Rh—CH_2 distances of each π-allyl unit differ — they are 212 and 225 pm. This points to an asymmetric π system that is intermediate between the

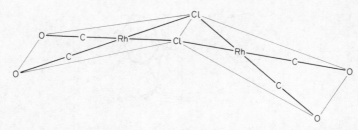

Figure 6.8 The dimeric structure of $[Rh(CO)_2Cl]_2$. The cause of the 'bent' bridge is unknown

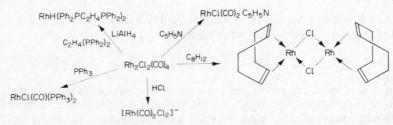

Figure 6.9 Some reactions of rhodium carbonyl chloride

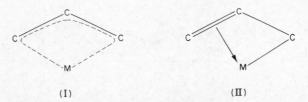

Figure 6.10 'Extremes' of metal–allyl bonding in rhodium allyls

two extremes indicated in figure 6.10. It is notable that in the asymmetric π allyls that are chlorine-bridged, the 'short' metal–carbon distances are *trans* to chlorine; this may be an example of *trans* influence determining the asymmetry.

On treatment of this compound with allyl magnesium chloride, $Rh(C_3H_5)_3$ is formed, which, from n.m.r. data, appears to be a symmetrical π allyl at low temperatures, but on warming exchange between nonequivalent sites occurs.

A large number of other π allyls are known; their n.m.r. behaviour is frequently complicated.

The other halide analogues of *trans*-$[RhCl(CO)(PPh_3)_2]$ may be prepared by metathesis with LiX in acetone. These readily undergo oxidative addition with, for example, MeI, the initial addition being followed by methyl migration to yield $RhXI(COMe)(PPh_3)_2$. The compounds *trans*-$[RhX(CO)(PPh_3)_2]$ are catalysts for the hydroformylation of alkenes, the probable intermediate being $RhH(CO)_2(PPh_3)_2$. Another example of the reactions of the bis-phosphine mono-carbonylchloride is with carbon disulphide, to form $[RhCl(CS)(PPh_3)_2]$, but the most important reaction is with, for example, hydrazine in alcohol, in the presence of triphenylphosphine, when $[RhH(CO)(PPh_3)_3]$ is produced. This is an extremely versatile catalyst for hydrogenation, hydroformylation and the isomerisation of alkenes. The stereoselectivity shown in the hydrogenation of alk-1-enes may be attributed to steric effects imposed by the bulky phosphines; in the hydrido carbonyl complex the rhodium atom is 36 pm above the P_3 plane and the Rh–H distance is 160 pm.

It seems that the active catalytic species is $RhH(CO)(PPh_3)_2$; initially the alkene co-ordinates to form an alkyl group and this may either be hydrogenated further to form an alkane or be transformed to an *acyl* group, which can then be converted into an aldehyde. Another example of the stereospecificity of this catalyst is that when alk-1-enes are used in the hydroformylation, the aldehyde product is 95 per cent straight chain.

Conjugated dienes and allene, by contrast, form stable adducts with both the rhodium and iridium systems. The π-allyl $Rh(\pi\text{-}C_3H_5)(CO)(PPh_3)_2$, for example, shows dynamic allylic behaviour, as evidenced by its n.m.r. spectrum.

Returning briefly to complexes of the type *trans*-$[RhCl(CO)(PR_3)_2]$, recent study has been made of the effects of bulky phosphines. ^{31}P n.m.r. evidence shows that a phosphine such as PBu^t_2R causes restricted rotation about the metal–phosphorus bond and thus the molecules tend to be 'locked' into specific configurations. Thus three conformers of *trans*-$[RhCl(CO)(PBu^t_2Et)_2]$ can be identified, two giving A_2X and the other ABX patterns (see figure 6.11). The ABX pattern is of course obtained from the compound with nonequivalent ^{31}P nuclei.

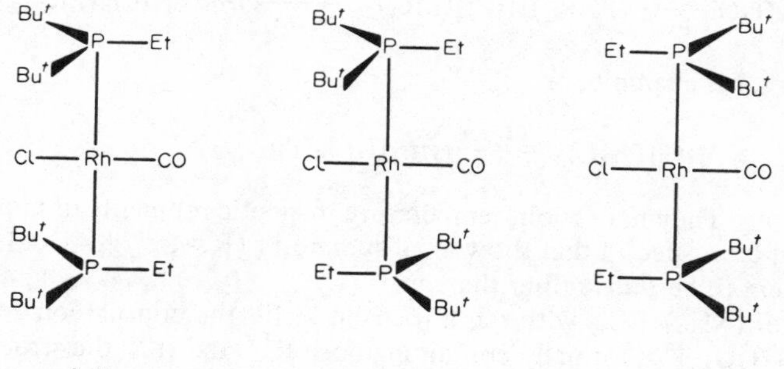

Figure 6.11 Conformers of the complex $Rh(CO)Cl(PBu^t_2Et)_2$ identified by low-temperature ^{31}P n.m.r.

6.5.5 Rhodium(0)

Electrochemical reduction of $[RhCl(R_3P)_3]$ in MeCN yields $Rh(R_3P)_4$ ($R_3P =$ Ph_3P, Ph_2MeP); the former is diamagnetic and thus probably a dimer, but the latter is paramagnetic (1.16 B.M.) and may be a mixture of forms.

6.6 Complexes of Iridium

These compounds present a slightly greater variety than those of rhodium, mainly because the +4 oxidation state is rather more stable and better represented than in the case of the lighter metal. However, nearly all the co-ordination chemistry of iridium is divided between the oxidation states of +3 (low-spin d^6) and +1 (π-bonded square planar or five-co-ordinated d^8). An important exception is the group of binary carbonyls, which formally contain Ir(0).

6.6.1 Iridium(IV)

Fairly stable halide complexes are formed with iridium(IV). Thus $Na_2[IrCl_6]$ can be obtained by chlorine oxidation of $Na_3[IrCl_6]$ and is sufficiently stable to give a free acid $(H_3O)_2IrCl_6.4H_2O$. It undergoes slow stepwise substitution by bromide ion in aqueous solution to give species from $[IrBrCl_5]^{2-}$ to $[IrBr_6]^{2-}$. Substitution by pyridine is possible to give $[IrCl_4(C_5H_5N)_2]$. As may be expected, $[IrF_6]^{2-}$ salts are known; the ion is sufficiently stable to give a nitrosonium salt $(NO^+)_2[IrF_6]^{2-}$. The hexahalo complexes are of considerable historical interest since the first experiment showing unequivocally the delocalisation of electrons from metallic orbitals onto ligands was performed on $IrCl_6^{2-}$ by e.s.r. in the early 1950s. These systems have been much studied by e.s.r. because of their simplicity; most recent results indicate that the unpaired electron in these $5d^5$ ions spends only about half its time in 5d orbitals. There is good evidence for the existence of fac- and mer- $[IrCl_3(H_2O)_3]^{n+}$ (n = 1 and 0) and cis- and trans-$[IrCl_4(H_2O)_2]^{n-}$ (n = 0,1) ions. The iridium(IV) species are, not surprisingly, strong oxidising agents.

Some iridium(IV) phosphine complexes are known; they may be prepared as follows

$$IrCl_3.xH_2O \xrightarrow[PR_3]{HCl} (PHR_3)[IrCl_4(PR_3)_2] \xrightarrow[CH_2Cl_2]{Cl_2} trans\text{-}[IrCl_4(PR_3)_2]$$

where R = Pr^n, for example; or

$$IrCl(Ph_3E)_3 \xrightarrow{Cl_2} trans\text{-}[IrCl_4(Ph_3E)_2]$$

where E = P, As. They have room-temperature magnetic moments of approximately 1.7 B.M., and e.s.r. spectra that show axial symmetry (R = Pr^n, $g_\perp$ = 2.43, $g_\parallel$ = 0.80), indicating them to be trans rather than cis.

Reaction of $(NH_4)_2IrCl_6$ with sulphuric acid yields the unusual ion $[Ir_3N(SO_4)_6(H_2O)_3]^{4-}$, formally containing both Ir^{IV} and Ir^{III}; the structure of the core (shown in a simplified form in figure 6.12) has bidentate bridging sulphate groups and terminal water molecules. The $(H_2O)_3Ir_3N$ unit is virtually planar. The related compound $K_{10}[Ir_3O(SO_4)_9].3H_2O$ can be prepared by boiling Na_2IrCl_6 with sulphuric acid and potassium sulphate.

6.6.2 Iridium(III)

The most stable oxidation state of the metal is iridium(III), which provides a wide variety of complexes, mainly of orthodox types. A plain aquo-ion has not been isolated, though aquation of $[IrCl_6]^{3-}$ does proceed as far as $[IrCl_2(H_2O)_4]^+$. Salts of the $[IrCl_6]^{3-}$ ion may be obtained by mild reduction of the corresponding $[IrCl_6]^{2-}$ salt. $K_3[IrBr_6]$ and $K_3[IrI_6]$ are both known, but the $[IrF_6]^{3-}$ ion is not

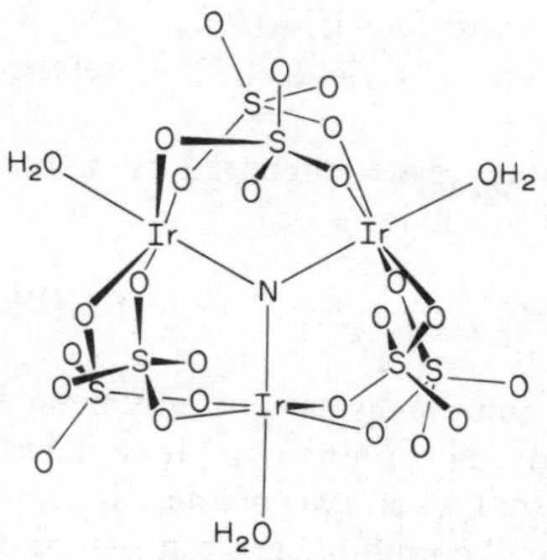

Figure 6.12 The polynuclear $[Ir_3N(SO_4)_6(H_2O)_3]^{4-}$ ion

known as yet. All the properties so far mentioned form an interesting contrast with Co(III), but the ammines of Ir(III) are rather similar to the well-known Co(III) compounds. Thus yellow $[Ir(NH_3)_5Cl]Cl_2$ is obtained on treatment of $IrCl_3$ with ammonia solution. It may be converted into the hexammine by hot aqueous ammonia, or into the monoaquo-complex $[Ir(NH_3)_5H_2O]^{3+}$ by aqueous alkali. The water in the last complex may be replaced by anions, such as NCS^-, NO_2^-, N_3. or CH_3COO^-, and the existence of ligand isomerism has been established in the first two cases.

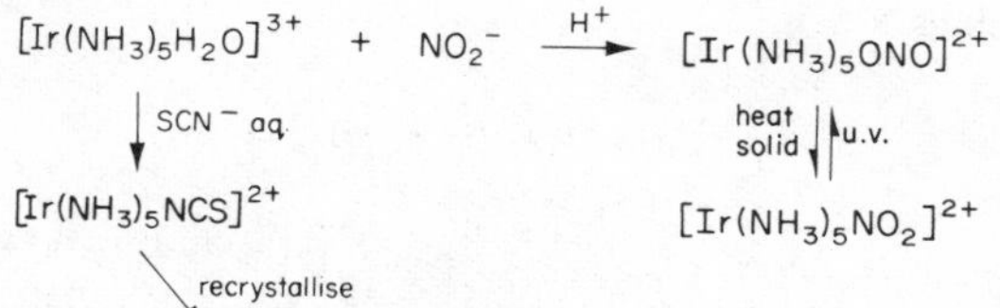

Ir(III) gives several pyridine complexes of an orthodox type, for example cis and trans-$K[IrCl_4py_2]$ are obtained by the action of pyridine on $K_3[IrCl_6]$. Other straightforward species are the orange tris(oxalato)iridate $K_3[Ir(C_2O_4)_3]$, made by the action of oxalate ion on $K_3[IrCl_6]$; the tris(acetylacetonate), made from Ir_2O_3 and acetylacetone; and $[Ir\,en_3]Cl_3$ made from ethylenediamine and $[IrCl_6]^{3-}$.

Phosphine complexes Iridium (III) has the usual affinity for phosphines, arsines and stibines possessed by heavy metals of the later groups in medium oxidation states. Thus $Na_3[IrCl_6]$, when treated with triethylphosphine in alcoholic solution, gives mer-$[IrCl_3(PEt_3)_3]$, while the diarsine o-$C_6H_4(AsMe_2)_2$ can occupy four co-ordination positions, giving $[IrCl_2(diars)_2]^+$. A rather less commonplace characteristic of iridium (III) is its ability to form stable hydrides and carbonyls, the +3 oxidation state being usually considered rather high for this type of compound to be stable, since for stability in carbonyls electron donation from the metal is necessary. One interesting reaction discovered by accident is a mer→fac conversion; it was found that a benzene solution of mer-$[IrCl_3(PEt_2Ph)_3]$ placed near a fluorescent light had been 74 per cent converted into the fac isomer in a week. This reaction has been found to be general for most mer-$IrX_3(PR_3)_3$ (X = Cl, Br) systems, usually going to approximately 100 per cent yield. In the presence of excess Y (SCN, N_3, NCO), fac-$IrY_3(PR_3)$ is formed.

Generally the iodide complexes are prepared by metathesis with LiI in acetone, while a convenient route for bromides is

$$Na_2IrBr_6 \xrightarrow[\text{MeEtCO}]{PR_3} mer\text{-}[IrBr_3(PR_3)_3]$$

The configurations of these complexes have usually been assigned with the aid of n.m.r. studies; dimethylphenyl phosphine complexes have been especially studied since it frequently happens that when two of these ligands are $trans$ in a complex, there is equally strong, 'virtual' coupling between the methyl protons and $both$ phosphorus atoms so that the ^{1}H n.m.r. resonance due to the methyl protons appears as a triplet (1:2:1). Thus mer-$[IrCl_3(PMe_2Ph)_3]$ gives methyl resonances at τ8.09 (intensity 2; 1:2:1 triplet, J[P–H] = 4.5 Hz) and at τ 8.75 (intensity 1; doublet, J[P–H] = 11.3 Hz). The former resonance comes from the methyl protons of the mutually $trans$ phosphines, and the latter from the third phosphine.

In a similar way $[IrCl_4(PMe_2Ph)_2]^-$ gives a triplet (τ 8.05, J[P–H] = 4.0 Hz) showing it to be $trans$. Virtual coupling does not $invariably$ $occur$ (it depends on J[P–P] being very large) but can be a very convenient aid in assigning configurations. The reactivity of the anion $trans$ to phosphine in mer-$[MX_3(ER_3)_3]$ (E = P, As) is generally greater than that of the other anionic ligands. Thus, reaction of mer-$[IrCl_3(ER_3)_3]$ with silver nitrate gives $[IrCl_2(NO_3)(ER_3)_3]$, which undergoes further reactions such as

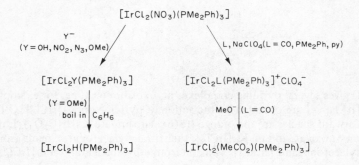

$Carbonyl$ and $other$ $complexes$ $with$ Ir–C $bonds$　　　Iridium (III) also forms σ bonds to carbon as, for example

$$mer\text{-}[IrCl_3(PR_3)_3] \xrightarrow[\text{THF}]{MeMgCl} fac\text{-}[IrMe_3(PR_3)_3]$$

where PR_3 = PEt_3, PEt_2Ph, for example, or, in a more exotic fashion

$$[IrHCl_2\{P(OPh)_3\}_3] \Big|\xrightarrow[\text{boil}]{\text{decalin}} [IrCl\{P(OPh)_3\}\{P(OPh)_2(o\text{-}OC_6H_4)\}_2]$$

This compound has an iridium–carbon σ bond as shown in figure 6.13. It provides an excellent example of σ-bond formation to an *ortho* carbon atom of a phenyl group.

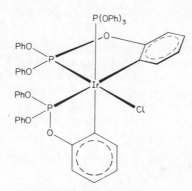

Figure 6.13 The complex $[IrCl\{P(OPh)_3\}\{P(OPh)_2(o\text{-}OC_6H_4)\}_2]$

Carbon monoxide displaces triphenylarsine from $[IrI_3(AsPh_3)_3]$, giving $[IrI_3(CO)(AsPh_3)_2]$, while the combined action of phosphine and CO on $[IrCl_6]^{3-}$ gives $[IrCl_3(CO)(PR_3)_2]$

$$H_3IrCl_6 + CO + PEt_3 \xrightarrow[\text{Cl}^-]{\text{MeOCH}_2\text{CH}_2\text{OH}} \textit{cis-} \text{ and } \textit{trans-}[IrCl_3(CO)(PEt_3)_2]$$

One CO ligand molecule is usually the maximum number that can be introduced.

Hydride complexes Iridium(III) continues and extends the tendency of rhodium(III) to form complex hydrides; these are octahedral, the remaining ligands usually being PR_3, AsR_3, Cl or CO. They may be made either conventionally, using a reducing agent such as $LiAlH_4$, or by hydride abstraction from the solvent

$$IrCl_3 + AsPh_3 \xrightarrow{\text{EtOH}} [IrCl_2H(AsPh_3)_3] + HCl + CH_3CHO$$

The last reaction exemplifies the inadequacy of ordinary ideas of oxidation and reduction in organometallic chemistry; the ethanol has undoubtedly been oxidised by $IrCl_3$ to acetaldehyde, but the iridium has remained in the same formal oxidation state as a conventional octahedral Ir(III) complex.

One difference between *fac-* and *mer-*$[IrCl_3(PR_3)_3]$ is that on boiling the *fac* isomer in alcohols for a few hours, $IrHCl_2(PR_3)_3$ is formed, while the *mer* isomer does not react. Far i.r. data for a large number of iridium(III) phosphine complexes containing halide and hydride ligands show interestingly a marked dependence of ν(Ir–Cl) on the *trans* ligand. Thus when Cl is *trans*, ν(Ir–Cl) occurs at about 300–320 cm^{-1}; when phosphine is *trans*, it is seen at about 260–280 cm^{-1}, and for *trans* hydride at approximately 250 cm^{-1}. Similar changes have been noted in platinum(II) complexes, and are likewise ascribed to the *trans* effect.

When iridium phosphine complexes are heated in alcohols, hydride complexes are frequently formed. One novel compound is $IrH_5(PEt_2Ph)_2$, prepared as follows

$$Ir(PEt_2Ph)_3Cl_3 \xrightarrow{LiBH_4} IrH_5(PEt_2Ph)_2$$

Analytical evidence is unreliable in determining the number of hydrogens in such hydrides; however, integration of the ^1H n.m.r signals gives $CH_2:CH_3:H = 8:12:4.8$, while the ^{31}P n.m.r. shows a sextet due to coupling with five equivalent hydrogens. It undergoes a number of reactions in which hydride is lost.

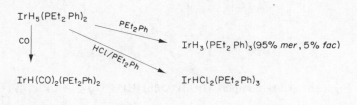

A related compound, $IrH_3(Ph_3P)_3$, catalyses the hydrogenation of aldehydes. $[IrHCl_2(PEt_2Ph)_3]$ is also prepared by reaction of *mer*-$[IrCl_3(PEt_2Ph)_3]$ with alcoholic KOH, the chlorine *trans* to phosphorus being displaced. The phosphine *trans* to the hydride is labile to substitution by, for example, alkylphosphines — another case of a *trans* effect at work.

A convenient pathway to $[IrHCl_2(PBu^t_2R)_2]$ (R = Me, Et) is

$$[IrCl_6]^{2-} \xrightarrow{PBu^t_2R/Pr^iOH} [IrHCl_2(PBu^t_2R)_2]$$

N.M.R. and far-i.r. spectra imply that the chlorines and phosphines are *trans,* while the high-field ^1H n.m.r. line is at $\tau \approx 60$, considerably outside the 'normal' range for hydrides bound to metals; this compound forms (presumably) six-co-ordinate adducts with dimethylsulphide or pyridine where the high-field n.m.r. line has a 'normal' τ value of approximately 25. The anomalous value in the former compounds may be due to the absence of a ligand *trans* to hydride, giving a system with a strong pseudo-tetragonal distortion, and concomitant low-lying excited states.

Hydrido-carbonyl-phosphine complexes also are obtained readily

$$IrH_3(PPh_3)_3 \xrightarrow{CO} IrH_3(CO)(PPh_3)_2$$

$$IrCl(CO)(PPh_3)_2 \xrightarrow{H_2} IrH_2Cl(CO)(PPh_3)_2$$

6.6.3 Iridium(II)

There are few complexes of this oxidation state. However, the reaction of o-$Bu^t_2P.C_6H_4.OMe$ with CO and $IrCl_6^{3-}$ in boiling propan-2-ol gives first an Ir(I) carbonyl, which is rapidly oxidised by air to the square planar, unequivocally Ir(II), complex *trans*-$[(o$-$Bu^t_2P.C_6H_4.O)_2Ir]$ which has $\mu = 1.76$ B.M.

6.6.4 Iridium(I)

Iridium(I) is a very versatile species. Being a low oxidation state of a third transition-series metal it should be restricted to covalent and π-bonding ligands and its co-ordination number is not expected to be greater than five, since the d^8 configuration leaves only five unoccupied orbitals for formation of σ bonds. It should thus adopt either a five co-ordinate structure or the four-co-ordinate square planar configuration, which although co-ordinatively unsaturated is specially stable for the d^8 configuration. In practice, the ligands that occur most frequently in the chemistry of this oxidation state are PPh_3 and its relatives; Cl, Br, I; CO and H. However, a number of unusual ligands, for example CS, also occur co-ordinated with Ir(I). The geometry adopted in practice is most often the square planar configuration, but many examples of five co-ordination are also found.

The square planar complexes are of the types $[IrL_4]^+$, IrL_3X or $[IrL_2X_2]^-$. Thus the action of CO at 150° on $IrCl_3$, $IrBr_3$ or IrI_3 gives the uncharged carbonyl halides $[IrX(CO)_3]$. The most versatile Ir(I) complex, $[IrCl(CO)(PPh_3)_2]$, may be obtained from the chloride

$$[IrCl(CO)_3] + 2PPh_3 \rightarrow [IrCl(CO)(PPh_3)_2] + 2CO$$

It is more conveniently obtained by the interesting reaction

$$IrCl_3 + Cl^- + PPh_3 \xrightarrow{\text{MeOC}_2\text{H}_4\text{OH}} IrCl(CO)(PPh_3)_2$$

Note that the CO ligand is obtained from the solvent; the possibility of the unexpected appearance of ligands such as CO, H, O or N has always to be borne in mind when investigating this type of co-ordination chemistry. The complex has a *trans* configuration (dipole moment = 3.90 Debye units), and its reactions provide a good illustration of Ir(I) chemistry. It reacts with many simple molecules, often by addition to give an Ir(III) complex.

Reactions of IrCl(CO)(PPh₃)₂   Table 6.2 gives some examples of conversion of $IrCl(CO)(PPh_3)_2$ into Ir(III) complexes.

Table 6.2

Reactant	Product
H_2	$IrH_2Cl(CO)(PPh_3)_2$
HCl	$IrHCl_2(CO)(PPh_3)_2$
Cl_2	$IrCl_3(CO)(PPh_3)_2$
CH_3I	$Ir(CH_3)ClI(CO)(PPh_3)_2$
$HgCl_2$	$IrCl_2(HgCl)(CO)(PPh_3)_2$
$SiHCl_3$	$IrHCl(SiCl_3)(CO)(PPh_3)_2$
$SiEtHCl_2$	$IrEtCl(SiHCl_2)(CO)(PPh_3)_2$
C_3H_5Br	$Ir(\sigma\text{-}C_3H_5)(CO)Cl(PPh_3)_2Br$

Often, however, the Ir(I) state is retained, but as a five-co-ordinated species, as in most of the examples of addition reactions of $IrCl(CO)(PPh_3)_2$ given in table 6.3.

Table 6.3

Reactant	Product
CO	$IrCl(CO)_2(PPh_3)_2$
O_2	$IrCl(CO)(O_2)(PPh_3)_2$
C_2F_4	$IrCl(C_2F_4)(CO)(PPh_3)_2$
SO_2	$IrCl(CO)(SO_2)(PPh_3)_2$
N_2H_4	$IrH(CO)(PPh_3)_3$
$PhCON_3$	$IrN_2Cl(PPh_3)_2$

Some of these complexes call for special comment. $[IrH(CO)(PPh_3)_3]$ is a good catalyst for the homogeneous hydrogenation of ethylene to ethane. The reaction probably proceeds as follows

$$IrH(CO)(PPh_3)_3 \xrightarrow{C_2H_4} Ir(C_2H_5)(CO)(PPh_3)_3 \xrightarrow{H_2} IrH(CO)(PPh_3)_3 + C_2H_6$$

$[IrCl(CO)(SO_2)(PPh_3)_2]$ has a square pyramidal structure; the sulphur dioxide binds only via sulphur.

The structure of the oxygen adduct $IrCl(O_2)(CO)(PPh_3)_2$ has been determined by X-ray methods (see figure 6.14). It can be regarded as either an Ir(I) or an Ir(III) complex depending on whether we consider the O_2 group to be uncharged O_2 (121 pm

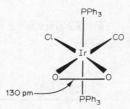

Figure 6.14 The co-ordination geometry in $IrCl(O_2)(CO)(PPh_3)_2$

separation when free) or O_2^{2-} (149 pm in peroxides). It may be considered as a five-co-ordinate adduct of Ir(I), with the O—O link weakened by d-π^* bonding as in ethylene complexes. One oxygen π^* orbital would be nonbinding, containing two electrons, and the other (in the O—Ir—O plane) would form a bonding M.O. with a filled Ir d orbital. The overall metal—O_2 bond strength is low, since the addition is reversible. Rather similar alternative formulations are possible in the case of the C_2F_4 adduct. The addition of SO_2 as an uncharged ligand is rather unusual and is worth notice, as is the addition of hydrogen, which takes place reversibly. It is also interesting that the Ir—C_2H_5 bond is formed on addition of $SiCl_2HEt$.

Reaction of $[IrCl(CO)(PPh_3)_2]$ with CS_2 and triphenylphosphine gives the dark violet ion $[Ir(CO)(\pi\text{-}CS_2)(PPh_3)_3]^+$. A thiocarbonyl analogue of $[IrCl(CO)(PPh_3)_2]$ has been obtained by treatment of $[IrCl(PPh_3)_3]$ with carbon disulphide, and also by

$$IrCl_3 \xrightarrow[DMF]{PPh_3} IrCl(CO)(PPh_3)_2 \xrightarrow{RCON_3} IrCl(N_2)(PPh_3)_2 \xrightarrow[PPh_3]{CS_2} IrCl(CS)(PPh_3)_2$$

Like the carbonyl analogue, it adds NO^+ thus

$$[IrCl(CS)(PPh_3)_2] \xrightarrow{NO^+BF_4^-} [IrCl(CS)(NO)(PPh_3)_2]^+ BF_4^-$$

With excess CO, $[Ir(CO)_2(CS)(PPh_3)_2]^+$ is formed.

Reaction with $PhCON_3$ appears to proceed via initial *cis* addition, followed by loss of PhNCO from the adduct to give the dinitrogen complex, from which the N_2 is readily displaced by acetylenes (giving $IrCl(RCCR)(PPh_3)_2$) or PPh_3 (giving $IrCl(PPh_3)_3$).

Considerable interest has been aroused by the question of whether the oxidative addition reaction to molecules such as $[IrCl(CO)(PPh_3)_2]$ ('Vaska's compound') proceeds via *cis* or *trans* addition, and, again, whether the initial addition product isomerises. In the solid state and in benzene solution, MeI adds *trans* to $[IrCl(CO)(PMe_2Ph)_2]$, but in the better-ionising solvent methanol a mixture of products is obtained. HX adds *cis* to $[IrCl(CO)(PPh_3)_2]$ in both the solid state and anhydrous benzene, but in moist solvents, a mixture of products results. Reactivity patterns of these complexes do depend on the nature of the phosphine: thus while $[IrCl(CO)(PPh_3)_2]$ will not add a third phosphine, the dimethylphenylphosphine analogue (much less sterically hindered) will. $[Ir(CO)(PMe_2Ph)_3]^+$ reacts with hydrogen rapidly to give $[IrH_2(CO)(PMe_2Ph)_3]^+$. It is notable that if the concentration of free phosphine is too great, the reactivity decreases, presumably owing to the formation of the inert $[Ir(CO)(PMe_2Ph)_4]^+$.

Iridium(I) hydrido-complexes The compound $[IrH(CO)_2(PPh_3)_2]$ has been widely studied; of special interest is its reaction with allenes, when it forms π allyls, which, like the rhodium systems (page 92) are dynamic. They react with CO to form a number of complexes, for example $IrR(CO)_2(PPh_3)_2$ and $Ir(COR)(CO)_2(PPh_3)_2$. The allyls here are σ in the solid state but π (dynamic) in solution. The behaviour in solution indicates that more than one species is present. Other carbonyl hydrides are prepared as follows

$$IrH_5(PEt_2Ph)_2 \xrightarrow{CO} [IrH(CO)_2(PEt_2Ph)_2]$$

$$IrCl(CO)(ER_3)_2 \xrightarrow[BH_4^-]{CO} [IrH(CO)_2(ER_3)_2] \qquad (ER_3 = PPh_3, PEtPh_2, AsPh_3)$$

The structure of the orthorhombic form of $[IrH(CO)_2(PPh_3)_2]$ has been determined (see figure 6.15). The molecule is a distorted trigonal bipyramid with one hydride and one phosphine in near-axial positions. The Ir—H distance is 164 pm. A mixture of forms exist in solution. The compound reacts with alk-1-ynes (RCCH) to yield $[Ir(C{\equiv}CR)H_2(CO)(PPh_3)_2]$, which eliminates hydrogen and forms $Ir(C{\equiv}CR)(CO)(PPh_3)_n$ ($n = 2, 3$); the bis complexes even in the solid state rapidly take up one mole of, for example, O_2, CO or SO_2.

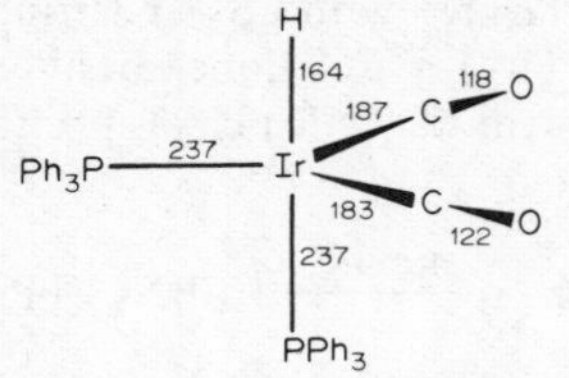

Figure 6.15 The structure of $IrH(CO)_2(PPh_3)_2$ (after M. Ciechanowitz, A.C. Skapski and P.G.H. Troughton, *Acta crystallogr.*, 22 (1969), S 172)

There is an Ir(I) hydride which is an isomeric form of Vaska's compound. Its structure involves internal metallation of a phosphine ligand

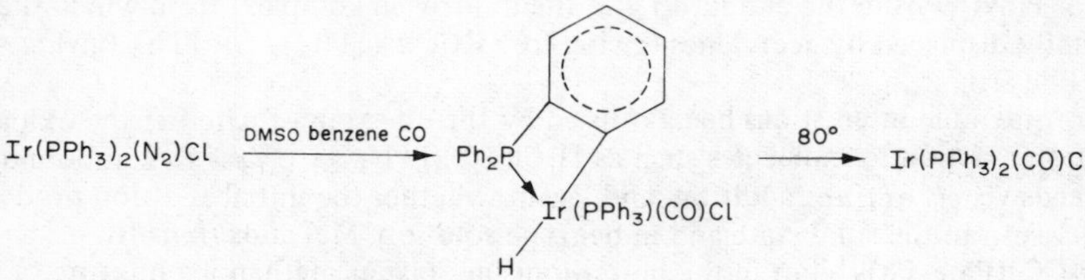

Other iridium(I) complexes The π bonding of the O_2 in dioxygen complexes is influenced by the other ligands; thus, in the series $[IrXO_2(CO)(PPh_3)_2]$ (X = Cl, I), O—O = 130 pm in the chloride and 151 pm in the iodide. The former compound is a reversible oxygen carrier, the latter is irreversible. The compound $[IrCl(CO)(O_2)(PPh_2Et)_2]$, also trigonal bipyramidal (if the midpoint of the O—O bond is regarded as a co-ordination site) has an O—O distance of 146 pm, and is a reversible oxygen carrier.

From the published crystallographic data, those adducts with O—O distances of between 130 and 145 pm are reversible carriers of oxygen while those with O—O distances of between 145 and 163 pm are not. While a simple qualitative view is that the longer (and the weaker) the O—O bond, the stronger is the metal—O_2 bond (both the halide and phosphine affect the O—O distance in the iridium(I) adducts) and that irreversible adducts have an O—O distance (145 pm upwards) compatible with 'peroxide' behaviour (O_2, 121 pm; O_2^-, 128 pm; O_2^{2-}, 149 pm), a more detailed explanation is not possible at present.

$[IrCl(CO)(O_2)(PPh_3)_2]$ has a prominent i.r. band at 850 cm^{-1}, assignable as ν(O—O); it undergoes two interesting reactions shown in the scheme below. The sulphate complex can be prepared in the two ways shown, although the reaction of $[IrCl(CO)(SO_2)(PPh_3)_2]$ with oxygen is slow.

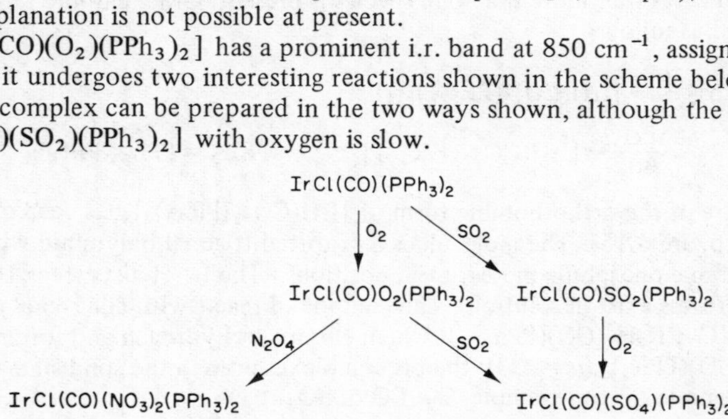

IrCl(PPh₃)₃; its reactions; other phosphine complexes $[IrCl(PPh_3)_3]$, analogous to the well-known catalyst $RhCl(PPh_3)_3$, is conveniently prepared from the cyclo-octene or cyclo-octa-1,5-diene complexes $IrCl(C_8H_{14})_2$ or $\{IrCl(C_8H_{12})\}_2$, by re-fluxing with Ph_3P. It may *not* be prepared by refluxing $IrCl_3$ and a large excess of Ph_3P in alcohol, the method used for the Rh analogue, since hydride complexes, $[IrHCl_2(PPh_3)_3]$ and $[IrH_2Cl(PPh_3)_3]$, are formed, again demonstrating the affinity of iridium for hydride ligands. Unlike the rhodium complex, $[IrCl(PPh_3)_3]$ is not a catalyst; this may be because the Ir–H bond is sufficiently strong for the hydrogen transfer to a co-ordinated olefin not to occur.

Some reactions of $[IrCl(PPh_3)_3]$ and related compounds are summarised in figure 6.16.

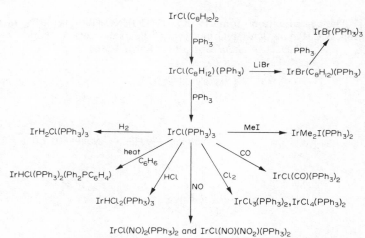

Figure 6.16 Reactions of some iridium (I) complexes

Treatment of $[IrCl(PPh_3)_3]$ with oxygen gives the dioxygen complex $[IrCl(O_2)(PPh_3)_2]$. This undergoes further reactions as follows

The cationic complex $[Ir(Ph_2PC_2H_4PPh_2)_2]^+$ may be obtained from $IrCl(CO)_3$ by complete substitution with the diphosphine. This complex will also add an additional ligand giving $[Ir(diphos)_2 L]^+$ where the added ligand L can be CO, PF_3, SO_2 or O_2.

Iridium nitrosyl complexes   The structures of a number of iridium nitrosyl complexes have been determined recently; indeed, the report in 1968 that

$[IrCl(NO)(CO)(PPh_3)_2]^+$ contained a 'bent' Ir–N–O bond was the first time that this phenomenom had been unequivocally authenticated by three-dimensional X-ray analysis (see figure 6.17). Preparations for this and other compounds are shown in figure 6.18. $[IrCl(CO)(NO)(PPh_3)_2]^+$ has i.r. absorptions at 2050 and

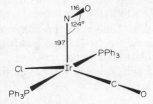

Figure 6.17 The co-ordination geometry in $[IrCl(CO)(NO)(PPh_3)_2]^+BF_4^-$ (after D.J. Hodgson, N.C. Payne, J.A. McGinnety, R.G. Pearson and J.A. Ibers, *J. Am. chem. Soc.*, **90** (1968), 4486)

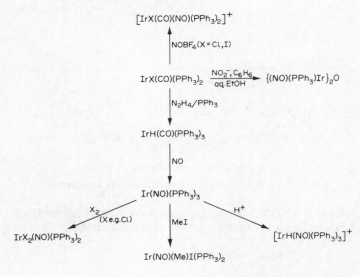

Figure 6.18 Preparation of some iridium nitrosyl complexes

1680 cm^{-1}, due to ν(C–O) and ν(N–O) respectively, the latter being in the region regarded as characteristic for NO$^+$. The iodide analogue is isostructural, indicating that, unlike the dioxygen adducts, the bonding is σ in character.

$[IrCl_2(NO)(PPh_3)_2]$ and *trans*-$[IrI(CH_3)(NO)(PPh_3)_2]^+$ are both square pyramidal, with Ir–N–O angles of 123° and 120°. $[Ir(NO)_2(PPh_3)_2]^+ClO_4^-$ and $[IrH(NO)(PPh_3)_3]^+ClO_4^-$ are, however, pseudotetrahedral with Ir–N–O angles of 164° and 175°. In the structure of the latter compound, the hydrogen atom was not located; an alternative description of the structure is distorted trigonal bipyramidal.

A rationale of this, at first sight, confusing information is that, in the compounds with Ir–$\hat{\text{N}}$–O angles of about 120° and Ir–N bond lengths of between 189 and 197 pm NO is bonding as NO⁻, while in those with Ir–$\hat{\text{N}}$–O angles of about 180° and Ir–N between 157 and 177 pm, it is bonded as NO⁺. This behaviour is analogous to that of NO_2^+ (linear) and NO_2^- (bent, 120°) (see table 6.4).

Table 6.4

	Ir–N(pm)	Ir–$\hat{\text{N}}$–O(°)	ν(N–O)(cm⁻¹)	Structure
[IrH(NO)(PPh₃)₃]⁺	168	175	1780	trig. bipy.
Ir₂O(NO)₂(PPh₃)₂	176	177	1758–1735	†
Ir(NO)(PPh₃)₃	167	180	1600	tetrahedral
[IrI(NO)(CO)(PPh₃)₂]⁺	189	125	1720	pyramidal
[IrCl(NO)(CO)(PPh₃)₂]⁺	197	124	1680	pyramidal
IrCl₂(NO)(PPh₃)₂	194	123	1560	pyramidal
IrI(CH₃)(NO)(PPh₃)₂	191	120	1525	pyramidal

† μ-oxo-bridged dimer with metal–metal bonding (Ir–Ir = 255 pm); approximately square planar co-ordination.

Of course, the formulations Ir(I)–NO⁺ and Ir(III)–NO⁻ are in a sense π-equivalent

In terms of ligand-field theory, it has been calculated for these d⁸ systems that d_{z^2} overlaps strongly with π *(NO) in the 'bent' geometry, resulting in a fairly stable molecular orbital. Thus the square pyramidal Ir(NO)(PPh₃)₂(CO)Cl, having the configuration $(d_{xy})^2$, $(d_{yz})^2$, $(d_{xz})^2(d_{z^2})^2$, $(d_{x^2-y^2})^0$ has a 'bent' Ir–N–O bond system; in the trigonal bipyramidal [IrH(NO)(PPh₃)₃]⁺ the configuration is $(d_{xz})^2$, $(d_{yz})^2(d_{x^2-y^2})^2$, $(d_{xy})^2(d_{z^2})^0$; nonoccupation of d_{z^2} means that no stabilisation is afforded by 'bending'.

It has often been accepted that there is an infrared criterion for NO bonding: that NO⁺ (MNO linear) absorbs in the region 1600–1850 cm⁻¹ and NO⁻ (MNO bent) absorbs at lower energy. Perusal of table 6.4 does not show such a correlation.

NICKEL

Metal: c.c.p.; m.p. 1452°; I_1: 7.63 eV; I_2: 18.15 eV; I_3: 35.16 eV
Oxides: NiO
Halides: NiX$_2$ (X = F, Cl, Br, I)

Oxidation State and Representative Compounds

	0	1	2	3	4
Typical donor atom/group	CO,PR$_3$	P,CN	O,N,Hal	Hal,As	F
Co-ordination number					
3		√ Ni(Ph$_3$P)$_2${N(SiMe$_3$)$_2$}			
4 tet	√ Ni(CO)$_4$	√ NiBr(PPh$_3$)$_3$	√√ [NiCl$_4$]$^{2-}$		
4 planar			√√√ [Ni(CN)$_4$]$^{2-}$ √√		
5 T.B.P.			√√ Ni(PPhMe$_2$)$_3$(CN)$_2$	√ NiBr$_3$(PEt$_3$)$_2$	
5 S.P.			√√ [Ni(Me$_3$AsO)$_5$]$^{2+†}$	√	
6			√√√ [Ni(OH$_2$)$_6$]$^{2+}$	√ [NiCl$_2$(diars)$_2$]$^+$	√ [NiF$_6$]$^{2-}$
7					

† More than one spin-state
√ Known; √√ Several examples; √√√ Very common

7 Palladium and Platinum

These two metals have a very rich and extensive chemistry which almost in its entirety consists of the study of co-ordination compounds. Like the heavy metals that are near to them in the periodic table, they form rather few polymeric salt-like compounds but give large numbers of molecular complexes with a variety of ligands.

Compared with the chemistry of nickel, which for convenient reference is summarised opposite, the usual trends towards higher oxidation number are observed, with palladium occupying an intermediate position between nickel and platinum in this respect. Thus nickel has to be forced into an oxidation state of +4 by the use of ligands such as oxide or fluoride, while Pd(IV) is stabilised by chloride ligands in $[PdCl_6]^{2-}$. However, Pd(II) is certainly the more stable state compared with Pd(IV). The +2 and +4 states of platinum are about equally stable and between them these two oxidation states provide a wealth of well-defined complexes which have been intensively investigated for many years and are still being studied with profit.

It is very noticeable that the increased value of the crystal-field splitting $10Dq$ that accompanies the larger effective nuclear charge of the heavier atoms affects the stereochemistry. Thus while the complexes of Ni(II) may be octahedral, tetrahedral or square planar, almost all those of Pd(II) and Pt(II) are square planar. Further, Pd(II) and Pt(II) may be distinguished in stability by the formation of isomeric complexes of the type *cis-* and *trans-*MA_2B_2; this is typical behaviour in the case of Pt(II) but Pd(II) does not normally form isomers. The stability of the square planar complexes relative to tetrahedral ones depends on the very high crystal-field stabilisation energy of the d^8 system for such an arrangement; in the case of Ni(II) this stabilisation needs to be magnified by the use of high-field ligands such as cyanide but for Pd(II) and Pt(II) no special ligands need be used to ensure the stabilisation of square planar geometry.

Similarly to their left-hand neighbours in the periodic table, palladium and platinum form complex hydrides and alkyls, those of platinum being very stable.

7.1 The Metals and their Aqueous Chemistry

The metals are lustrous, high melting (Pd, 1552°; Pt, 1769°), moderately soft and rather inert, both being used in jewellery.† They both adopt cubic close packing.

---

† Platinum apparently has a limited use as small-arms ammunition: 'I shoot the Hippopotamus With bullets made of platinum, Because if I use leaden ones His hide is sure to flatten 'em.' (H. Belloc, *The Bad Child's Book of Beasts*.)

Palladium as a powder dissolves readily in hot concentrated nitric acid. This reagent does not affect platinum which does, however, dissolve slowly in aqua regia and is attacked by nonmetals such as chlorine, sulphur and phosphorus.

Platinum forms no aquo-ions, so strong is its tendency to form hydroxy-species. Alternatively, if suitable ligand anions, such as chloride, are available in an aqueous reaction mixture, it complexes with them. Palladium will, however, form a conventional hydrated perchlorate $[Pd(OH_2)_4](ClO_4)_2$ when $PdCl_2$ is treated with $AgClO_4$; most other anions complex with the metal.

Palladium readily and reversibly forms an interstitial nonstoichiometric hydride $PdH_{0.7}$; both metals are widely used as hydrogenation catalysts.

7.2 Halides

Binary halides MX_n are formed by palladium in the +2 and +4 oxidation states, and by platinum in the +2, +4, +5 and +6 oxidation states. They are listed here in the following table.

Table 7.1

	+2	+4	+5	+6
Pd	F, Cl, Br, I	F	–	–
Pt	Cl, Br, I	F, Cl, Br, I	F	F

Preparation is often by direct reaction between the metal and a halogen, some examples being

$$Pt + Cl_2 \xrightarrow{\ 500°\ } PtCl_2$$

$$Pt + 2Cl_2 \xrightarrow{\ 250°\ } PtCl_4$$

$$Pt \xrightarrow[\text{(2) heat}]{\text{(1) } BrF_3} PtF_4$$

$$Pt(\text{hot wire}) + 3F_2 \longrightarrow PtF_6$$

$$2PtCl_2 + 5F_2 \xrightarrow{\ 350°\ } 2PtF_5 + 2Cl_2$$

These halides, of course, have a variety of structures and reactivities. Thus the thermally stable dichlorides $PtCl_2$ and $PdCl_2$ have structures that are apparently dependent on the precise conditions of preparation, but X-ray diffraction has established the occurrence of Pt_6Cl_{12} (as shown in figure 7.1) and of isostructural Pd_6Cl_{12}. $PdCl_2$, however, usually takes the form of a chlorine-bridged coplanar chain.

Rather surprisingly, PtF_2 is unknown, but PdF_2 is quite stable and has an octahedrally co-ordinated structure. It is therefore paramagnetic. There are no known

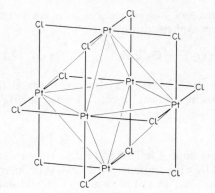

Figure 7.1 The structure of Pt_6Cl_{12}

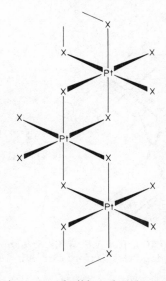

Figure 7.2 Structure of platinum tetrahalides, showing *cis* bridging in the solid state

tetrahalides of palladium except PdF_4, which emphasises that Pd(IV) is rather a highly oxidising state. The tetrahalides of platinum are all known and have fair thermal stability.

$PtCl_4$ is isostructural with $PtBr_4$ and α-PtI_4 and appears to have the structure shown in figure 7.2.

The +5 oxidation state of platinum is rare, but PtF_5 is obtainable as a low-melting ($80°$) red polymeric solid. It easily disproportionates

$$2PtF_5 \longrightarrow PtF_6 + PtF_4$$

The PtF_6 obtained in this way, or by direct combination, is a low-melting ($61°$) volatile red solid. It is an avid electron-seeker

$$PtF_6 + O_2 \longrightarrow O_2{}^+PtF_6{}^- \qquad (I_1 \text{ for } O_2, \text{ 12.1 e.v.})$$

It attacks even dry glass and is instantly hydrolysed by water. The molecule is octahedral. $O_2{}^+PtF_6{}^-$ has been shown to contain octahedral $[PtF_6]^-$ ions.

7.3 Oxides

The anhydrous oxides are limited to PdO and PtO_2. They are respectively made by direct combination and by dehydration of $PtO_2.xH_2O$ (from hydrolysis of $Na_2[PtCl_6]$). PdO shows the effect of the high crystal-field stabilisation energy of the d^8 configuration, since it has a square planar co-ordination (figure 7.3). A hydrated oxide $PtO.H_2O$ may be obtained by hydrolysis of Na_2PtCl_4 and subsequent drying of the product; however, further drying provokes disproportionation into Pt and PtO_2.

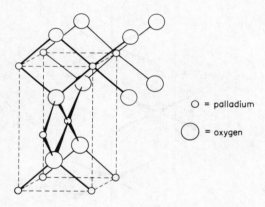

O = palladium

◯ = oxygen

Figure 7.3 The structure of PdO (after A.F. Wells, *Structural Inorganic Chemistry,* Clarendon Press, Oxford (3rd edn, 1962)

7.4 Some other Binary Compounds

On heating, palladium and platinum metals combine with the heavier elements in Groups V and VI to give a variety of binary compounds, many of which are non-stoichiometric. Others, however, are stoichiometric and their structures are therefore simpler. Of these latter, PtS is isostructural with PdO, and PdS is similar. These sulphides are obtainable from aqueous Pt(II) or Pd(II) solutions, respectively, on treatment with H_2S. PtS_2 has the octahedrally co-ordinated CdI_2 structure; the structure of PdS_2 is unknown but may possibly be of a $Pd^{2+}(S_2)^{2-}$ type. PtP_2 has the pyrites structure (see figure 7.4), thus placing the platinum in the +4 state; the stoichiometry alone is, of course, no guide to the oxidation state of the metal in this type of compound.

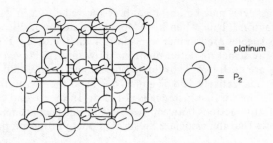

Figure 7.4 The pyrites structure of PtP_2, containing P_2 groups

7.5 Complexes of Palladium and Platinum

7.5.1 Palladium(0) and Platinum(0)

In this low oxidation state, π-bonding ligands are necessary for the formation of stable complexes. The best-known π-bonding ligand, carbon monoxide, does not, however, give $Pd(CO)_4$, which would be the palladium analogue of nickel carbonyl. Explanations of the absence of this compound and of its platinum analogue have been offered, mainly in terms of the difference in ground-state electronic structure between $Ni(3d^8 4s^2)$, $Pd(4d^{10})$ and $Pt(5d^9 6s^1)$. The phenomenon does not, however, extend to analogous complexes with phosphorus ligands, as described below, so it is possible that the π bond, relatively important for complexed CO, is weaker in Pd(0) and Pt(0) complexes than in those of Ni(0); while the σ bond, relatively more important for complexed phosphines, retains its strength along the series Ni(0), Pd(0), Pt(0).

Bidentate tertiary phosphines and arsines give complexes $[Pd(di-L)_2]$. Either the corresponding Pd(II) complex may be reduced or direct reaction used

$$[Pd(diars)_2]^{2+} \xrightarrow[\text{H}_2\text{O/EtOH}]{\text{NaBH}_4} [Pd(diars)_2]$$

$$Pd + 2 \enspace \text{[benzene ring with} -PEt_2 \text{ and } -PEt_2 \text{]} \xrightarrow[\text{N}_2]{180°} [Pd\{C_6H_4(PEt_2)_2\}_2]$$

The orange products are fairly stable thermally but are rapidly attacked by air. They presumably have tetrahedral (strictly, D_{2d}) co-ordination.

The co-ordination number of Pd(0) and Pt(0) is not restricted to four, and these ions resemble the neighbouring d^{10} ions Ag(I) and Au(I) in showing preferences for lower co-ordination. Thus in a two-stage reaction

$$cis-[PtCl_2(PPh_3)_2] \xrightarrow[\text{EtOH}]{\text{KOH}} trans-[PtHCl(PPh_3)_2]$$

$$\downarrow_{\text{PPh}_3}^{\text{KOH}}$$

$$Pt(PPh_3)_3 + Pt(PPh_3)_4$$

Similar palladium species may be obtained; here the value of x in PdL_x depends largely on steric factors. Thus when $L = PMe_3$, PdL_4 does not dissociate, but when $L = PBu^n_3$ the tris-species predominates in solution, and with $L = P(cyclo\text{-}C_6H_{11})_3$, the bis-species. X-ray structure determinations show that $Pd\{P(cyclo\text{-}C_6H_{11})_3\}_2$ and $Pd(PBu^t_2Ph)_2$ are both two-co-ordinated monomers in the solid state. The $P\text{--}\widehat{Pd}\text{--}P$ angles are respectively $158.4°$ and $175.7°$, and the reason for the large deviation from $180°$ in the first instance is not clear. In the second compound, there are close (273 pm) interactions between two *ortho* hydrogen atoms and the metal atom. N.M.R. shows that the associated restricted phenyl rotation persists in solution.

The crystal structure of $Pt(Ph_3P)_3$ shows that the Pt atom lies at the centre of an equilateral triangle, co-ordinated by three phosphorus atoms; the Pt atom is 10 pm from the P_3 plane (Pt--P, 226 pm). On treatment of $Pt(Ph_3P)_3$ with acids HX (X = HSO_4, for example) the species $[PtH(PPh_3)_3]^+$ results. By reaction with H_2O_2 and KOH, '$Pt(PPh_3)_2$' is formed

$$[PtH(PPh_3)_3]^+ + H_2O_2 + OH^- \longrightarrow Pt(PPh_3)_2 + Ph_3PO + 2H_2O$$

The bis-species is also produced by photolysis of $Pt(PPh_3)_2(C_2O_4)$ or by refluxing ethanolic $Pt(PPh_3)_2(CO_3)$ in the dark. There is evidence that it is in fact a Pt(II) compound produced by elimination of benzene.

The structure of $Pt(PPh_3)_4$ has not been reported but gas-phase electron-diffraction results indicate that $Pt(PF_3)_4$ is tetrahedral (as expected) with Pt--P = 223 pm. $Pt(Ph_3P)_4$ dissociates in solution to form predominantly the tris-complex (a little of the bis-species may be formed). Some analogous compounds of other phosphines are known, for example $Pt(Ph_2MeP)_4$. However, n.m.r. study has shown that below about 300 K, $Pt(Ph_2MeP)_3$ is the predominant species in solution; above this temperature rapid exchange occurs. Here a $Pt(Me_2PhP)_4$ and $Pt(Me_2C_6F_5P)_4$ do not appear to dissociate under these conditions; the enhanced stability of the tetrakis-species may be attributed, partially at least, to steric factors. The triethyl-phosphine complexes $Pt(PEt_3)_3$ and $Pt(PEt_3)_4$ are more reactive than the triphenyl-phosphine analogues. The coupling constant $J[^{31}P\text{--}^{195}Pt]$ is greater for the tris-complex, indicating a greater s character in the Pt--P bond; in fact, the tris-species reacts with chlorobenzene to give square planar *trans*-$PtClPh(PEt_3)_2$, a nucleophilic reaction unknown for the triphenylphosphine analogue.

Reactions of Pd(0) and Pt(0) phosphine complexes $Pt(PPh_3)_4$ (and the Pd analogue) react readily with molecular oxygen in benzene to give the adduct $M(Ph_3P)_2O_2$, which is the starting material for a large number of reactions. The structure of the platinum compound $Pt(PPh_3)_2O_2.1.5C_6H_6$ determined by X-ray diffraction, is given in figure 7.5, which shows the co-ordination geometry to be planar, although

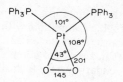

Figure 7.5 The benzene solvate of the $Pt(PPh_3)_2(O_2)$ molecule (after T. Kashiwagi, N. Kasai and M. Kakudo, *Chem. Commun.* (1969), 743)

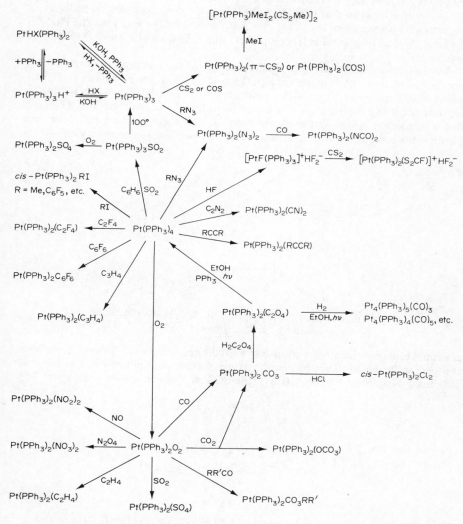

Figure 7.6 Triphenylphosphine–platinum (0) complexes and their reactions

it is necessarily not 'square'. In the chloroform solvate $[Pt(PPh_3)_2O_2].2CHCl_3$, O–O is 150 pm, both distances being compatible with the fact that the two compounds are irreversible oxygen carriers (see page 102). Recent study has been made of the electronic binding energies in a series of complexes $(PPh_3)_2PtL_2$ ($L_2 = (PPh_3)_2$, C_2H_4, CS_2, O_2, Cl_2), where the oxidation state formally changes from Pt(0) to Pt(II), using the E.S.C.A. (electron spectroscopy for chemical analysis) technique. While the phosphorus 2p binding energies (and Pt–P bond lengths) remain virtually constant, the Pt $4f_{7/2}$ binding energy increases from about 71.7 eV in $(PPh_3)_4Pt$ to about 73.4 eV in $(PPh_3)_2PtCl_2$ (for Pt metal it is 71.2 eV). It has been shown that (i) $(PPh_3)_4Pt$ may be classified as a Pt(0) complex and (ii) in $(PPh_3)_2PtO_2$, the $4f_{7/2}$ energy is close to that in $(PPh_3)_2PtCl_2$, so that this compound is really a platinum(II) peroxo-complex. This idea is supported by the fact that $Pt(PPh_3)_2O_2$ is planar, typical of platinum(II) complexes, while those of platinum(0) tend to be tetrahedral.

We may now proceed to a more general survey of the reactions of triphenyl-phosphine–platinum (0) complexes (see figure 7.6). Both the palladium and platinum tetrakis-complexes react with cyanogen to afford *cis*-$M(PPh_3)_2(CN)_2$, this reaction involving the breaking of a C–C bond. By way of contrast, $RhCl(PPh_3)_3$ forms an adduct. Allene and acetylene form air-stable adducts too; the allene complex has the structure shown in figure 7.7. The co-ordination geometry about

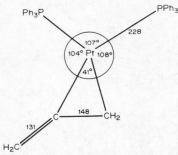

Figure 7.7 The allene complex $Pt(PPh_3)_2(CH_2:C:CH_2)$ (after M. Kadonaga, N. Yasuoka and N. Kasai, *Chem. Commun.* (1971), 1597)

platinum is planar; the novel feature is the bonding of the allene – in most complexes it is perpendicular to the co-ordination plane. Carbon disulphide forms an iso-structural π complex, $Pt(PPh_3)_2(CS_2)$ (the COS analogue is known – see figure 7.8).

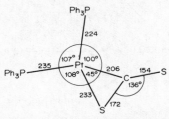

Figure 7.8 The co-ordination geometry of $Pt(PPh_3)_2(CS_2)$ (after R. Mason and A.I.M. Rae, *J. chem. Soc. (A)* (1970), 1767)

This complex reacts with methyl iodide to give $[PtMeI_2(CS_2Me)(PPh_3)_2]$. On heating in chloroform, the carbon oxysulphide complex rearranges as follows

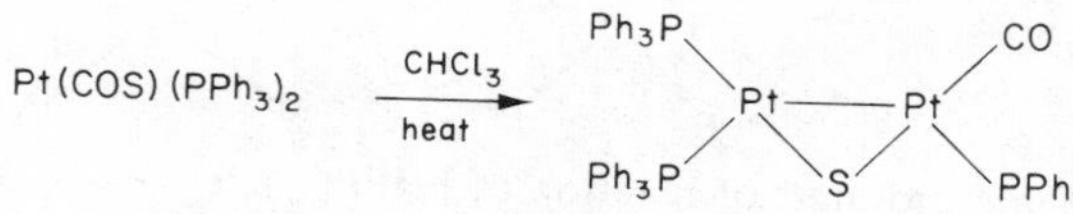

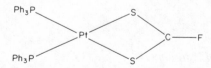

$Pt(PPh_3)_2(C_2O_4)$ forms from reaction of the carbonate with oxalic acid; on u.v. irradiation in ethanol, $Pt(PPh_3)_2$ is formed from it, while in the presence of excess phosphine, $Pt(PPh_3)_4$ is obtained.

An unusual type of reaction occurs on treatment of $Pt(PPh_3)_4$ with organic azides RN_3 (R = Me, Et, Ph, etc.) where *trans*-$Pt(PPh_3)_2(N_3)_2$ results from the dealkylation, without any gas evolution. This compound reacts with CO to give the cyanate complex

$$Pt(PPh_3)_2(N_3)_2 \xrightarrow{\ CO\ } Pt(PPh_3)_2(NCO)_2 + 2N_2$$

Reactions of a more normal kind occur with fluoroalkyl iodides (R_FI) to give $M(PPh_3)_2R_FI$ (M = Pt, Pd) involving σ-bonded fluorocarbon, with C_2F_4 to give

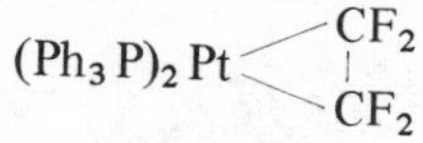

$$(Ph_3P)_2Pt\underset{CF_2}{\overset{CF_2}{<}}$$

and with hexafluorobenzene to yield $Pt(PPh_3)_2(C_6F_6)$; this latter compound appears to contain two σ bonds to neighbouring carbons in the benzene ring. $Pt(PPh_3)_4$ reacts with anhydrous HF to afford $[PtF(PPh_3)_3]^+HF_2^-$; on treatment with carbon disulphide, $[Pt(PPh_3)_2(FCS_2)]^+HF_2^-$ is formed. The cation has the structure shown in figure 7.9, again involving planar co-ordination. Sulphur dioxide gives $Pt(PPh_3)_3SO_2$ as a benzene solvate; on heating to 100° $Pt(PPh_3)_3$ is formed.

Figure 7.9 The structure of the cation $[Pt(Ph_3P)_2(FCS_2)]^+$

Reaction with H_2S or RSH is of interest from the aspect of the poisoning of platinum catalysts in industrial processes. The yellowish compounds produced are of the form $Pt(PPh_3)_2SRR'$. The ^1H n.m.r. spectra of the adducts where R,R' = H are of some interest: on bubbling H_2S into benzene solutions of $Pt(PPh_3)_n(n = 2$ or 3) three signals are detected, one at about $\tau 8.0$ (A) and two others at about $\tau 11.5$ (B) and 19 (C), interpretable in terms of the presence of two species (see figure 7.10). Sulphur reacts directly with the platinum (0) complexes to give compounds of the type $PtS_4(PPh_3)_2$ and $Pt_4S_8(PPh_3)_4$. This reaction is irreversible, the final products being very inert.

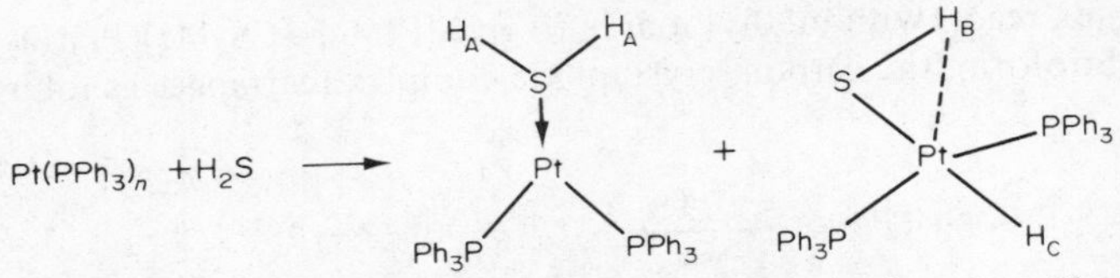

Figure 7.10 Probable products of reaction of $Pt(PPh_3)_n$ (n = 2 or 3) with H_2S

Reactions of $Pt(PPh_3)_2O_2$ The dioxygen adduct, $Pt(PPh_3)_2O_2$ undergoes a variety of reactions; this and the other platinum compounds we have considered have been much more extensively studied than the palladium analogues, largely because of the greater stability of the platinum compounds; nevertheless the palladium compounds generally react in an analogous way.

Addition reactions of the dioxygen adduct are shown in figure 7.6. Thus addition of NO gives $Pt(PPh_3)_2(NO_2)_2$, NO_2 gives $Pt(PPh_3)_2(NO_3)_2$ and SO_2 gives $Pt(PPh_3)_2(SO_4)$. The reaction with CO_2 can yield either the carbonate $Pt(PPh_3)_2CO_3$ or the peroxycarbonate $Pt(PPh_3)_2(OCO_3)$, the latter being stable in the presence of oxygen. The structure of the carbonate is shown in figure 7.11. Aldehydes and ketones react with the dioxygen adduct to give compounds containing the following ketone-peroxy-chelate ring

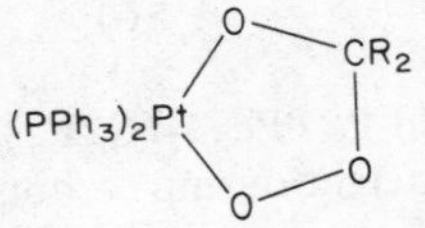

Interestingly, ethylene *displaces* the co-ordinated oxygen to form $(PPh_3)_2Pt(C_2H_4)$; the PtP_2C_2 unit is nearly planar (Pt–P 227 pm, Pt–C 211 pm, C–C 143 pm), again implying it to be a platinum(II) complex rather than platinum(O).

7.5.2 *Palladium(I) and Platinum(I)*

These are not fully authenticated yet. A recent report gives the following sequence

$$\left\{\begin{matrix} PdX_2\ (X = Br, I) \\ \text{or} \\ PdCl_2(PhCN)_2 \end{matrix}\right. \xrightarrow[C_6H_6]{\text{excess } Bu^tNC} PdX_2(Bu^tNC)_2 \quad (X = Cl, Br, I)$$

$$1 : 1 \text{ ratio} \longrightarrow PdX(Bu^tNC)_2$$

$$(C_3H_5)_2Pd_2Cl_2 \xrightarrow{NaC_5H_5} (C_3H_5)Pd(C_3H_5) \xrightarrow{Bu^tNC} Pd(Bu^tNC)_2$$

These compounds appear to be quite stable. They are diamagnetic and may be

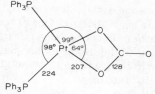

Figure 7.11 The structure of Pt(PPh₃)₂(CO₃) (after F. Cariati, R. Mason, G.B. Robertson and R. Ugo, *Chem. Commun.* (1967), 408)

dimers; they react with some other donor molecules thus

PdX(Bu'NC)₂ ⟶ (PPh₃, X = I) ⟶ PdX(PPh₃)(Bu'NC)₂

⟶ (NO, X = Cl, Br) ⟶ *trans*-[PdX(NO₂)(Bu'NC)₂]

There is evidence that the reaction

$$Pt(PPh_3)_3 + EtI \longrightarrow \textit{trans}\text{-}PtEtI(PPh_3)_2$$

proceeds by a free-radical mechanism involving a Pt(I) intermediate such as PtI(PPh₃)₂.

7.5.3 Palladium(II) and Platinum(II)

The chemistry of Pd(II) and Pt(II) is quite extensive. The two ions resemble each other fairly closely in their properties, the principal difference being the greater lability of the Pd(II) complexes. The compounds are nearly all square planar and may be divided into three overlapping classes: (a) classical complexes of such ligands as ammonia and halides; (b) σ-bonded alkyl and hydride complexes; and (c) olefin and acetylene complexes.

The red tetrachloropalladate (II) and tetrachloroplatinate (II) ions may be obtained by the action of halide ion on the dichlorides. They may then be successively substituted by ligands such as Br, I, CN, PR₃, NH₃ or SR₂ to give a large variety of species, which in the case of Pt usually exist in alternative isomeric forms. The pattern of substitution is mainly determined by the *trans effect,* whose existence was first noted in 1920 by Chernaev, a member of the Russian school of co-ordination chemists who did much fundamental work on platinum chemistry.

The trans effect The *trans* effect has the result that in a substitution reaction where an incoming ligand might enter in two or more alternative positions with consequent displacement of the ligand occupying the position chosen, the position chosen is *trans* to that ligand present which ranks highest in the *trans* effect series. This series is an empirical sequence of ligands commonly found bonded to Pt(II), which is the ion that illustrates the effect most strikingly. Such a series runs as follows

$$H_2O \langle OH^- \langle NH_3 \langle Cl^- \langle Br^- \langle I^- \langle NO_2^- \langle SCN^- \langle C_6H_5^- \langle Me^- \langle SC(NH_2)_2 \approx H^- \approx$$

$$R_3P \langle CN^- \approx CO \approx NO \approx C_2H_4$$

The reader will notice the approximate similarity to the nephelauxetic series, which is probably a measure of σ *and* π bonding. One explanation of the *trans* effect is in terms of π bonding and is best described diagrammatically (see figure 7.12). The region of the complex *trans* to the strongly π-bonding ligand is rendered electrostatically positive relative to the other positions and the incoming nucleophilic ligand L′ thus tends to substitute there.

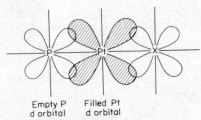

Empty P   Filled Pt
d orbital   d orbital

Figure 7.12 π-type interaction postulated in platinum phosphine complexes

The *trans* effect is a kinetic phenomenon. The explanation just given does not necessarily involve any thermodynamic weakening of the *trans* ligand-platinum bond; there is, however, considerable structural and spectroscopic evidence for such weakening and the capability of a ligand to weaken the bond *trans* to it is known as its *trans influence*. The kinetic *trans* effect produced by a ligand L also depends on the stabilisation of the activated complex that is formed during the displacement reaction. This stabilisation has been suggested to be due to strong Pt—L bonding using the Pt 6p orbital.

Much recent work suggests that the *trans* influence of a ligand is dominated by the nature of its σ-bonding orbital (that is, by partial rehybridisation of metal σ orbitals in response to changes in the character of ligand orbitals used in σ bonding) and that electronegativity and π-acceptor effects have less influence on the *trans* σ-bond. The *trans* influence parallels the kinetic *trans* effect; extended Hückel M.O. calculations suggest that the influence is not so much due to weakening of the $Pt(p_\sigma)$ bond as to weakening of Pt(6s) and $(d_{x^2-y^2})$ interactions with the *trans* ligand. Perhaps the most spectacular example of the failure of the π-mechanism is the failure to account for the high position of H^- in the *trans*-effect series; such influence seems only explicable on the basis of a σ mechanism of some kind.

There is much structural evidence for the *trans* influence. Thus in *trans*-$PtCl_2(PEt_3)_2$, Pt—Cl is 232 pm, but in *trans*-$PtHCl(PhEt_2Ph)_2$ and *cis*-$PtCl_2(PEt_3)_2$, Pt—Cl is 242 pm. Thus it may be seen that PR_3 and H^- both have higher *trans* influences than Cl. Similarly, in *trans*- and *cis*-$PtCl_2(NH_3)_2$, Pt—N is 201 pm and Pt—Cl 233 pm in the former and 205 and 232 pm, respectively, in the latter. While Pt—P bond lengths are sensitive to the nature of the substituents on phosphorus, overall comparison of X-ray data shows that Pt—P bond lengths are shorter in *cis*-$Pt(PR_3)_2X_2$ compared with the corresponding *trans* complexes. Far-infrared studies also show *trans* influence. In *trans*-$PtCl_2L_2$, Pt—Cl is insensitive to the nature of L (L = NH_3, 330 cm^{-1}; L = PMe_3, 326 cm^{-1}, while in *cis*-$PtCl_2L_2$ there is a marked influence (L = NH_3, 326 cm^{-1}; L = PEt_3, 303 and 281 cm^{-1}). On the other hand evidence for the importance of π bonding in square planar complexes is suggested by the anomalous stability of some *cis* complexes relative to the *trans* isomer. Thus, in contrast to the corresponding palladium complex, *trans*-$[PtCl_2(PPr^n_3)_2]$ in benzene solution partially isomerises into the *cis* compound

$$trans\text{-}[PtCl_2(PPr^n_3)_2] \xrightleftharpoons{\text{benzene}} cis\text{-}[PtCl_2(PPr^n_3)_2]$$

The variation between the Pt and the Pd complexes, and the marked variation of equilibrium position along the series P, As, Sb, Bi refute the possibility that the *cis* form might be mainly stabilised by solvation effects rather than by electronic ones. The appearance of a substantial proportion of the *cis* form, which on account of the energy associated with its high electric dipole moment must otherwise be unstable relative to the *trans* form, is most naturally explained by the greater degree of π bonding possible in a *cis* configuration, where three metal d orbitals can π bond, rather than a *trans*, where only two can (see figure 7.13).

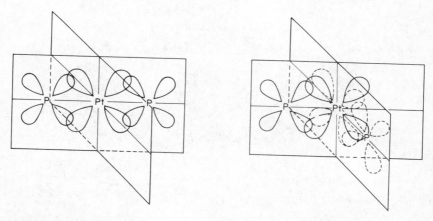

Figure 7.13 Postulated d_π-d_π bonding in *trans*- and *cis*-phosphine complexes $Pt(R_3P)_2X_2$ (after L.M. Venanzi, *Chem. Brit.*, **4** (1968), 162)

Trans influence has been studied by n.m.r. spectroscopy. The n.m.r. evidence arises from the proposal that the magnitude of the ^{31}P–^{195}Pt coupling constant forms a measure of the s character of the Pt–Pσ-bond (see table 7.2). These n.m.r. data show that the difference in s character in the Pt–P bond between *cis* and *trans* isomers is virtually the same in Pt(II) and Pt(IV) systems. In Pt(II) complexes with, formally, dsp^2 hybridisation, the s character is about a quarter, while in Pd(IV) complexes, formally d^{2}sp^3, it should be about a sixth. Thus the ratio $J[Pt(IV)-P]$: $J[Pt(II)-P] \approx 0.67$. The degree of π bonding is expected to be less in the Pt(IV) systems owing to the reduction in d_{Pt}-d_P interaction due to the contraction of d orbitals consequent upon the increase in formal charge (4 against 2). Some *cis* and *trans*-PtX_2L_2 (X = Cl, Br; L = $C_{12}H_{25}{}^{15}NH_2$) also show a 1.2 times larger ^{15}N–^{195}Pt coupling in the *cis* complexes compared with the *trans*. Here, of course, there is no π component of the Pt–N bond and thus an effect purely σ in nature is operative.

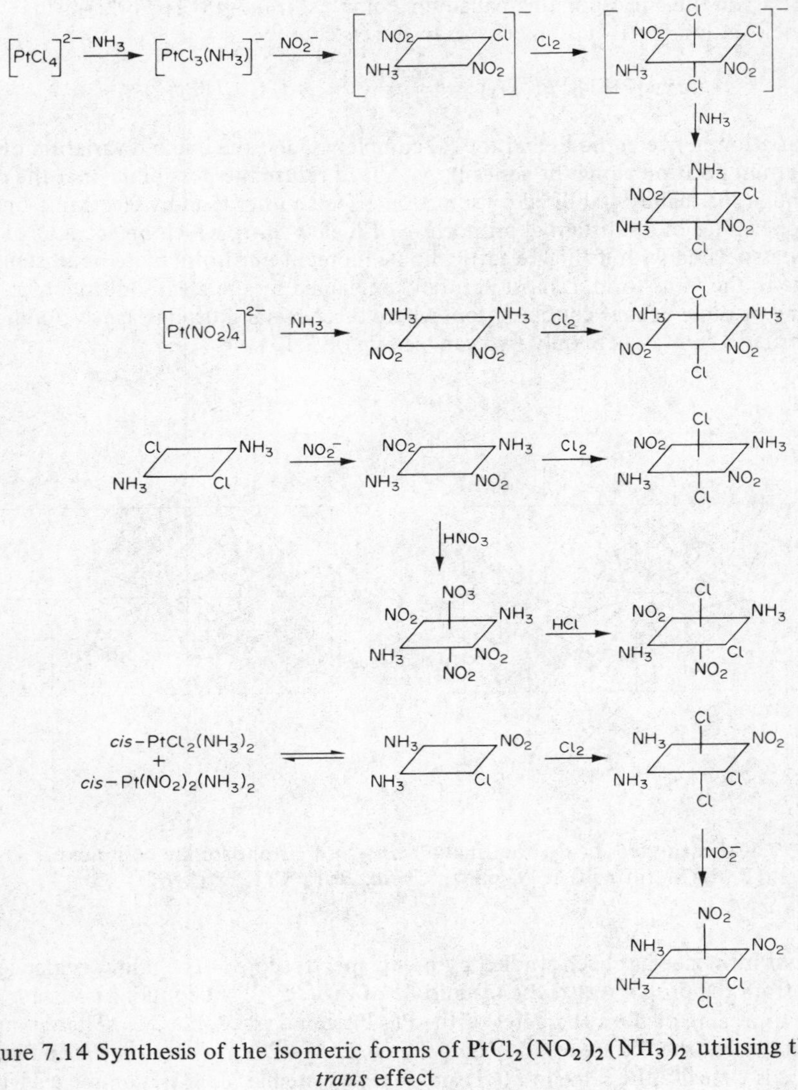

Figure 7.14 Synthesis of the isomeric forms of $PtCl_2(NO_2)_2(NH_3)_2$ utilising the *trans* effect

Many workers at present interpret the *trans* influence in terms of a σ-bonding mechanism. Ligands that form strong σ bonds to platinum thus have a strong *trans* influence (for example PR_3, H^-). It cannot be said, however, that π-bonding effects are unimportant.

The syntheses of all five isomeric forms of the Pt(IV) complex $PtCl_2(NO_2)_2(NH_3)_2$ provide examples of the practical use of the *trans* effect (see figure 7.14). In many

Table 7.2 Platinum–phosphorus coupling constants
from ^{31}P n.m.r. spectra†

Complex	$J[^{195}Pt-^{31}P]$ (Hz)
cis-PtCl$_2$(PBu$_3$)$_2$	3508
cis-PtI$_2$(PBu$_3$)$_2$	3372
trans-PtCl$_2$(PBu$_3$)$_2$	2380
trans-PtI$_2$(PBu$_3$)$_2$	2200
cis-PtCl$_2${P(EtO)$_3$}$_2$	5698
cis-PtI$_2${(P(EtO)$_3$}$_2$	5472
cis-PtCl$_4$(PBu$_3$)$_2$	2070
trans-PtCl$_4$(PBu$_3$)$_2$	1462

† A. Pidcock, R.E. Richards and L.M. Venanzi, *J. chem. Soc. A.* (1966), 1707

of the individual steps of these syntheses it is the *trans* effect that decides the stereochemical nature of the product. Other principles that also are relevant are: (a) the oxidation of Pt(II) to Pt(IV) proceeds with retention of configuration; (b) anionic ligands are much more readily substituted than uncharged ones.

Phosphine complexes and others The platinum(II) complexes PtCl$_2$(PR$_3$)$_2$ may be *cis* or *trans*; the palladium analogues are all *trans,* probably due to a greater rate of *cis* → *trans* isomerisation in this case. A convenient preparation (applicable to many complexes of Pd and Pt) is to shake an aqueous mixture of the phosphine with K$_2$PtCl$_4$. The initial product is [Pt(PR$_3$)$_4$] [PtCl$_4$], which on heating gives a mixture of the *cis* and *trans* complexes. The mixed complexes may be extracted with petrol to remove the more soluble yellow *trans* isomer, yielding the white *cis* complex. Alternatively, the *cis* → *trans* conversion may be accomplished by heating the *cis* form above its melting point for about one hour.

The bispyridine complexes may be readily prepared

$$K_2PtCl_4 \xrightarrow[H_2O]{py} \textit{cis-}PtCl_2py_2 \text{ (yellow)}$$

py, H$_2$O

$$Pt(py)_4Cl_2 \text{ (colourless solution)}$$

HCl, evaporate

$$\textit{trans-}PtCl_2py_2 \text{ (pale yellow)}$$

Certain *cis* complexes such as *cis*-[PtCl$_2$(PPh$_3$)$_2$] (prepared by refluxing K$_2$PtCl$_4$ with Ph$_3$P in xylene) are very stable; here the conversion may be accomplished by irradiation of a benzene solution of the *cis* complex with a mercury lamp. The configuration of the complexes may be deduced with the aid of dipole moment, n.m.r. and i.r. measurements.

Cationic complexes are also formed

$$\textit{cis-}\text{PtCl}_2(\text{PMe}_2\text{Ph})_2 \xrightarrow[\text{PMe}_2\text{Ph}]{\text{MeOH}} [\text{PtCl}(\text{PMe}_2\text{Ph})_3]^+ \text{ and } [\text{Pt}(\text{PMe}_2\text{Ph})_4]^{2+}$$

Hydrido-complexes are conveniently prepared by reduction of *cis*-PtX$_2$(PR$_3$)$_2$ with aqueous hydrazine or base/alcohol mixture

$$\textit{cis-}\text{PtX}_2(\text{PR}_3)_2 \longrightarrow \textit{trans-}\text{PtHX}(\text{PR}_3)_2$$

Ethylene reacts reversibly with these species to form an alkyl complex

$$\text{C}_2\text{H}_4 + \textit{trans-}\text{PtHX}(\text{PR}_3)_2 \underset{\text{temperature}}{\overset{\text{pressure}}{\rightleftharpoons}} \textit{trans-}\text{Pt}(\text{C}_2\text{H}_5)\text{X}(\text{PR}_3)_2$$

Studies on the thermal decomposition of a bromide, employing deuteration, show that the product PtHBr(PEt$_3$)$_2$ contains hydrogen originating from both α and β-hydrogens of the ethyl group implying a transition state having symmetrically placed carbon atoms.

Other interesting reactions are

$$\textit{trans-}\text{PtHX}(\text{PEt}_3)_2 \xrightarrow{\text{R}_3\text{SiH}} \textit{trans-}\text{PtX}(\text{SiR}_3)(\text{PEt}_3)_2 + \text{H}_2$$

$$\textit{cis-}\text{PtCl}_2(\text{PBu}^n{}_3)_2 \xrightarrow{\text{R}_3\text{SiH}} \textit{trans-}\text{PtClH}(\text{PBu}^n{}_3)_2 + \text{R}_3\text{SiCl}$$

$$\textit{cis-}\text{PtCl}_2(\text{PMe}_2\text{Ph})_2 \xrightarrow{\text{R}_3\text{SiH, NEt}_3} \textit{trans-}\text{PtCl}(\text{SiR}_3)(\text{PMe}_2\text{Ph})_2 + \text{Et}_3\text{NHCl}$$

$$\textit{trans-}\text{PtHX}(\text{PEt}_3)_2 \xrightarrow{\text{HX}} \textit{cis, cis, trans-}\text{PtH}_2\text{X}_2(\text{PEt}_3)_2$$

Compounds of the type [Pt(SnCl$_3$)$_5$]$^{3-}$ and [PtCl$_2$(SnCl$_3$)$_2$]$^{2-}$ result from reaction of SnCl$_2$ and PtCl$_2$ in methanol and hydrochloric acid; the former is a rare example of five-co-ordinate Pt(II). Reaction in acetone yields the ion [Pt$_3$Sn$_8$Cl$_{20}$]$^{4-}$, a Pt$_3$Sn$_2$ cluster, which has been shown to be [{Pt(SnCl$_3$)$_2$}$_3$(SnCl)$_2$]$^{4-}$.

Compounds involving platinum–carbon bonds A number of stable methyl platinum(II) compounds exist; for convenience they are considered with the analogous Pt(IV) methyls (page 132). The analogous neopentyls and 2-silaneopentyls are also stable, as are alkenyls.

$$\textit{cis-}\text{PtCl}_2\text{L}_2 \xrightarrow{\text{LiCH}_2\text{SiMe}_3} \textit{cis-}\text{Pt}(\text{CH}_2\text{SiMe}_3)_2\text{L}_2$$

$$(\text{L} = \text{PMe}_2\text{Ph}, \text{AsPh}_3; \text{L}_2 = \text{bipy})$$

$$\textit{cis-}\text{Pt}(\text{PPh}_3)_2(\text{PhC}{\equiv}\text{CPh}) \xrightarrow[(\text{X = Cl, CF}_3\text{CO}_2)]{\text{HX}} \textit{trans-}\text{Pt}(\text{PPh}_3)_2\text{X}(\text{PhC}{=}\text{CPhH})$$

Insertion reactions also occur

$$trans\text{-}PtClR(PEt_3)_2 \xrightarrow{\text{CO}} trans\text{-}PtCl(COR)(PEt_3)_2$$
$$(R = Me, Et)$$

Platinum(II) and palladium(II) can form σ bonds to carbon by means of internal-metallation reactions in phosphine and phosphite complexes (as do rhodium and iridium)

$$trans\text{-}PtX_2L_2 \xrightarrow[\text{reflux}]{\text{MeOCH}_2\text{CH}_2\text{OH}} trans\text{-}PtHXL_2 + PtXL(PBu^t_2.C_6H_4)$$

$(L = PBu^t_2Ph; X = Cl, Br, I)$

This last compound has the structure shown in figure 7.15. The metallation process,

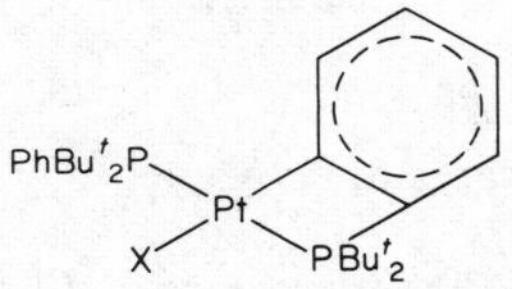

Figure 7.15 An example of platinum–carbon bond formation via an internal-metallation reaction

which competes with hydride formation, is in this instance favoured in the order I ⟩ Br ⟩ Cl. It is also favoured by bulky phosphines. Two related reactions are as follows

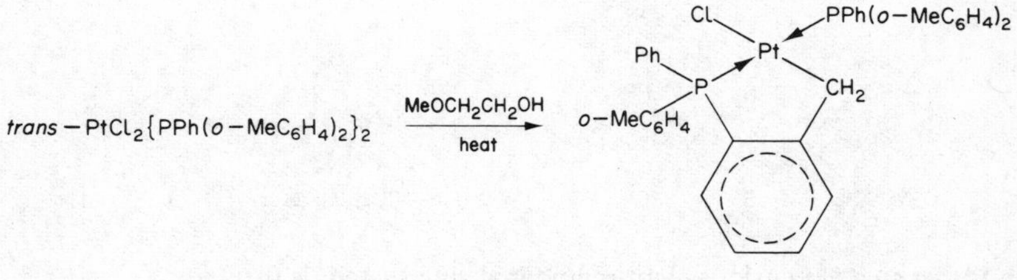

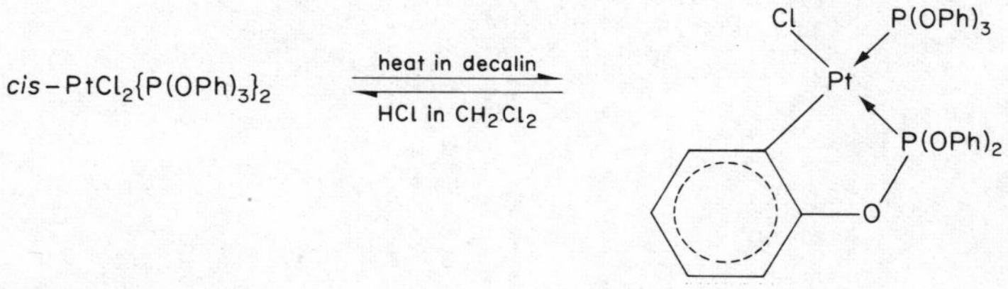

A compound believed to contain only σ or π bonds from carbon to the platinum atom is prepared as follows

There are two Pt–phenyl σ bonds and two Pt–olefin co-ordinate links.

Binuclear compounds and isomerism In addition to the mononuclear Pd(II) and Pt(II) complexes, a variety of binuclear compounds are known. Many anionic ligands are capable of forming a bridge to another metal ion and can be considered to use one of their otherwise nonbonding lone pairs for this. The situation may be expressed as resonance between canonical forms

These compounds may be prepared directly

$$2PtCl_2 + 2PPr^n_3 \xrightarrow{200°} Pt_2Cl_4(PPr^n_3)_2$$

Both the bridging and the terminal anions may be replaced by other anions such as SR^-, PR_2^- or SCN^-, and fairly extensive isomerism is observed. In the following example the sequence in which the substituting reagents are used determines which isomer is formed

With thiocyanate, a curious eight-membered ring system is formed

Addition of further ligand easily breaks these bridges

$$Pt_2Cl_4(PPr_3)_2 \xrightarrow{2PPr_3} 2 \; trans\text{-}PtCl_2(PPr_3)_2$$

Reaction of the dimeric tertiary phosphine complexes with CO gives carbonyl complexes in good yield

$$Pt_2Cl_4(PR_3)_2 \xrightarrow{CO, \; 1 \; atm., \; 20°} cis\text{-}[PtCl_2(CO)(PR_3)]$$

An interesting and unusual case of isomerism in the crystalline state is provided by $PdI_2(PMe_2Ph)_2$. When a methanol solution of $trans\text{-}PdCl_2(PMe_2Ph)_2$ is treated with sodium iodide and allowed to crystallise, both red needles (98 per cent) and yellow plates (2 per cent)form; the latter turn red when subjected to heat. The red form contains a $trans\text{-}PdI_2P_2$ square unit, but with a short Pd—I contact in an axial position, giving square pyramidal geometry. The yellow form is again square planar, but with two hydrogens from phenyl rings occupying the 'axial' positions.

Two groups capable of *ligand* isomerism are thiocyanate and dimethyl sulphoxide, $Me_2SO.K_2[Pd(SCN)_4]$ contains entirely S-bonded thiocyanates, but in the phosphine complex $Pd(NCS)(SCN)(Ph_2PCH_2CH_2PPh_2)$ X-ray analysis shows that one thiocyanate is N-bonded and the other S-bonded. It appears that both electronic and steric effects determine the mode of co-ordination. The thiocyanato-type would be expected for Class B metals and isothiocyanato bonding for Class A metals: evidently palladium (II) must lie close to the borderline. In both $trans\text{-}PdCl_2(Me_2SO)_2$ and $cis\text{-}Pd(NO_3)_2(Me_2SO)$ (also a rare authenticated case of monodentate nitrato-ligands), the donor is sulphur, but $cis\text{-}[Pd(Me_2SO)_4]^{2+}$ has two O-bonded and two S-bonded sulphoxides (according to X-ray analyses), while there is infrared evidence that the presence of the bulky neopentyl group in $[Pd\{(Me_3CCH_2)_2SO\}_4]^{2+}$ leads to entirely O-bonded groups, undoubtedly because of steric effects.

Complexes of olefins and other unsaturated compounds While the general theme is considered in chapter 9, it is pertinent to mention here some palladium and platinum complexes with olefins, since their behaviour is in many respects similar to that of the other types of complex.

Reaction of ethylene and ethanolic chloroplatinate yields $K[Pt(C_2H_4)Cl_3]$, Zeise's salt; it is best made by dissolving a ground mixture of K_2PtCl_4 and $SnCl_2$ in 1M HCl, deoxygenating and passing ethylene until the red solution becomes yellow. Vacuum evaporation affords the complex. An ethanolic solution of this complex reacts with HCl to give the orange chloro-bridged dimer $\{Pt(C_2H_4)Cl_2\}_2$, which, unlike the precursor, is unstable in moist air, though it is stable in concentrated H_2SO_4.

The anion of Zeise's salt has the structure shown in figure 7.16. It is notable

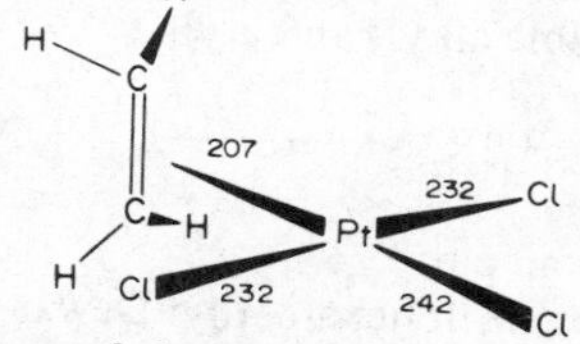

Figure 7.16 The structure of the anion of Zeise's salt $[Pt(C_2H_4)Cl_3]^-$

that the C–C bond is perpendicular to the $PtCl_3$ plane. Bonding in metal–olefin complexes is thought of in terms of both filled ligand $\pi \to$ metal σ bonding and 'back bonding' from filled metal d to empty ligand π^* orbitals (see figure 7.17). The ethylene molecule, however, does appear to undergo considerable rotational

Figure 7.17 Bonding in olefin complexes

oscillation about *both* the C–C axis and the platinum–olefin axis.

Zeise's salt reacts readily with donors such as cyanide or 2,2'-bipyridyl to evolve ethylene; the reaction does, however, appear to be an indirect process in the latter case

$$[PtCl_3(C_2H_4)]^- \xrightarrow{\text{bipy}} [PtCl(C_2H_4)(bipy)]^+ + 2Cl^-$$

$$\downarrow Cl^-$$

$$[PtCl_3 bipy] + C_2H_4$$

An important reaction of $[Pd(C_2H_4)Cl_3]^-$ is that it may be hydrolysed to form acetaldehyde; this is related to the industrial Smidt Wacker process for converting olefins into aldehydes.

Related to these compounds are the diolefin complexes such as the cyclo-octa-1, 5-diene complexes $(C_8H_{12})MCl_2$ and the dicyclopentadiene complex $PtCl_2(C_{10}H_{12})$; their structures are shown in figure 7.18(a) and (c). Reaction with alcohol (ROH)

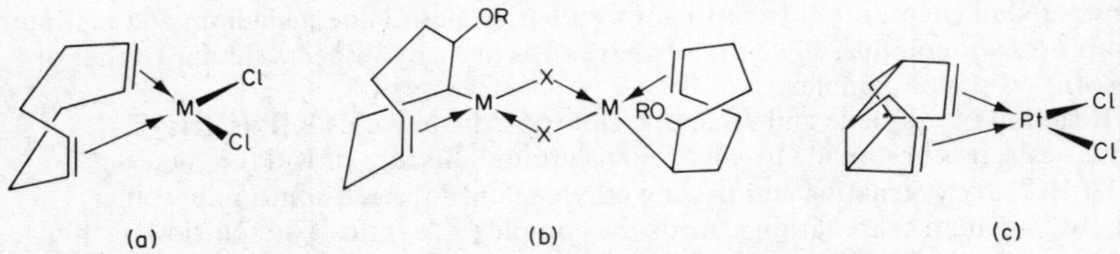

(a)　　　　　(b)　　　　　(c)

Figure 7.18 Some diolefin complexes

in the presence of weak base gives the stable binuclear species shown in figure 7.18(b). The palladium and platinum compounds differ sharply in that platinum forms the more stable alkyls and aryls.

$$C_8H_{12}PtI_2 + 2RMgI \longrightarrow C_8H_{12}PtR_2 + 2MgI_2$$

where R = Me, Et, Ph, for example.

Reaction of 'Zeise's dimer' with hexamethyl-Dewar-benzene yields a complex of that ligand, which in turn reacts with methoxide ion to give an allylic derivative.

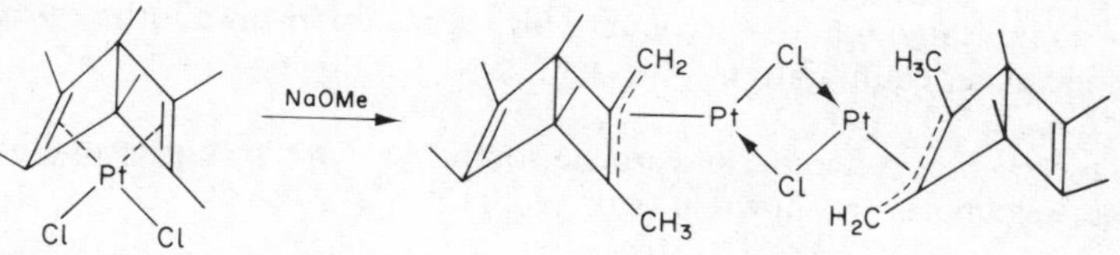

An analogous complex of the unsubstituted Dewar-benzene with palladium chloride has also been obtained

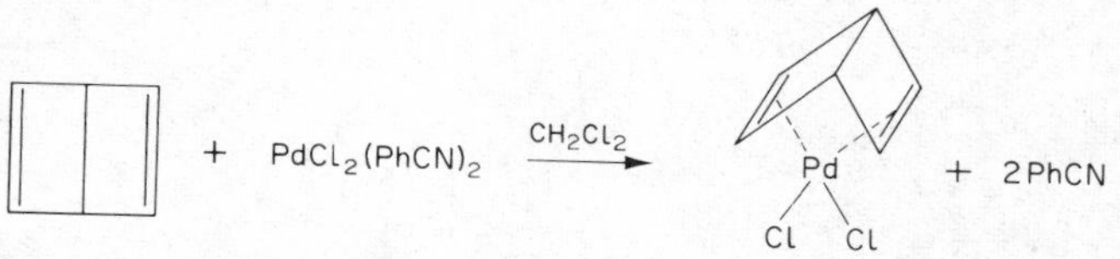

The tetraphenyl cyclobutadiene complex $Ph_4C_4PdCl_2$ is conveniently prepared from $PhC{\equiv}CPh$ and $PdCl_2$ in alcoholic HCl; this type of coupling does not occur generally with alkynes, and complexes of the type $[Pt_2Cl_4(RCCR')_2]$ and $K[PtCl_3(RCCR')]$ are formed (by analogous reactions to those used for Zeise's complexes) where $R = Bu^t$, $R' = Et$, Pr^i, for example. In many cases, these complexes are too labile to be isolated. Cyclo-octatetraene reacts directly with $PtX_4{}^{2-}$ to give $(C_8H_8)PtX_2$.

Allyls The structure of the dimeric π-allyl palladium (II) chloride differs from the alkenes in that the bonding is somewhat asymmetric (see figure 7.19). The central

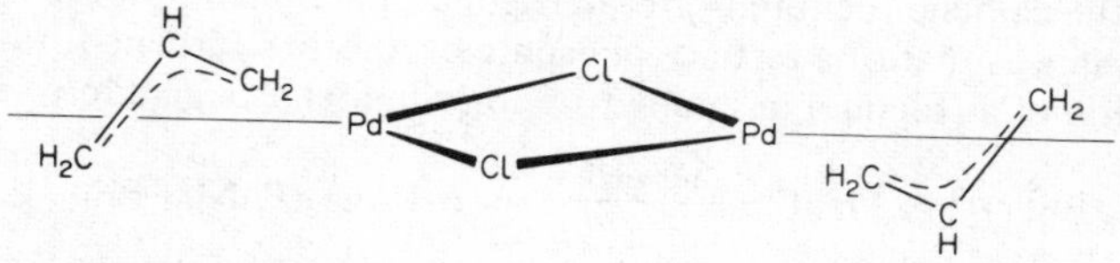

Figure 7.19 The dimeric π-allyl palladium (II) chloride complex

carbons are 'tipped' slightly away from the Pd–Pd axis; the n.m.r. spectrum is of the A_2B_2X variety, compatible with π-allyl behaviour; however, in the presence of donor ligands such as R_3P, Me_2SO, an A_4X signal is obtained; presumably the equivalence of the terminal hydrogens here results from rapid exchange. The compound is conveniently prepared from $PdCl_2$ and allyl alcohol. One other convenient route to the synthesis of π-allylic halide complexes uses acetate ion as a proton acceptor

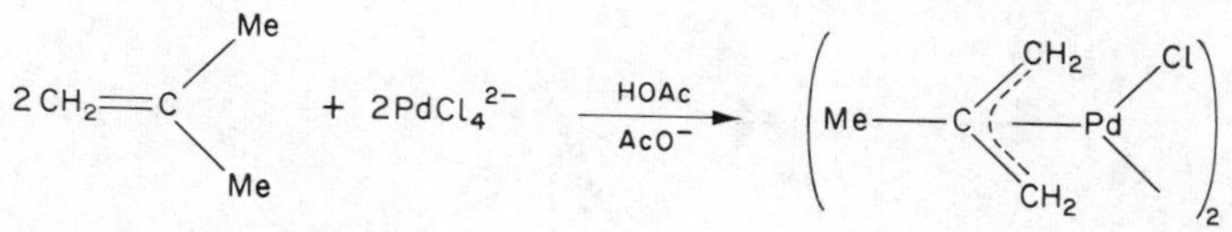

The biallyl, prepared as follows

$$PdCl_2 \xrightarrow[\text{ether, } -20°]{C_3H_5MgCl \text{ (excess)}} Pd(\pi\text{-}C_3H_5)_2$$

has quite complicated n.m.r. behaviour, which seems to involve both syn–anti exchange and isomer equilibration.

Carbene complexes A number of carbene complexes have been prepared in recent years; some examples are shown in figure 7.20.

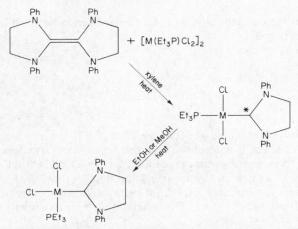

Figure 7.20 Preparation of carbene complexes of Pt or Pd

The crystal structure shows the platinum–carbon distance to be 200 pm and the geometry at the carbene (denoted*) to be trigonal; the *trans* influence of carbene is about the same as that of a tertiary phosphine. Another carbene-forming reaction utilises addition of an alcohol or amine to a co-ordinated isocyanide

$$\textit{trans-}[Pt_2Cl_4(PEt_3)_2] + PhNC \xrightarrow{\text{benzene}} \textit{cis-}[PtCl_2(PhNC)(PEt_3)]$$

$$\downarrow \text{EtOH}$$

$$\textit{cis-}[PtCl_2\{C(NHPh)OEt\}(PEt_3)]$$

The resulting carbene complex has the structure shown in figure 7.21 with a planar

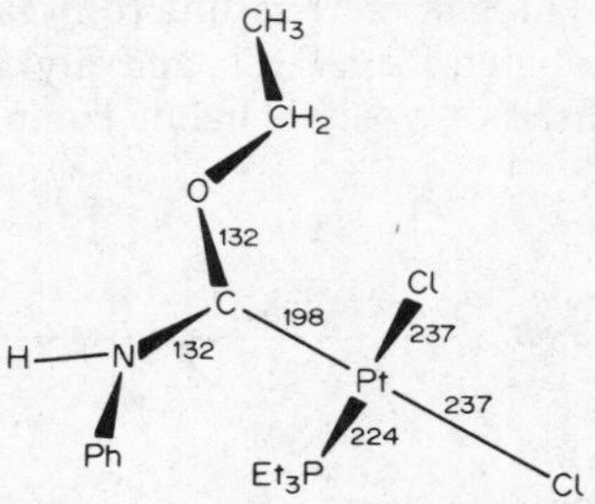

Figure 7.21 Structure of the carbene complex *cis*-[PtCl₂{C(NHPh)(OEt)}(PEt₃)] (after E.M. Badley, J. Chatt, R.L. Richards and G.A. Sim, *Chem. Commun* (1969), 1321)

OCNPt unit and short C—N and C—O distances. Recent ^{13}C n.m.r. data show that carbene complexes may be considered as metal-stabilised carbonium ions, since the carbene carbon gives a resonance some 300 p.p.m. upfield from tetramethylsilane, indicating a marked deshielding comparable to that for carbonium ions. Also, X-ray studies indicate that in all carbene complexes the π-system based on the sp^2 hybridised carbon atom shows little interaction with the metal.

A compound only recently shown to be of a carbene type is Chugaev's salt, first reported in 1915. A bright orange salt $PtC_8H_{15}N_6Cl$ separates when excess methylisocyanide, followed by hydrazine, is added to aqueous K_2PtCl_4. It is a 1:1 electrolyte, and n.m.r. data indicates it to have the cyclic structure shown in figure 7.22; the palladium analogue is also known.

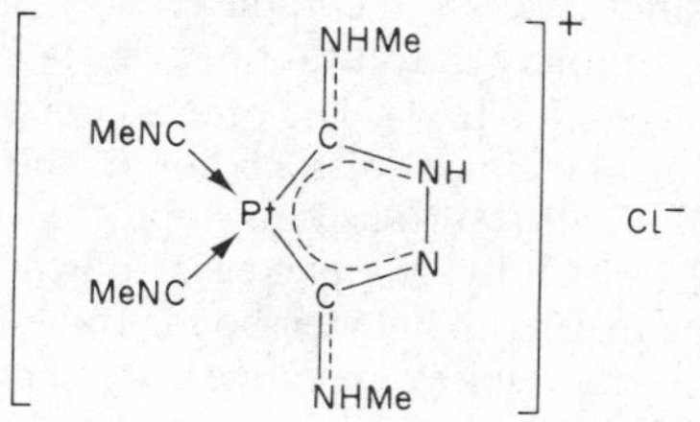

Figure 7.22 The cyclic structure assigned to Chugaev's salt

Cyclopropane reacts with chloroplatinic acid in acetic anhydride to give the brown $[C_3H_6PtCl_2]_n$; on treatment with pyridine the brown monomeric $[C_3H_6PtCl_2py]$ with *trans* chlorines and pyridines is formed. On refluxing in $CHCl_3$–CCl_4, the yellow $[C_3H_6PtCl_4py_2].CCl_4$ is formed. This has the structure shown in figure 7.23(a); the formulation as an ylide (b) rather than a carbene (c) is

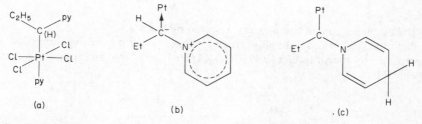

(a)   (b)   . (c)

Figure 7.23 (a) The structure of $[C_3H_6PtCl_4py_2]$; (b) ylide formulation; (c) carbene formulation

supported by the fact that on thermal decomposition propene is formed quantitatively, and that the stereochemistry about the σ-bonded carbon is not trigonal planar, as expected for (c).

7.5.4 Palladium(III) and Platinum (III)
No complexes in this oxidation state have been unequivocally established.

7.5.5 Palladium (IV) and Platinum (IV)

Pd (IV) and Pt (IV) complexes, having a d^6 metal ion in a fairly high oxidation state, will have a very considerable ligand-field stabilisation in the octahedral configuration, and this is accordingly the configuration adopted. The few Pd(IV) complexes are much less stable than those of Pt(IV), which are very stable indeed. Preparation is often by oxidation of square planar complexes, which in the case of platinum proceeds with retention of configuration

$$cis-PtCl_2(PPr_3)_2 \xrightarrow{Cl_2} $$

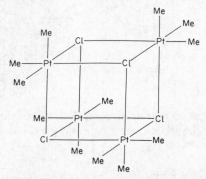

Despite their apparent kinetic stability, complexes such as $PtBr_2Cl_2(PEt_3)_2$ will interchange halogen fairly rapidly in solution to give statistical quantities of complexes $PtBr_xCl_{4-x}(PEt_3)_2$ ($x = 0$ to 4). The process possibility involves catalysis by traces of $PtX_2(PEt_3)_2$ (from incipient dissociation of the Pt(IV) complex) involving a halogen-bridged $Pt(IV)-X-Pt(II)$ transition step.

A series of PtX_6^{2-} (X = F, Cl, Br) complexes is known. N.Q.R. data indicates the ionic character of the platinum–halogen bonds to be in the region 0.30–0.53, in moderate agreement with the results of e.s.r measurements on the analogous Ir(IV) complexes. The acid $(H_3O)_2[PtCl_6]$ may be obtained by dissolution of platinum in aqua regia followed by repeated evaporation with hydrochloric acid. A variety of complexes, some employed by Werner in his elucidation of co-ordination theory, may be obtained from $PtCl_6^{2-}$ by substitution with ammonia or other ligands

$$K_2[PtCl_6] \xrightarrow{\text{liq. } NH_3} [Pt(NH_3)_6]Cl_4$$

$$K_2[PtCl_6] \xrightarrow{\text{en}} [Pt(en)_3]Cl_4$$

$$K_2[PtBr_6] \xrightarrow{NH_3} [PtBr_4(NH_3)_2]$$

The ethylenediamine complex was resolved by Werner.

Platinum (IV) has an unusually strong tendency to form bonds to carbon, as exemplified by the structures of $Pt(CH_3)_3Cl$ (see figure 7.24) and its acetylacetonato-derivative.

Figure 7.24 The tetrameric structure of $PtMe_3Cl$

A convenient synthesis is

$$12R_2Hg + 4\ PtCl_4 \xrightarrow{\ Me_2CO,\ -78°\ } (PtR_3Cl)_4 + 12RHgCl$$

where R = Me, Et.

Triethylplatinum chloride has also recently been shown to be a tetramer. In the acetylacetonato-derivative $Pt(CH_3)_3CH_3COCHCOCH_3$ an unusual type of acetylacetonato-bridge provides Pt—C bonds and at the same time brings about an octahedral configuration (see figure 7.25).

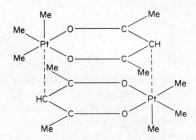

Figure 7.25 The dimeric structure adopted by $PtMe_3(acac)$

In the adduct $Me_3Pt(acac)(bipy)$, the bonding is also rather exotic; there are 3 Pt—Me σ bonds, a bidentate 2,2′-bipyridyl and an acetylacetone bound σ to carbon, six-co-ordination thus being achieved. The following interrelationship exists

$$\{Me_3Pt(acac)\}_2 \xrightarrow{\ bipy\ } Me_3Pt(acac)(bipy)$$

KI | Tl acac   Tl acac | KI

$$(Me_3PtI)_4 \xrightarrow{\ bipy\ } Me_3PtI(bipy)$$

The balance between C- and O,O-bonding in the acetylacetonates is very delicate; taking Pt(II) examples, $K[Pt(acac)Cl_2]$ is O,O-bonded, $Na_2[Pt(acac)_2Cl_2].2H_2O$, is probably C-bonded and in $K[Pt(acac)_2Cl]$, one acetylacetone is O,O bonded and the other C-bonded. In view of this, it is hardly surprising that the force constants for Pt—C and Pt—O stretching modes in platinum (acac) complexes have been found to be very similar.

Treatment of $K[Pt(acac)_3]$ with HCl gives a green-yellow compound $C_{10}H_{14}O_3PtCl_2$ (n.m.r.: methyls, $\tau\,7.8$ (6H, $J[Pt-H] = 34$ Hz), $\tau\,8.3$ (6H, $J[Pt-H] = 3$ Hz) and olefinic resonances at $\tau\,5.5$ (2H, $J[Pt-H] = 84$ Hz). Condensation of two acetylacetones has taken place. The structure of the compound is shown in figure 7.26.

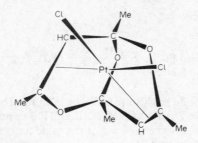

Figure 7.26 The structure of $C_{10}H_{14}O_3PtCl_2$, the condensation product of
KPt(acac)$_3$ with HCl (after D. Gibson, C. Oldham, J. Lewis, D. Lawton, R. Mason
and G.B. Robertson, *Nature, Lond.*, **208** (1965), 580)

7.6 Organo-compounds of Pt(II) and Pt(IV)

Figure 7.27 shows reactions employed to synthesise both platinum (II) and (IV)
complexes, as well as some related compounds. The configuration of many of these
complexes has been assigned with the aid of n.m.r. and dipole-moment measurements. Thus *trans*-[PtMeX(PMe$_2$Ph)$_2$] reacts with X_2 (X = Cl, Br) in MeOH to give
PtMeX$_3$(PMe$_2$Ph)$_2$; this complex has a dipole moment very similar to that of the
precursor — thus X_2 adds *trans*, and the methyl proton resonances show 'virtual'
coupling — hence the phosphines are *trans*. The configuration of this is structure
(I) of figure 7.28. Reaction of the complex with excess MeLi gives PtMe$_4$(PMe$_2$Ph)$_2$.
The methyl proton resonance is a doublet, showing the phosphines to be *cis*; this
and other evidence support configuration (II) of figure 7.28; the complex may also
be prepared from *cis*-[PtCl$_4$(PMe$_2$Ph)$_2$] and MeLi. Addition reactions generally
proceed with retention of configuration: thus *trans*-[PtIMe(PMe$_2$Ph)$_2$] reacts with
MeI to give PtI$_2$Me$_2$(PMe$_2$Ph)$_2$ (structure (III) of figure 7.28). Oxidative addition
of MeCOX (X = Cl, Br) also occurs; thus

$$cis\text{-PtMe}_2(\text{AsMe}_2\text{Ph})_2 \xrightarrow{\text{MeCOX}} \text{PtMe}_2(\text{COMe})\text{X}(\text{AsMe}_2\text{Ph})_2 \quad \text{(IV)}$$

N.M.R. studies of many of these complexes have been of value in elucidation of the
trans effect. Typically, $J[^{195}\text{Pt}-^1\text{H}]$ shows a sharp dependence on the *trans* ligand,
as can be seen by examination of the methyl resonances, reflecting the

trans-ligand	Cl	PR$_3$	Me
J(Hz)	67–73	56–59	44

σ-character of the Pt—C bond. Within the series *trans*-[PtMeL(PMe$_2$Ph)$_2$]$^+$,
$J[^{195}\text{Pt}-^1\text{H}]$ increases in the order PMe$_2$Ph ⟨ PPh$_3$ ⟨ CO ⟨ AsPh$_3$ ⟨ py, which is
compatible with the view that good σ donors exert their *trans* influence by
reducing the platinum 6s character in the Pt—CH$_3$ bond.

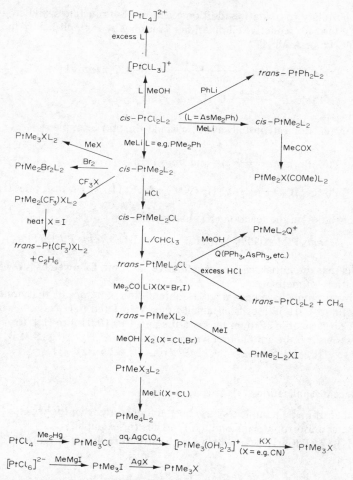

Figure 7.27 Some methyl-platinum complexes and related compounds

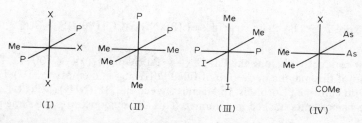

Figure 7.28 Configuration of some methyl-platinum (IV) complexes

The corollary to oxidative-addition reactions, but much less studied, is what might be termed 'reductive elimination'. This may be exemplified by the reactions (L = PMe$_2$Ph or AsMe$_2$Ph)

$$PtMe_2(COMe)XL_2 \xrightarrow[\text{quantitatively}]{\text{heat}} Me_2CO + \textit{trans-}PtMeXL_2$$

$$\textit{cis-}PtMe_4L_2 \xrightarrow{\text{heat}} C_2H_6 + \textit{cis-}PtMe_2L_2$$

However, reactions are not always so clean as this; for example

$$PtMeBr_3L_2 \xrightarrow{\text{heat}} \textit{cis-} \text{ and } \textit{trans-}PtBr_2L_2 + MeBr$$

$$PtMe_3ClL_2 \xrightarrow{\text{heat}} C_2H_6 \,(90\%) + MeCl\,(10\%) + \textit{trans-}PtMeClL_2$$

Kinetic studies on the reaction (X = halogen)

$$\textit{trans-}PtMeX(PPh_3)_2 + HCl \longrightarrow CH_4 + \textit{trans-}PtClX(PPh_3)_2$$

suggests that the initial reaction is oxidative addition, followed by slow 'reductive elimination' of methane.

Many attempts have been made to obtain PtMe$_4$; it appears that none have been successful. π-C$_5$H$_5$)PtMe$_3$ is obtained from PtMe$_3$I and C$_5$H$_5$Na in benzene–THF at 20°C; it is volatile at room temperature and 1 Pa (0.01 mm Hg). Crystallographic study shows it to be a monomer with a *pentahapto-*C$_5$H$_5$† ring. N.M.R. data are: τ 4.39 (5; J[Pt–H] 5.8 Hz); τ 9.12 (9; J[Pt–H] 82.5 Hz).

7.7 Recent Applications

A number of metal complexes have been under study recently because they show promise as anisotropic semiconductors. Thus for example, Pd(NH$_3$)$_2$Cl$_2$.Pd(NH$_3$)$_2$Cl$_4$, which has the structure shown in figure 7.29, displays

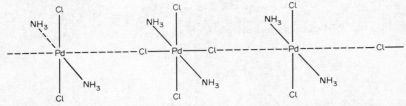

Figure 7.29 The chain structure of [Pd(NH$_3$)$_2$Cl$_2$] [Pd(NH$_3$)$_2$Cl$_4$]

its greatest conductivity along the Pd–Cl ––– Pd chain ($\sigma_\parallel$: $\sigma_\perp \approx 300:1$). The mechanism of this may be in overlap of filled Pd(II) d$_{z^2}$ and empty Pd(IV) d$_{z^2}$ orbitals with chlorine p$_z$ orbitals. In Magnus's green salt, [Pt(NH$_3$)$_4$] [PtCl$_4$], there are square planar units stacked in columns. This also demonstrates the same pheno-

† The *hapto* notation (Greek 'ηαπτεω, to fasten)

menon ($\sigma_\parallel : \sigma_\perp \gtrsim 100:1$). $6p_z$–$5d_z$ overlap is invoked here to give bands of molecular orbitals capable of this behaviour, the band gap being of the order of 5 eV.

It may be mentioned finally that certain complexes of the platinum metals, including *cis*-[Pt(NH$_3$)$_2$Cl$_2$] and *cis*-[Pt(NH$_3$)$_2$Cl$_4$] act as anti-tumour agents; they inhibit tumours caused by certain varieties of leukaemia and sarcoma viruses. The *trans* isomers are inactive. Both *cis* and *trans* isomers are known to bind to the DNA bases, but the *cis* isomers have a much greater capacity to inhibit DNA synthesis, and this may be relevant to their anti-tumour action.

COPPER

Metal: c.c.p.; m.p. 1084°; I_1: 7.72 eV; I_2: 20.29 eV; I_3: 36.83 eV
Oxides: Cu_2O, CuO
Halides: CuX (X = Cl, Br, I), CuX_2 (X = F, Cl, Br)

Oxidation State and Representative Compounds

	1	2	3
Typical donor atom/group	Hal, CN	O, N, Hal	O^{2-}, F
Co-ordination number			
2	$[Cu(NH_3)_2]^+$ √		
3	$[Cu(CN)_2]_n^-$ √ $Cu(Ph_3P)_2(NO_3)$ √	$CuCl_2$ (gaseous) √	
4 tet		$[CuCl_4]^{2-\dagger}$ √√	$Cu(S_2CNBu^n_2)Br_2$ √
4 planar		$[Cu(imidazole)_4]I_2^\dagger$ √√	
5 T.B.P.		$[CuCl_5]^{3-}$ √√	
5 S.P.		$[Cu\{NH_2(CH_2)_3NH_2\}_2OH_2]^{2+}$ √√√	
6		$[Cu(NO_2)_6]^{4-\dagger}$ √√√ √	√? $[CuF_6]^{3-}$
8		$CaCu(AcO)_4 \cdot 6H_2O$ √	

† Usually distorted: the dividing line between planar and octahedral co-ordination is indefinite.
? Suspected; √ Known; √√ Several examples; √√√ Very common.

8 Silver and Gold

These two metals are superficially the best known of those considered in this book, on account of their use in jewellery and as reserve currency metals, and their use, now passing, in coins. Considered from the chemical point of view, their behaviour is fairly straightforward compared with many of the other 4d and 5d metals. Silver has one particularly stable oxidation state, the d^{10} ion Ag(I), though several compounds of Ag(II) are known, as are a few Ag(III) compounds. In the case of gold, the d^{10} unipositive state is again stable but the heavier metal shows a marked increase in the stability of the +3 oxidation state compared with silver, so much so that Au(III) compounds are numerous and well-defined. Au(II), however, is extremely uncommon since it is sandwiched between the two very stable states Au(I) and Au(III) and is thus liable to immediate disproportionation. This situation is the converse of that obtaining for the oxidation states of copper, where Cu(II) is the most stable state, especially in aqueous solution or with oxygen or nitrogen ligands, but Cu(I) is stabilised by ligands having some π-acceptor character, such as phosphines, while Cu(III), although examples are known, is very rare. The d^8 ions in the +3 oxidation state usually adopt a square planar four-co-ordinate array of ligands, the resulting crystal-field stabilisation energy being very considerable and probably a major factor in the stability of these compounds. The effect of chemical environment on the relative stability of the various oxidation states of copper, silver and gold forms an interesting study.

The types of compound formed by silver and gold can be summarised very briefly as follows.

Table 8.1

Ag(I)	Linear and tetrahedral co-ordination; π-bonding ligands and halides; olefin complexes	Au(I)	Linear and tetrahedral co-ordination; π bonding ligands and halides
Ag(II)	Six-co-ordinate compounds with amine ligands	Au(II)	Very few; stabilised by 'suspect' ligands
Ag(III)	Square planar compounds with amine and fluoride ligands	Au(III)	Square planar co-ordination; π-bonding, halide and alkyl ligands
		Au(V)	Fluoro-complex only

8.1 The Metals, their Occurrence and Extraction

Silver, as is well known, is a soft metal with a colourless lustre, which has an attractive pearly quality. However, if unprotected, it forms a tarnish of sulphide within a few days in urban atmospheres. It is malleable and fairly soft, and its melting point is 960°. Silver dissolves in concentrated nitric acid.

Gold is not usually seen in a pure state (24-carat), since owing to its great softness it is alloyed with copper when used for jewellry. Although it has a pleasing yellow lustre in bulk, fine particles are purple. The melting point is 1063°. It is unattacked by the atmosphere or by single acids but dissolves in aqua regia, giving this reagent its name.

Gold and silver occur native, but silver also occurs as chloride or sulphide, and is often associated with copper ores. Gold typically occurs as minute particles in quartz but erosion also leads to the fairly widespread presence of small quantities of gold powder in rivers. Extraction from the quartz matrix is by action of cyanide and peroxide, when the metal is oxidised to the $[Au^I(CN)_2]^-$ ion. Gold is an intrinsically rare element, its abundance being about 5 mg per ton in the earth's crust and 1 mg per ton in seawater (from which it can be recovered but not economically).

8.2 Hydrated Ions

Silver, as Ag(I), is the only stable hydrated ion. Even this does not hydrate readily, but $AgClO_4$, AgF and $AgNO_3$ are readily soluble in water to give conducting solutions. It is evident that the hydration energy of the Ag^+ ion ($r = 113$ pm, intermediate between Na^+ and K^+) will be fairly small, and owing to its tendency to form strong covalent bonds with polarisable ligands silver salts usually crystallise in an anhydrous form. Ag^+ ion can be oxidised to Ag^{2+} ion by ozone but this ion is unstable in aqueous solutions except in strongly acidic ones; thus in 4M $HClO_4$, E_o for Ag^{2+}/Ag^+ is 2.0 V. Ag^{2+} ion may also be obtained by dissolution of AgO $(Ag^IAg^{III}O_2)$ in acid. The Ag^{3+} aquo-ion is unknown.

Gold forms no aquo-ions. The instability of the Au^+ aquo-ion, in contrast with Ag^+, is associated with the disproportionation $3Au(I) \xrightarrow{H_2O} 2Au + Au(III)$, the Au(III) state being much more stable than either Ag(II) or Ag(III). The behaviour partly resembles that of copper

$$2Cu(I) \xrightarrow{H_2O} Cu + Cu(II)$$

Thus aurous† chloride AuCl at once disproportionates in water into metallic gold and auric chloride. The auric aquo-ion is unknown.

8.3 Oxides

These metals have little affinity for oxygen ligands, and their oxides are relatively unstable. The known oxides are Ag_2O, AgO, Au_2O and Au_2O_3.

Black insoluble Ag_2O may be obtained by the action of alkali on silver nitrate solution followed by drying, but traces of water and alkali metal ion tend to remain

---

† As gold has only two common oxidation states, the -ous, -ic, nomenclature can be retained.

as impurity. It decomposes into silver and oxygen above 160°, and is slightly soluble in sodium hydroxide and soluble in nitric acid. The co-ordination of the Ag(I) ions by oxygen is tetrahedral, as in the isostructural Cu_2O. Fusion of K_2O with Ag_2O under argon gives the mixed oxide KAgO (others, such as CsAuO, are known). These compounds AMO contain M_4O_4 units involving linear co-ordination of M (Ag or Au).

Anodic oxidation of silver or the action of aqueous persulphate on Ag(I) gives AgO, a black substance, stable to about 100°. This is diamagnetic, and neutron diffraction shows two types of silver ion, one with linear and one with square planar co-ordination. The compound is thus $Ag^I Ag^{III} O_2$, but on dissolution in acid it gives a solution of Ag(II) ions. However, in alkaline periodate, the Ag(I) – Ag(III) distribution is retained, the two species being separated

$$4AgO + 6KOH + 4KIO_4 \longrightarrow 2K_5H_2[Ag^{III}(IO_6)_2] + Ag_2O + H_2O$$

A higher oxide approximating to Ag_2O_3 is the product of continued anodic oxidation of silver. It is probably a genuine but impure oxide of Ag(III).

Aurous oxide Au_2O is obtained by the action of alkali on $[AuBr_2]^-$. The grey solid evolves oxygen above 200°. Auric oxide, Au_2O_3, a brown substance, is obtained by alkaline precipitation from a Au(III) solution followed by drying. It reverts to Au_2O above 160°. Both the gold oxides, when freshly precipitated and hydrous, are soluble in alkali.

8.4 Halides

Silver and gold form the following halides.

Table 8.1

Ag(I)	F	Cl	Br	I	Au(I)		Cl	Br	I
Ag(II)	F				Au(III)	F	Cl	Br	I
Ag(III)	none								

Silver(I) fluoride AgF is a colourless solid, m.p. 435°. When anhydrous, it has the NaCl structure. However, it forms a tetrahydrate and dihydrate and is exceedingly soluble (14M) in water; this behaviour is very different from that of most other silver salts. On treatment of AgF solution with silver, a subfluoride Ag_2F is formed whose structure includes Ag–Ag bonds. Silver chloride, m.p. 449°, and silver bromide also have the NaCl structure; the iodide is trimorphic. The deepening colour along the series AgCl, AgBr, AgI is not uncommon in semicovalent metallic halides, the movement of absorption to longer wavelength reflecting the increasing ease of the charge-transfer process $Ag^+ \ldots X^- \to Ag \ldots X$. The lattice energies of AgCl, AgBr and AgI are respectively 5 per cent, 7 per cent and 9 per cent higher than the theoretical (that is, based on an electrostatic model) values. These values though small are significant and indicate some covalent bonding.

Silver difluoride is obtained by the action of fluorine on silver at 200°. It is dark brown and paramagnetic, the moment being low, and is thermally stable, having m.p. 690°. It is isomorphous with CuF_2, forming monoclinic crystals, and thus exhibits the tetragonal Jahn–Teller distortion of a d^9 ion. There are no other higher halides of silver.

The compounds AuX (X = Cl, Br, I) are made by gently heating the corresponding trihalides. They are insoluble in water, which converts them rapidly (AuCl) or slowly (AuI) into metallic gold and complex Au(III) halides. They decompose into the elements at moderate temperatures ($\approx 200°$).

The auric halides are dimeric Au_2X_6 (X = Cl, Br, I). The chloride and bromide are made by direct combination, while the iodide is made by the reaction of KI and $AuCl_3$ in cold aqueous solution in theoretical proportions (under different conditions aurous iodide is obtained instead). These halides have the planar structure characteristic of a d^8 ion (see figure 8.1).

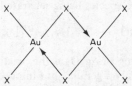

Figure 8.1 The structure of the gold halides Au_2X_6 (X = Cl, Br, I)

Auric fluoride may be obtained by the action of BrF_3 on gold, followed by heating to 300°. It is diamagnetic (low-spin d^8) and contains a nearly tetragonally distorted octahedron of fluoride ions around gold (Au–F = 2 of 191 pm, 2 of 204 pm and 2 of 269 pm). It is decomposed by water.

8.5 Other Compounds of Ag(I) and Au(I) and their Complexes

In their +1 oxidation states, silver and gold resemble each other fairly closely. Any differences are mainly caused by the ready oxidation, absent for silver, of Au(I) to Au(III) and by the slight preference of Ag(I) for higher co-ordination numbers than are stable for Au(I).

These two ions adopt a linear, two-co-ordinate configuration or a tetrahedral configuration with about equal readiness. Silver(I) is just occasionally octahedral, as in AgF, AgCl, AgBr and Ag_2MoO_4, but this co-ordination polyhedron has not apparently been established for Au(I). Also, silver(I) may possibly adopt a three-co-ordinate structure in certain complexes $AgI(PRR'_2)_2$.

Silver nitrate $AgNO_3$, whose crystal structure was determined as recently as 1966 despite the commonplace nature of this compound, has complicated co-ordination polyhedra with about ten oxygen atoms surrounding each silver atom. This ionic structure can be related to the large radius of Ag(I) (113 pm) and the electronegative character of oxygen.

The linear covalent situation is well demonstrated by the cyanides. Both AgCN and AuCN have linear polymeric chains $\rightarrow$ M–CN $\rightarrow$ M–CN $\rightarrow$ in contrast to the NaCl structure of NaCN; the otherwise unnecessarily low co-ordination number indicates very considerable covalent character. Silver thiocyanate has a polymeric structure rather similar to that of AgCN

$$Ag–S–C–N \rightarrow Ag–S–C–N \rightarrow Ag–$$

the angle Ag–$\hat{S}$–C being 165°. Both AgCN and AuCn readily dissolve in KCN solution to give the linear complex ion $[NC–M–CN]^-$. The diammine cation

$[Ag(NH_3)_2]^+$ is also quite stable. Other complex anions are $[AgCl_2]^-$ and $[AuCl_2]^-$. The latter and its analogue $[AuBr_2]^-$ may be made by phenylhydrazine (for the chloride) or acetone (for the bromide) reduction of $NBu_4[AuX_4]$. They are decomposed by water. $Ag(imidazole)_2 NO_3$ also exemplifies the linear two-co-ordination of Ag(I), with Ag–N distances of 212 and 213 pm; the nearest oxygen is as much as 296 pm away. The pyrazine complex $Ag(pyrazine)NO_3$ contains nonlinear

$$-Ag-N\underline{}N-Ag-$$

units in infinite chains, having Ag–N = 221 pm and N–$\widehat{Ag}$–N angles of 160° (see figure 8.2).

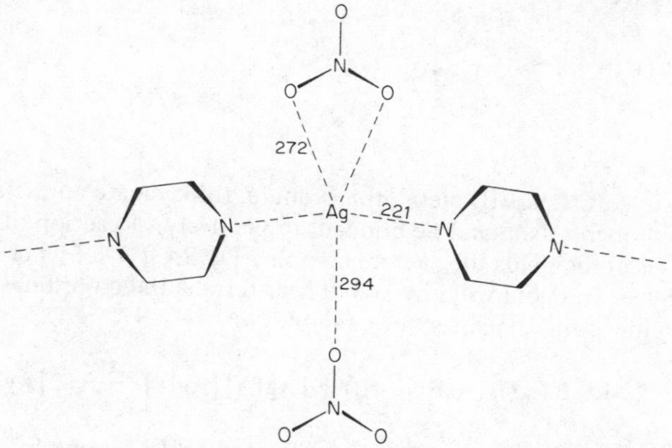

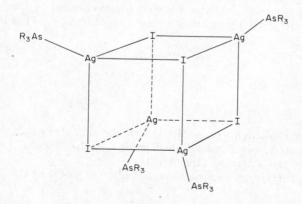

Figure 8.2 Nearest neighbours of the Ag^+ ion in the complex $Ag(pyrazine)(NO_3)$

It is to be expected that these filled d-shell ions in a rather low oxidation state will bond strongly with phosphines and arsines, and many compounds R_3PMX and R_3AsMX (X = Cl, Br, I or SCN) are known. In these, however, silver forms four-co-ordinated tetramers (see figure 8.3) while gold forms two-co-ordinated

Figure 8.3 Structure adopted by silver (I) arsine complexes

monomers. The gold–phosphine compounds are especially stable and after melting can be distilled under reduced pressure.

A range of triphenylphosphine aurous complexes can be prepared by the following scheme; triphenylphosphine reduces Au(III) to Au(I)

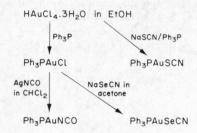

Au(I) being a 'soft' (class B) metal, the cyanate, thiocyanate and selenocyanate complexes are probably N, S and Se bonded, respectively. Reaction of Ph_3PAuCl with argentous acetate yields the acetate complex $Ph_3PAu(OAc)$ whose structure is unknown. Silver and gold will give stable near-tetrahedral co-ordination with chelate diphosphines and diarsines; for example

$$2AgI + 2o\text{-}Et_2P.C_6H_4.AsEt_2 \ (=L) \rightarrow [AgL_2][AgI_2] \xrightarrow{\ 2L\ } 2[AgL_2]I$$

At the 1:1 metal:ligand ratio, no three-co-ordinate [AgIL] species is formed, but an ionic dimer having four- and linear two-co-ordination. Silver(I) is also found tetrahedrally co-ordinated with sulphur in $AgAlS_2$, which has the zinc blende structure.

The typical π-bonding ligand CO also combines with AuCl giving crystalline AuCl(CO), which is, however, at once decomposed by water; the silver analogue is unknown. The formation of two stable linear covalent bonds by Ag(I) and Au(I) is characteristic of d^{10} ions in this portion of the periodic table. The ions Zn(II), Cd(II) and Hg(II), especially the latter, form many two-covalent compounds such as $HgPh_2$. In Group III, Tl(III) recurs to this theme with $[TlMe_2]^+$. Thus the group of metals

	Cu	Zn	
Pd	Ag	Cd	
	Au	Hg	Tl

show this behaviour, where Au(I) and Hg(II) form the most stable compounds. While the cause must be associated with favourable s, p_z, d_{z^2} bonding, it is not at present possible to give a detailed explanation.

Silver(I) and gold(I) form covalent bonds to carbon of moderate stability. Thus the reaction

$$AgCl + PhMgBr \rightarrow AgPh + MgBrCl$$

proceeds at $-18°$ to give silver phenyl as an unstable solid. In the case of gold, stable compounds may be obtained using phosphine ligands. Thus phenyl lithium and $AuCl(PPh_3)$ give $PhAuPPh_3$; $MeAuPPh_3$ and $Me_3SiCH_2AuPPh_3$ are also known. The C_6F_5 group often forms stable metal–carbon σ bonds and the fairly stable compound $LiAg(C_6F_5)_2$ may be obtained by the reaction of AgCl with LiC_6F_5 in ether. The compound AuC_6F_5 is rather unstable with respect to decomposition to $C_6F_5.C_6F_5$ but is stabilised by formation of the adduct $Ph_3PAuC_6F_5$.

A further example, which again shows the preference of Au(I) for linear σ-bonded two-co-ordination, is provided by the cyclopentadienyl compound $Ph_3PAuC_5H_4Me$. The relative magnitude of the P–H coupling constants in the proton n.m.r. spectra of this compound in 2-methyltetrahydrofuran solution between room temperature and $-100°$ indicates rapid interconversion ('fluxion') between σ-bonded isomers such as (I) and (II) rather than a π-bonded system as in (III) (see figure 8.4).

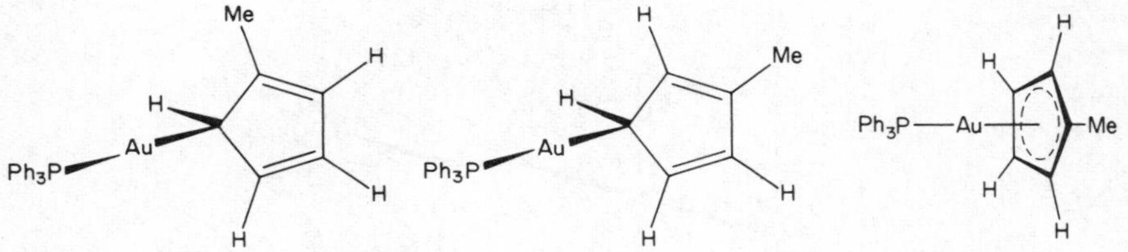

Figure 8.4 Possible isomers of $Ph_3PAuC_5H_4Me$: (I) and (II) are σ-bonded; (III) is π-bonded

Silver forms an explosive acetylide Ag_2C_2 said to be more sensitive than mercury fulminate. It may be made from ammoniacal silver nitrate and acetylene; corresponding Cu and Au compounds are known (in addition, the action of aqueous ammonia on Au_2Cl_6 gives a solid product which on rubbing produces an expensive explosion!).

Silver(I) forms complexes with unsaturated hydrocarbons; gold(I) hardly does so at all. Polymeric complexes (LAgX), for example, are formed with L = cyclo-octatetraene, $X = NO_3$ and with L = benzene, $X = ClO_4$. The relative reactivity of Ag and Au here contrasts with the behaviour of palladium and platinum, where the heavier metal co-ordinates more strongly with unsaturated hydrocarbons. The silver–olefin bonds tend to be somewhat unsymmetrical, but are doubtless of the same general nature as in the platinum complexes discussed previously (see figure 8.5).

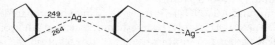

Figure 8.5 Unsymmetrical π bonding in the complex $AgClO_4.C_6H_6$

When $MeAuPMe_3$ reacts with $F_3CC{=}CCF_3$ under the influence of ultraviolet light, one product is the σ-alkylene complex(I) (figure 8.6) rather than a π-olefin complex. Also formed in this reaction is an insertion product $cis\text{-}Me_3PAuC(CF_3):C(CF_3)Me$. Both these compounds arise by reductive elimination from an intermediate(II) (figure 8.6), which has been isolated and characterised by X-ray analysis. Interestingly, it contains one Au(I) and one Au(III) atom (the presence of both oxidation states is clearly shown by the [197]Au Mössbauer spectrum). Furthermore, an isomer of (II), which is *cis* at the square planar Au(III) atom, has also been isolated from this reaction.

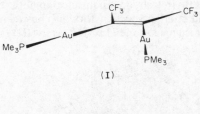

(I)

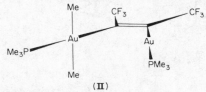

(II)

Figure 8.6 Products of the reaction between $MeAuPMe_3$ and $F_3C.C{=}C.CF_3$ (after C.J. Gilmore and P. Woodward, *Chem. Commun.* (1971), 233 and J.A.J. Jarvis, A. Johnson and R.J. Puddephatt, *Chem. Commun.* (1973), 373)

Gold(I) forms remarkably stable metal–metal bonds, and in this property it resembles its d^{10} neighbour Hg(II). Thus $AuCl(PPh_3)$ will split the osmium triangle in osmium dodecacarbonyl to give a linear complex

$$Os_3(CO)_{12} + AuCl(PPh_3) \longrightarrow Ph_3PAuOs_3(CO)_{12}Cl$$

and will react with $Na[Co(CO)_4]$ to give $Ph_3PAuCo(CO)_4$. The structure of the latter compound involves, as expected, nearly linear co-ordination of gold ($Co\widehat{-Au}-P$ = 177.5°; Au–Co = 250 pm). The isoelectronic iron compound $Ph_3PAuFe(CO)_3NO$ is also known.

There are some remarkable gold cluster compounds such as $Au_{11}(SCN)_3(PPh_3)_7$ and the similar compound $Au_{11}I_3\{P(p\text{-}FC_6H_4)_3\}_7$. The structure of the second is illustrated in figure 8.7. The central gold atom is co-ordinated with ten Au–X groups (X = I or $P(p\text{-}FC_6H_4)_3$; the gold atoms are separated by distances that indicate mutual bonding (260–319 pm).

The cluster compound $Au_9(NO_3)_3\{P(p\text{-}tolyl)_3\}_8$ may be prepared by borohydride reduction of $Au(NO_3)\{P(p\text{-}tolyl)_3\}$ in ethanol; a central Au atom is co-ordinated to eight $-AuPR_3$ groups. A minor product of this reaction is $[Au_6\{P(p\text{-}tolyl)_3\}_6](BPh_4)_2$, in the cation of which six gold atoms appear at the

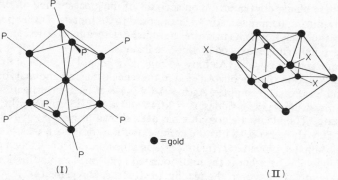

Figure 8.7 The structure of the gold cluster complexes $Au_9(NO_3)_3\{P(p\text{-tolyl}_3)\}_8$ (I) (p-tolyl groups omitted) and $Au_{11}I_3\{P(p\text{-FC}_6H_4)_3\}_7$ (II) (core only shown)

vertices of an octahedron, neighbours being 293–309 pm apart. Each gold atom is bonded to one $P(p\text{-tolyl})_3$ group. Octahedra of silver atoms are also known and occur in $Bu_4NAg_3I_4$ and $Ag(S_2CNPr_2)$, the Ag–Ag distances being respectively 320 and 290–297 pm (Ag–Ag in metallic silver = 289 pm).

The bonding in compounds of this type can be rationalised in terms of molecular orbitals that extend over the metal cage. These orbitals, of which the fully symmetric A_{1g} is the most important, are derived from those components of the metal s and p orbitals (the d orbitals are probably ineffective) that are not required for bonding to the peripheral phosphine (or other) ligands.

There is an interesting example of apparent intermetallic bonding in $Au_2(S_2CNR_2)_2$, the structure of which is shown in figure 8.8. The Au–Au separation is only 276 pm, compared to 288 pm in metallic gold; to accommodate this

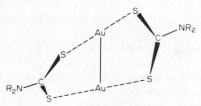

Figure 8.8 The structure of the dimeric gold (I) dithiocarbamates

short Au–Au distance, one of the dithiocarbamates is twisted relative to the other by 16.5°; the S–Au–S bonds are linear, as expected for Au(I), and this 'twist' seems to have been introduced to accept the mutual interaction of the gold atoms. A Raman band at 185 cm^{-1} has been assigned to an Au–Au stretching mode.

8.6 Complexes of Silver(II)

Ag(II) is a d^9 ion and should therefore adopt a tetragonal configuration with four short and two long bonds, as Cu(II) does on account of Jahn–Teller stabilisation, or even a square planar configuration on account of the greater value of Δ, the crystal-field splitting parameter, for 4d as opposed to 3d metals. This expectation is realised in those cases where structures have been determined. The complexes mainly have amine, amine–oxygen or fluoride ligands.

The complexes $[Ag\,py_4]^{2+}$, $[Ag\,bipy_2]^{2+}$ and $[Ag\,phen_2]^{2+}$ can be obtained by persulphate oxidation of aqueous Ag^+ ion in presence of ligand and may be precipitated as perchlorates. The complex $Ag(bipy)_2(NO_3)_2$ has been investigated by electron spin resonance; the g-values, $g_\perp = 2.045$ and $g_\parallel \approx 2.13$ confirm the Ag(II) oxidation state. The picolinate complex has been shown to have a square planar structure (I, X = H; figure 8.9), but the complex with pyridine-2,6-dicarboxylic acid has a six-co-ordinate structure(II, figure 8.9); one ligand molecule being a doubly charged anion while the other is the undissociated acid. In the pyridine-2,3-dicarboxylic acid complex, however, the picolinate structure is regained (I, X = CO_2H).

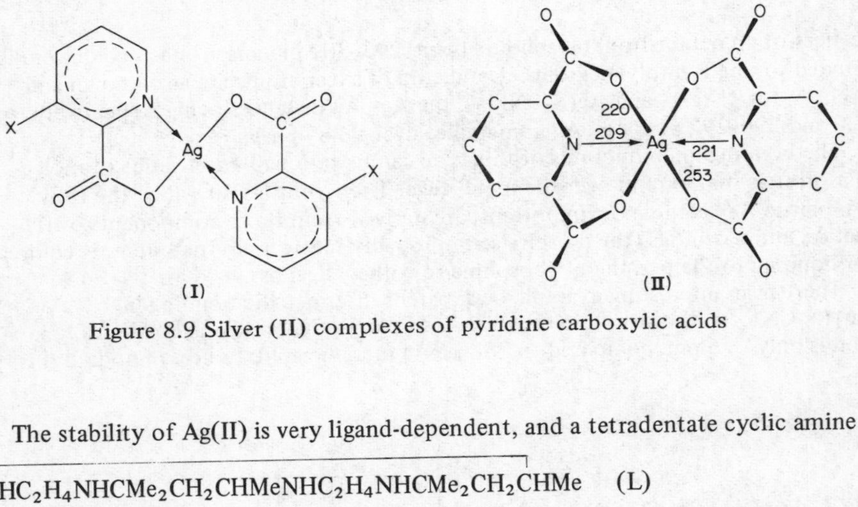

Figure 8.9 Silver (II) complexes of pyridine carboxylic acids

The stability of Ag(II) is very ligand-dependent, and a tetradentate cyclic amine

$$NHC_2H_4NHCMe_2CH_2CHMeNHC_2H_4NHCMe_2CH_2CHMe \quad (L)$$

will even cause Ag(I) to disproportionate in water solution (but not in acetonitrile) into Ag metal and the yellow Ag(II) complex of the tetra-amine

$$AgClO_4 + L \xrightarrow{CH_3CN} [Ag^IL]ClO_4 \xrightarrow{H_2O} [Ag^{II}L](ClO_4)_2 + Ag + L$$

Fluoro-complexes may be obtained by fluorination of a mixture of Ag_2SO_4 with alkaline earth carbonates, for example

$$BaCO_3 + Ag_2SO_4 \xrightarrow{F_2,\ 800°} Ba[AgF_4]$$

The violet complexes are square planar (Ag–F = 205 pm). They have magnetic moments of 1.7-2.0 B.M., consistent with their d^9 square planar status.

8.7 Complexes of Gold(II)

Gold(II) complexes are rare, but the presence of such species may be verified by e.s.r. spectroscopy or magnetic measurements. Thus Au(II) bis(dialkyldithiocarbamates) have been detected in solution by these means. The best characterised gold (II) complex is that of maleonitriledithiolate (MNT, *cis*-$S_2C_2(CN)_2{}^{2-}$) formed by reduction of the gold(III) complex or by the following reaction

$$(PhCH_2)_2SAuX + \{S(CH_2Ph)_2\}_2AuX_3 \xrightarrow[Bu^n{}_4NBr]{LiMNT} (Bu^n{}_4N)_2[Au(MNT)_2]$$

where X = Cl, Br. This complex is quite stable; the room-temperature magnetic moment is 1.85 B.M. The e.s.r. spectrum yields the *g*-values 1.978, 2.006 and 2.017, while the isotropic hyperfine splitting (^{197}Au has $I = 3/2$) is 4.04 millitesla (40.4 gauss); this indicates that delocalisation of the unpaired electron is considerable. The π-carbollyl complex $(Et_4N)_2[(\pi\text{-}3\text{-}1,2\text{-}B_9C_2H_9)_2Au]$ is also known; it is prepared by reduction of the auric analogue with sodium amalgam.

The apparent Au(II) complexes $AuX(S_2CNBu_2)$ (X = Cl, Br, I) are in fact Au(I) – Au(III) compounds $[Au(S_2CNBu_2)_2]^+[AuX_2]^-$.

8.8 Complexes of Silver(III)

If potassium salts are used instead of barium in the fluorination process that gives $[Ag^{II}F_4]^{2-}$, the complex fluoride $K[Ag^{III}F_4]$ is obtained. It is yellow and diamagnetic. The structure of the diamagnetic ethylene-bisbiguanide complex $Ag(ebbg)(ClO_4)_3$ has been determined (see figure 8.10); it is square planar as expected for a diamagnetic d^8 complex. It is prepared by persulphate oxidation of aqueous Ag^+ in the presence of ligand, a method that gives Ag(II) complexes of bipyridyl and related ligands. It is not clear what determines the relative stability of Ag(II) and Ag(III) in this situation.

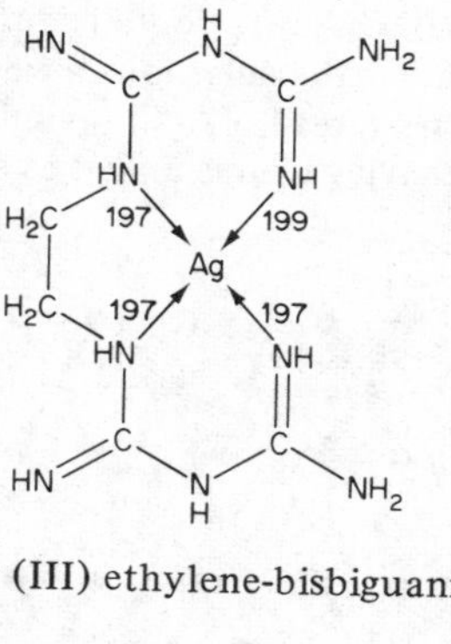

Figure 8.10 The silver (III) ethylene-bisbiguanide complex ion

8.9 Complexes of Gold(III)

The square planar, diamagnetic auric state has a stability approaching that of its d^8 neighbour, Pt(II), with which it also compares by forming σ bonds to carbon with similar ease. It does not, however, form bonds with olefins to such an extent. Analysis of the electronic spectra of several square planar AuX_4^- complexes (X = SCN, Cl, Br, for example) yields the energy sequence $b_{1g}(x^2 - y^2) \rangle b_{2g}(xy) \rangle e_g(xz, yz) \rangle a_g(z^2)$.

A convenient starting point for the preparation of auric complexes is chloro-auric acid, $HAuCl_4$, formed by dissolving gold in aqua regia and subsequently eva-porating with hydrochloric acid. This acid can be obtained as a solid hydrate $HAuCl_4.4H_2O$, shown by X-ray investigation to contain planar $[AuCl_4]^-$ and $H_5O_2^+$ ions, and can be converted into neutral salts, for example $K[AuCl_4]$. This latter is in turn converted by substitution, into other complex ions $[AuX_4]^-$ (X = F, Br, I, CN, SCN, NO_3). The last of these provides a rare authenticated example of nitrate ion acting as an essentially monodentate species with Au–O = 199 pm (the next nearest oxygen atoms are 287 pm distant from Au).

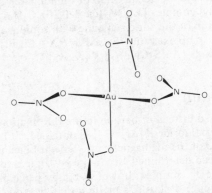

Figure 8.11 The structure of the $[Au(NO_3)_4]^-$ ion (after C.D. Garner and S.C. Wallwork, *J. chem. Soc. (A)* (1970), 3093)

The apparent Au(II) complex $CsAuCl_3$ contains square planar $[AuCl_4]^-$ and linear $[AuCl_2]^-$ ions linked so that the auric ions have (4 + 2) tetragonal co-ordina-tion. Similarly $Cs_2AuAgCl_6$ is $Cs_2(AuCl_4)(AgCl_2)$ while $(NH_4)_6Au_3Ag_2Cl_{17}$ is in fact $(NH_4)_6(Ag_2Cl_5)(AuCl_4)_3$ having chains linked as shown in figure 8.12.

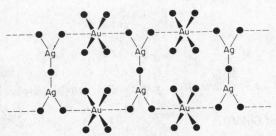

Figure 8.12 The chain structure of $(NH_4)_6(Ag_2Cl_5)(AuCl_4)_2$ (one $[AuCl_4]^-$ ion lies parallel to and between Ag_2Cl_5 ions in adjacent chains (after J.C. Bowles and D. Hall, *Chem. Commun.* (1971), 1523)

The action of amine, phosphine or arsine ligands on Au_2Cl_6 readily gives a variety of complexes, the phosphine and arsine complexes, however, having a tendency for reduction to the aurous state. Four-co-ordination, as in $[Au(diars)_2](ClO_4)_3$ and $[AuBr_3(PMe_3)]$, is not always preserved in these compounds. Thus $[AuCl_3(2,2'-biquinolyl)]$ has been shown to have an irregular five-co-ordinate structure and $[AuI_2(diars)_2]I$ is six-co-ordinate. Auric complexes may also be prepared by oxidation of aurous species

$$AuCl(PMe_3) + Br_2 \rightarrow AuBr_2Cl(PMe_3)$$

8.10 Gold(III) Organo-compounds

These are interesting compounds, first made by Pope and Gibson as long ago as 1907. They are fairly stable and have an extensive chemistry. They are distinguished from other transition-metal organo-compounds by the absence, in many instances, of π-bonding ligands. It is usual for the presence of this type of ligand to be required for stability, as for example in $PdMe_2(PEt_3)_2$ (though the platinum(IV) trimethyls, such as $Pt_4Me_{12}Cl_4$, provide another exception to this generalisation).

Although all three types AuR_3, AuR_2X and $AuRX_2$ (R = alkyl, X = anionic ligand) are known, AuR_2X is much the most stable — in fact, among the stablest of σ alkyls known. Most of the known compounds are alkyls rather than aryls, and this seems to reflect a real difference in stability and not merely lack of exploration. In the several cases that have been specifically checked by X-ray analysis, the gold is four-co-ordinate and square planar, with dimerisation or other polymerisation taking place if necessary to achieve this configuration. It is practically certain that all these compounds have this geometry. Trimethylgold may be prepared in ether solution by the reaction

$$Au_2Br_6 + 6MeLi \xrightarrow{\text{ether, }-65^\circ} 2AuMe_3 . Et_2O + 6LiBr$$

It decomposes at -35°, but forms a stable complex $AuMe_3(PPh_3)$ on addition of triphenylphosphine. The trimethyl may be converted into the dimethyl monohalide by treatment with HCl or Au_2Br_6

$$[AuMe_3(Et_2O)] \begin{array}{c} \xrightarrow{\text{HCl}} [Au_2Cl_2Me_4] + MeCl \\ \xrightarrow{Au_2Br_6} [Au_2Br_2Me_4] \end{array}$$

The triphenyls are unknown, but reaction of $AuCl_3$ with C_6F_5MgCl yields $Au(C_6F_5)_3$ (at 100° and 65 Pa (0.5 mm Hg), it readily yields Au and $C_6F_5 . C_6F_5$), which again forms a stable PPh_3 adduct. Bis(pentafluorophenyl) thallium(III) bromide may be conveniently used in the preparation of gold pentafluorophenyls thus

$$(C_6F_5)_2TlBr + Ph_3PAuCl \xrightarrow{\text{benzene}} cis\text{-}(C_6F_5)_2AuCl(PPh_3) + TlBr$$

This gold complex is shown by X-ray diffraction to be essentially square planar (see figure 8.13).

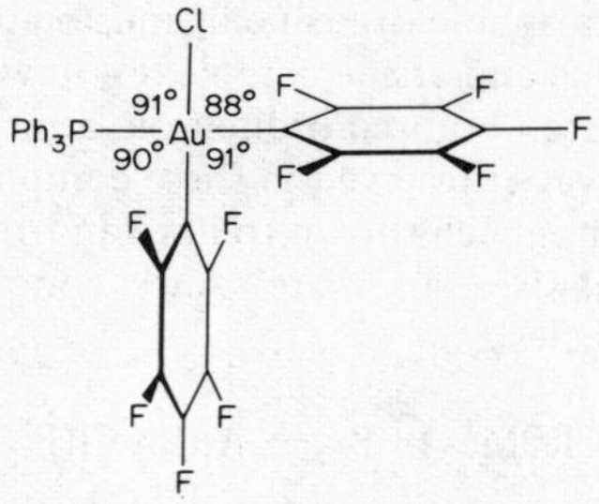

Figure 8.13 The co-ordination geometry in *cis*-$(C_6F_5)_2$AuCl(PPh$_3$) (after R.W. Baker and P. Pauling, *Chem. Commun.* (1969), 745)

The bromide and iodide are obtained by metathesis. Reduction of AuCl$(C_6F_5)_2$PPh$_3$ with alcoholic hydrazine yields the Au(I) complex Au(C_6F_5)PPh$_3$.

Dibromotetraethyldigold, a planar dimer (see figure 8.14), may be obtained by a Grignard reaction in pyridine solution

$$\text{Au}_2\text{Br}_6 \xrightarrow{\text{pyridine}} \text{AuBr}_3\text{py} \xrightarrow{\text{EtMgBr}} \text{Au}_2\text{Br}_2\text{Et}_4$$

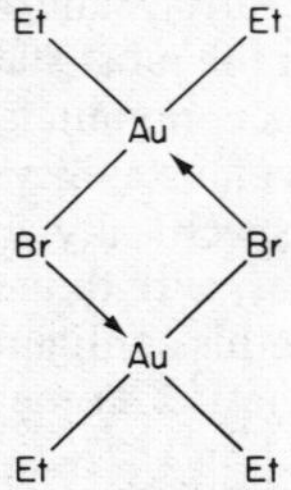

Figure 8.14 The structure of dibromotetraethyldigold

Many other compounds Au$_2$X$_2$R$_4$ (R = alkyl, X = Cl, Br, SCN, CN, ½SO$_4$, ½C$_2$O$_4$) are known. They are unaffected by stannous chloride, showing that the Au(III) state has very considerable stability. In the case of the cyanides, obtained by the action of silver cyanide on the bromides, a dimeric structure is unfavourable, because the cyanide ion has a lone pair on the nitrogen atom which is linearly directed, M–C≡N: →. Hence a tetrameric structure is adopted, confirmed by X-ray diffraction in Au$_4$(CN)$_4$Pr$^i_8$ (see figure 8.15).

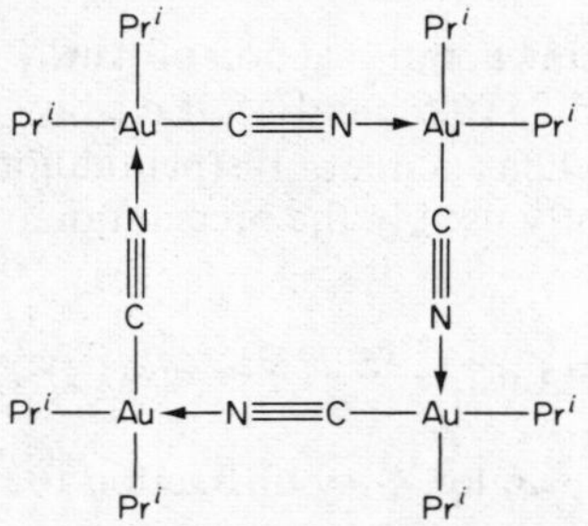

Figure 8.15 The structure of Au$_4$(CN)$_4$Pr$^i_8$

Oxy-acids will also displace the halide ion

$$[Au_2Br_2Et_4] \xrightarrow{Ag_2SO_4} Au_2SO_4Et_4$$

$Au_2I_2Me_4$ and related compounds have been employed in a number of syntheses.

$$Au_2Cl_2Me_4 \xrightarrow[\text{cyclopentane}]{L} AuClMe_2 . L \qquad (L = py, PPh_3, AsPh_3)$$

$$Au_2Br_2Et_4 \xrightarrow{NH_3} AuBrEt_2NH_3$$

$$Au_2I_2Me_4 \xrightarrow{Ph_3P} AuIMe_2PPh_3 \xrightarrow{MeLi} AuMe_3 . PPh_3$$

$[AuEt_2 en]Br$ (from en)

$$Au_2I_2Me_4 \xrightarrow{AgNO_3/HNO_3}$$

$$\text{'AuMe}_2^{+}\text{'}$$

$$\downarrow (1) NaOH; (2) 2M\ HNO_3$$

$$Au_2Cl_2Me_4 \xleftarrow{HCl} \{Au(OH)Me_2\}_4 \xrightarrow[HClO_4,\ PPh_3]{MeOH} [AuMe_2(PPh_3)_2]ClO_4$$

$Au_2I_2Me_4$ may readily be converted into the other dimethylhalides by treatment with $AgNO_3$ followed by NaX. $Au(N_3)Me_2$ adopts a dimeric structure in which each azide bridges via only one nitrogen atom; $Au(OH)Me_2$ is a tetramer, with bridging hydroxyls completing a puckered eight-membered ring (see figure 8.16). In aqueous solution, the dimethylgold cation appears, from Raman and n.m.r. data, to exist as the *cis*-$[AuMe_2(OH_2)_2]^{+}$ ion; $v(Au-CH_3)$ is assigned at 590 cm^{-1} and $v(Au-OH_2) \approx 400$ cm^{-1}. The structure of the sulphato-complex, which is soluble in benzene and in water, is uncertain.

Many of these compounds are quite volatile; a golden mirror may be deposited on glassware by heating the exterior of a flask containing the vapour of $[Au_2Br_2Et_4]$.

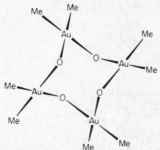

Figure 8.16 The puckered structure of $\{Au(OH)Me_2\}_4$ (after G.E. Glass, J.H. Konnert, M.G. Miles, D. Britton and R.S. Tobias, *J. Am. chem. Soc.,* **90** (1968), 1131)

The monoalkyl (and monoaryl) auric derivatives are intermediate in stability between the trialkyls and the dialkyls. The monoalkyls are made as follows

$$Au_2Br_2Et_4 \xrightarrow{\;Br_2\;} Au_2Br_4Et_2$$

They are dimeric and have a *cis* configuration, as demonstrated by the dipole moment (5.5 D) of the propyl compound. Of the two possible such configurations, $Et_2AuBr_2AuBr_2$ rather than *cis*-EtBrAuBr$_2$AuEtBr is possibly indicated by the reactions

$$Au_2Br_4Et_2 \xrightarrow{\;KBr\;} K[AuBr_4]^- + Au_2Br_2Et_4$$

$$\downarrow en$$

$$[Au\,en_2]Br_3 + [AuEt_2\,en]Br$$

$Au_2Br_4Et_2$ is reduced by stannous chloride and is decomposed into EtBr and AuBr by gentle heat.

Gold(I) bromide forms a 1:1 complex with the phosphine-olefin ligand $Ph_2P.o\text{-}C_6H_4.CH:CH_2$ (L), which is oxidised to $AuBr_3L$ by bromine. The product is, however, a monobenzyl derivative of AU(III) and does not contain the $AuBr_3$ group (see figure 8.17).

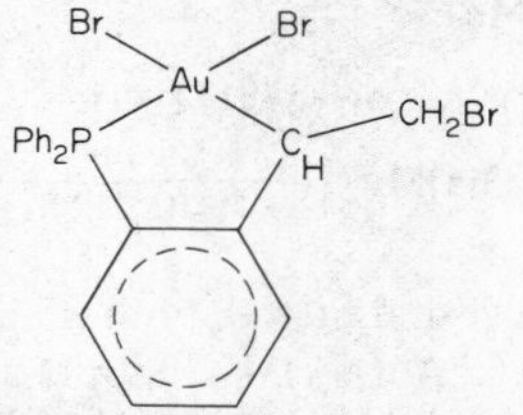

Figure 8.17 The structure of the complex $AuBr_2(Ph_2P.o\text{-}C_6H_4.CH.CH_2.Br)$

The reaction of Au_2Cl_6 with benzene gives 1,2,4,5-tetrachlorobenzene, HCl being evolved, but if the reaction is stopped prematurely, yellow $Au_2Cl_4Ph_2$ is obtained. It resists hydrolysis by water but is reduced by stannous chloride. $Au_2Cl_4Ph_2$ is stabilised by adduct formation and in a series of such adducts $[AuCl_2PhL]$ where $L = SPr^n_2$, py, PPh_3, the thermal stability and resistance of L to a displacement reaction both increase in that sequence.

The bond lengths in cis-$[AuCl_2(Ph)SPr^n_2]$ show the existence in Au(III) chemistry of a *trans* influence exerted by the phenyl group: Au—Cl = 238 pm (*trans* to phenyl), 227 pm (*cis*).

The monoaryl anions $[AuRX_3]^-$ (X = Cl, Br) may be made from arylhydrazines. For example

$$2NBu^n_4[AuCl_4] + p\text{-}NO_2C_6H_4NHNH_2.HCl \rightarrow NBu^n_4[AuCl_3(p\text{-}NO_2C_6H_4)] +$$

$$N_2 + 4HCl + NBu^n_4[AuCl_2]$$

8.11 Gold (V)

Although the d^6 configuration would tend to stabilise Au(V), this high oxidation state occurs only in complex fluorides. The action of XeF_6 on AuF_3 at 400° gives $[Xe_2F_{11}]^+[Au_VF_6]^-$, in which the octahedral anion has Au—F = 186 pm. Solid CsF converts this salt into $Cs[AuF_6]_4$.

9 Metal Complexes Containing π-bonding Ligands

Here we consider the compounds whose stability is believed to be connected with π bonding. If it is not yet true that they are as the sand that is upon the sea-shore in multitude, there are certainly very many and their influence on the study of inorganic chemistry has been revolutionary.

Strangely enough, well-defined examples of this class of compound have been known from what are, for the chemist, early times and for a period they remained as curious anomalies in bonding theory. Thus in 1830 Zeise prepared a solid crystalline ethylene complex, $K[PtCl_3(C_2H_4)]$ (Zeise's salt); in 1890 Mond and Langer prepared the volatile liquid nickel carbonyl $Ni(Co)_4$; and from 1921–9 Hein and co-workers described a number of chromium 'polyphenyls' which were in fact π-bonded complexes. The discovery (a fortuitous one) that seems to have detonated the comparatively recent explosive increase in our knowledge of this area of chemistry was that of ferrocene, $Fe(C_5H_5)_2$, by Kealy and Pauson and by Miller, Tebboth and Tremain in 1951. Ferrocene, for an organometallic, was almost absurdly stable, both to heat and to chemical reagents, and compelled attention. There also appeared at about this time a variety of significant new types of compound, among which may be mentioned as representative examples $PtHCl(PEt_3)_2$, a very stable covalent transition-metal hydride; $Cr(C_6H_6)_2$, which extended the idea of 'sandwich' complexes to six-membered rings; $Pd(C_3H_5)(C_5H_5)$, a π-allyl complex; and $Mn(CH_3)(CO)_5$, containing a σ-bonding alkyl group.

The principal correlating feature of all these compounds is that their bonding can be well described from a covalent, molecular-orbital standpoint, which involves, in addition to the metal-ligand σ bonds, d-orbital participation in metal-ligand σ bonds.

An octahedral complex, such as $W(CO)_6$, provides a convenient example on account of its high symmetry. Such octahedral π-bonded complexes are numerous and stable and so are also of much practical interest. The bonding may be treated either with or without the use of group theory and both approaches will be outlined here: the group-theoretical approach is of course the neater and more concise though possibly not the more instructive.

9.1 Bonding in an Octahedral Metal Carbonyl

The carbon monoxide molecule, if described in terms of sp hybridisation at C and O, has orbitals (as shown in figure 9.1), leading to two σ-symmetry lone pairs, one C–O σ bond and one empty C–O antibonding σ orbital (not shown). In addition there are two filled π bonds, one in the yz and one in the xz plane, formed by overlap of the two p_x orbitals and the two p_y orbitals. There are also two empty anti-

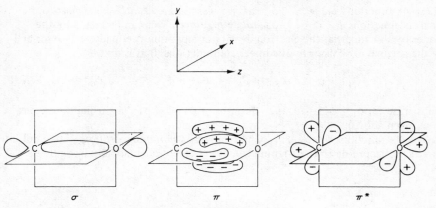

Figure 9.1 Orbitals of CO

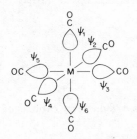

Figure 9.2 The sp carbon lone pairs as wave functions $\psi_1, \psi_2 \ldots \psi_6$

bonding π* orbitals arising from this overlap; these are four-lobed orbitals, one lying in each of the two planes.

We now assert that instead of describing the six carbon sp lone pairs by six individual wave functions $\psi_1, \psi_2, \ldots \psi_6$ (figure 9.2) we can equally well describe them by simple linear combinations

$$\phi_1 = \frac{1}{\sqrt{6}} (\psi_1 + \psi_2 + \psi_3 + \psi_4 + \psi_5 + \psi_6)$$

$$\phi_2 = \frac{1}{\sqrt{2}} (\psi_1 - \psi_6)$$

$$\phi_3 = \frac{1}{\sqrt{2}} (\psi_2 - \psi_4)$$

$$\phi_4 = \frac{1}{\sqrt{2}} (\psi_3 - \psi_5)$$

$$\phi_5 = \frac{1}{2} (\psi_2 - \psi_3 + \psi_4 - \psi_5)$$

$$\phi_6 = \frac{1}{2\sqrt{3}} (2\psi_1 + 2\psi_6 - \psi_2 - \psi_3 - \psi_4 - \psi_5)$$

These six functions $\phi_1, \phi_2, \ldots \phi_6$, *taken together*, are *exactly equivalent* to the original six functions $\psi_1, \psi_2, \ldots \psi_6$, *taken together*. This can be tested by the reader by showing that the electron density at any point is equally well described by the original orbitals or by the linear combinations, that is

$$\psi_1{}^2 + \psi_2{}^2 + \ldots + \psi_6{}^2 = \phi_1{}^2 + \phi_2{}^2 + \ldots + \phi_6{}^2$$

(substitute for $\phi_1{}^2, \phi_2{}^2, \ldots \phi_6{}^2$ in terms of $\psi_1, \psi_2, \ldots \psi_6$ using the expressions given above).

The six functions $\phi_1, \phi_2, \ldots \phi_6$ are known as *symmetry orbitals*. They fall into three sets, according to their symmetry, as indicated schematically in figure 9.3.

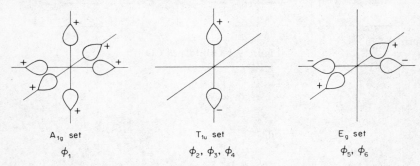

A_{1g} set
ϕ_1

T_{1u} set
ϕ_2, ϕ_3, ϕ_4

E_g set
ϕ_5, ϕ_6

Figure 9.3 The three types of symmetry orbitals of the set $\phi_1, \phi_2 \ldots \phi_6$

The terms A_{1g}, T_{1u} and E_g arise from group theory and are here used as convenient labels to describe the symmetry properties of the various orbitals. These sets can all be matched in symmetry with appropriate metal orbitals (figure 9.4).

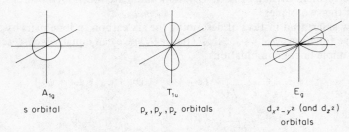

A_{1g}
s orbital

T_{1u}
p_x, p_y, p_z orbitals

E_g
$d_{x^2-y^2}$ (and d_{z^2})
orbitals

Figure 9.4 The metal orbitals that form groups matching in symmetry with $\phi_1 - \phi_6$ carbonyl symmetry orbitals

Where a metal orbital and a ligand symmetry orbital belong to the same symmetry set, bonding can occur. Thus the metal s orbital (A_{1g} symmetry) overlaps with the ligand symmetry orbital ϕ_1 (also A_{1g}) and bonding can occur because the overlap integral is not zero

$$\oint \phi_1 \psi_s d\tau \neq 0$$

where ψ_s is the wave function of the s orbital. Thus a molecular bonding orbital of

A_{1g} symmetry arises together with a molecular antibonding orbital usually denoted A_{1g}^*. This process can be expressed as a simple schematic energy-level diagram as

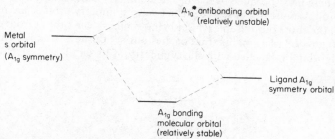

Figure 9.5 Schematic energy-level diagram showing the formation of a bonding and an antibonding molecular orbital from two other orbitals of the correct symmetry

in figure 9.5. When we extend this idea to consider all six ligand symmetry orbitals bonding with the six metal orbitals of appropriate symmetry, the octahedral molecular-orbital energy-level diagram is obtained (see figure 9.6). It is of very

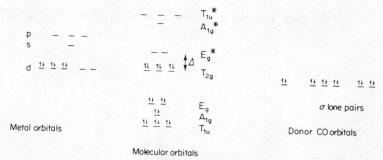

Figure 9.6 σ-bonding molecular-orbital diagram for an octahedral $M(CO)_6$ system, neglecting π bonding

considerable importance. Note that the metal d_{xy}, d_{yz} and d_{xz} orbitals cannot bond with any of the ligand symmetry orbitals and hence appear as nonbonding orbitals of unchanged energy (their energy would, however, be somewhat changed by electrostatic forces arising essentially from the ligand electrons and will be more radically changed by the π bonding to be considered shortly).

The six bonding M.O.s are filled by the six lone pairs of electrons from the carbon sp orbitals giving reasonably strong bonds contributing much to the stability of the complex.

Next, the π-bonding situation must be considered. Figure 9.7 shows one of the three mutually perpendicular π bonds formed between the T_{2g} metal orbitals d_{xy}, d_{yz} and d_{xz} and the π* antibonding orbitals. It can be seen that the symmetry of this combination of ligand orbitals is the same as that of the d orbital (they both have T_{2g} symmetry).

Note that there are originally twelve π^* ligand orbitals oriented as indicated in figure 9.8. These must produce twelve ligand symmetry orbitals since the original number of orbitals must be conserved. The twelve symmetry orbitals are in four sets of three, the three orbitals in each set being similar. One orbital from each set is shown in figure 9.9 as a combination of the four π^* orbitals from which it is derived; each orbital is labelled with the symmetry of its set (T_{2g}, etc.). Note

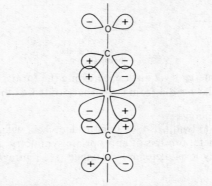

Figure 9.7 One of the three mutually perpendicular π bonds formed by $T_{2g} - \pi^*$ overlap

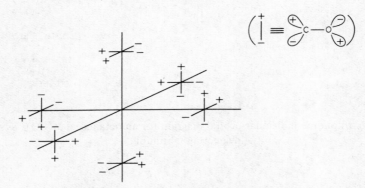

Figure 9.8 Orientation of π^* ligand orbitals

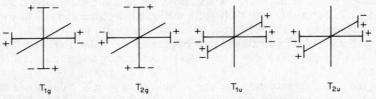

T_{1g} T_{2g} T_{1u} T_{2u}

Figure 9.9 Combination of π^* ligand orbitals produces four sets (of three) symmetry orbitals. Here one example from each set is shown

carefully that while the T_{2g} set can bond with the orbitals d_{xy}, d_{yz} and d_{xz} as we have seen, the T_{1g} and T_{2u} sets cannot bond with any s, p or d orbital and must remain unfilled. T_{1u} does have the same symmetry as the p orbitals and would probably slightly stabilise the σ bonds already formed between the p orbitals and ϕ_2, ϕ_3, ϕ_4; this interaction may be neglected, however.

To summarise the π-bonding situation, the twelve π* orbitals from the six CO ligand molecules produce twelve ligand symmetry orbitals. Three only of the latter have the same symmetry (T_{2g}) as the tungsten d_{xy}, d_{yz} and d_{xz} orbitals, which are filled with electrons in the d^6 W(0) 'ion' in $W(CO)_6$ and other π-bonded complexes. Bonding, therefore, occurs, arising from overlap of the d orbitals with the T_{2g} ligand symmetry orbitals. The bonding electrons are in a sense partially donated from the filled metal d_{xy}, d_{yz}, d_{xz} orbitals to the empty ligand π* orbitals and the process is known as *metal-to-ligand π bonding,* or, in a widely used but inelegant phrase, *back-bonding.*

Effects of π-bonding on the octahedral M.O. diagram   The most obvious effect is a marked stabilisation of part of the bonding system. Figure 9.10 shows the establishment of the π-bonding orbitals in an octahedral complex such as $W(CO)_6$. The T_{2g} molecular orbitals are now available as a well-stabilised alternative to the metal d_{xy}, d_{yz}, d_{xz} orbitals.

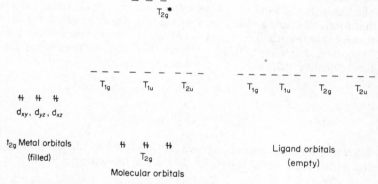

Figure 9.10 π-bonding orbitals in an octahedral complex such as $W(CO)_6$

As a matter of detail, but an important one, the T_{2g} M.O.s resemble the metal d orbitals more than they resemble the ligand T_{2g} π* symmetry orbitals. This is because they are closer in energy to the metal d orbitals than to the ligand orbitals. It is thus correct to think of the T_{2g} M.O.s as modified metal d orbitals, such that the associated electron density is partially delocalised onto the ligands. Conversely, the T_{2g}* M.O.s, although unoccupied, are 'located' mainly on the ligands.

We now have sufficient information to draw the complete energy-level diagram for $W(CO)_6$ or similar complexes. This diagram (figure 9.11) will be qualitative in that numerical values of energy levels cannot be obtained from the simple symmetry arguments just employed. Spectral data, however, often allow us to obtain numerical estimates of energy levels of molecular orbitals in specific compounds.

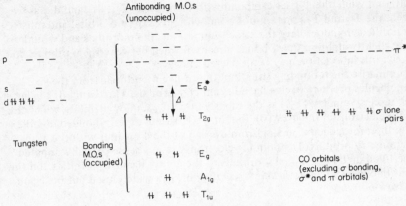

Figure 9.11 The complete energy-level diagram

9.2 Discussion of the Molecular-orbital Diagram

This full M.O. diagram has a number of very important consequences. The first is that a ligand such as carbon monoxide is enabled to form co-ordination compounds that it would not form otherwise. There is every reason to believe that without π bonding, the CO molecule would not form stable carbonyls at all, since it is a feeble σ-donor that forms no stable adducts with elements having few available d electrons. The additional π stabilisation is obtained in two ways. Firstly it is a direct consequence of the stabilisation of the filled T_{2g} levels, but secondly the concurrent metal-to-ligand electron drift induces a small electric dipole

$$\overset{\delta^+}{M} \longrightarrow \overset{\delta^-}{CO}$$

which counteracts the unfavourable negative charge placed on the metal (unfavourable because metals are electropositive) by the σ-bond lone-pair donation

$$\overset{\delta^-}{M} \longleftarrow \overset{\delta^+}{CO}$$

and hence augments and strengthens the σ-bond system. The result is the formation of reasonably strong σ and π bonds and an overall electric dipole which dipole-moment studies indicate is almost zero.

The second consequence of the M.O. diagram is that the value of Δ depends on the efficiency of the bonding. Δ is the difference in energy between the T_{2g} and E_g^* levels and is exactly analogous to the electrostatic Δ obtained by the crystal-field treatment. This latter treatment assumes that the d orbitals take no part in the bonding, which is purely electrostatic, but are split into two sets by the electrostatic effect of the six ligands considered as point negative charges. *These two sets exactly correspond with the two sets T_{2g} and E_g^* that arise from M.O. theory.* However, the M.O. treatment shows that the more σ bonding there is, the higher is the energy of E_g^*; and the more   bonding, the lower is the energy of T_{2g}. Hence, on the M.O. theory, Δ arises partly from σ bonding and partly from π bonding. It is significant that $M(CO)_6$ (M = Cr, Mo, W), which are very stable carbonyls in which the σ and π bonding are both presumably very effective, are colourless: Δ is so large that electronic transitions are in the ultraviolet.

9.2.1 The Eighteen-electron (E.A.N.) Rule

The third consequence of the octahedral M.O. diagram is of the utmost use in rationalising the existence of known compounds and predicting the possible existence of others as yet unknown. This is the *eighteen-electron rule* or *effective atomic number (E.A.N.) rule*. It is exceedingly simple: there are nine stable orbitals in the M.O. diagram; these must be filled for stability of the molecule, *and so there must be exactly eighteen bonding electrons.* Thus in $W(CO)_6$ there are 6 outer electrons in the W(0) atom $(5d^4 6s^2)$ and 6 x 2 electrons in the six lone pairs of the six CO ligands; the total is 18. It turns out that the rule applies to tetrahedral and trigonal bipyramidal configurations also, though this does not follow from the material mentioned so far in this chapter. Very many examples of molecules obeying this rule appear later along with a few exceptions. Some examples are

$$\text{Fe(CO)}_5 \; : \quad 8(\text{Fe}) + 5 \times 2 \; (5\text{CO}) \qquad\qquad\qquad = 18$$

$$\text{OsH}_2(\text{diphos})_2 \; : \quad 8(\text{Os}) + 2 \times 1 \; (2\text{H}) + 2 \times 4 \; (2 \text{ diphos}) = 18$$

$$\text{Re(CO)}_3(\text{C}_5\text{H}_5) \; : \quad 7(\text{Re}) + 3 \times 2 \; (3\text{CO}) + 5 \; (\pi\text{-C}_5\text{H}_5) \quad = 18$$

$$\text{OsHBr(CO)(PPh}_3)_3 \; : \quad 8(\text{Os}) + 1(\text{H}) + 1(\text{Br}) + 2(\text{CO}) +$$
$$3 \times 2 \; (3\text{PPh}_3) = 18$$

$$\text{Pd}(\pi\text{-C}_3\text{H}_5)(\pi\text{-C}_5\text{H}_5) \; : 10(\text{Pd}) + 3(\pi\text{-C}_3\text{H}_5) + 5(\pi\text{-C}_5\text{H}_5) \quad = 18$$

The rule is of importance in predicting structures. Taking $Mo(C_5H_5)_4$ as an example, Mo has six d electrons so $Mo(\pi\text{-}C_5H_5)_4$ would give us a total count of twenty-six. However, $Mo(\pi\text{-}C_5H_5)_2 (\sigma\text{-}C_5H_5)_2$ totals eighteen $(6 + 2 \times 5 + 2 \times 1)$ and this is the structure adopted.

The rule is *not* obeyed in a number of cases. One group of exceptions is Rh(I), Ir(I), Pd(II) and Pt(II), where a sixteen-electron rule applies. This is due to the stability of the square planar d^8 configuration. Another is exemplified by $V(CO)_6$ (seventeen electrons) where addition of one electron yields the very stable species $V(CO)_6^-$ (Nb and Ta form only $M(CO)_6^-$, incidentally).

9.3 The Bonding in an Octahedral Complex (Using Group Theory)

For completeness, and in order to show the brevity and elegance of the treatment, we now use group theory to investigate the bonding in $W(CO)_6$ or any octahedral π-bonding complex. We shall assume that the reader has a basic knowledge of group theory, although it is quite possible that such a reader is already familiar with this particular application. From the σ-bonding ligand orbitals $\psi_1, \psi_2, \ldots \psi_6$ (figure 9.2) we obtain a reducible representation

	E	C_3	C_2	C_4	$C_4{}^2$	i	S_4	S_6	σ_h	σ_d
Γ_σ	6	0	0	2	2	0	0	0	4	2

Comparison with the O_h group-character table, reproduced below, shows that $\Gamma\sigma = A_{1g} + E_g + T_{1u}$. The ligand orbitals thus form σ bonds with the metal orbitals listed on the right-hand side of the O_h table having symmetry A_{1g} (s orbital), E_g (d_{z^2}, $d_{x^2-y^2}$) and T_{1u} (p_x, p_y, p_z). From the π^* carbon monoxide orbitals (figure 9.1), we obtain the reducible representation.

	E	C_3	C_2	C_4	C_4^2	i	S_4	S_6	σ_h	σ_d
Γ_π =	12	0	0	0	−4	0	0	0	0	0

The components of Γ_π are T_{1g}, T_{1u}, T_{2g}, T_{2u} because

	E	$8C_3$	$6C_2$	$6C_4$	$3C_2$	i	$6S_4$	$8S_6$	$3\sigma_h$	$6\sigma_d$	
T_{1g} =	3	0	−1	1	−1	3	1	0	−1	−1	
T_{1u} =	3	0	−1	1	−1	−3	−1	0	1	1	from O_h table
T_{2g} =	3	0	1	−1	−1	3	−1	0	−1	1	
T_{2u} =	3	0	1	−1	−1	−3	1	0	1	−1	
Γ_π =	12	0	0	0	−4	0	0	0	0	0	(adding)

The metal orbitals of T_{2g} symmetry ($5d_{xy}$, $5d_{yz}$, $5d_{xz}$) therefore bond with the T_{2g} symmetry orbital. There are no T_{1g} metal orbitals; the T_{2u} orbitals are 4f orbitals and are too stable to bond, while the T_{1u} orbitals ($6p_x$, $6p_y$, $6p_z$) are empty. Three bonding orbitals only (the three components of T_{2g}) are thus formed.

The following is a brief outline of the mechanics of obtaining the reducible representations Γ_σ and Γ_π. For justification of the procedure, consult a specialised text.

For each symmetry operation (E, no operation; C_3, rotation through 120°; σ, plane of symmetry; etc.) possessed by the molecule and listed on top of the octahedral O_h character table (see figure 9.12) we add up the number of ligand orbitals unmoved by the operation and write this number under the symbol for the operation. If any orbital is changed in sign, for example

count −1. For rotation of a *pair* of orbitals through 120°, without translation, count −1.

9.4 General Considerations

After the foregoing general theory, which should provide a basis for consideration and discussion of specific compounds of the π-bonding type, we now consider the various classes of compounds formed by the second- and third-row transition metals.

	I	$6C_4$	$3C_2$	$6C_2$	$8C_3$	i	$6S_4$	$3\sigma_h$	$6\sigma_d$	$8S_6$	
A_{1g}	1	1	1	1	1	1	1	1	1	1	$x^2+y^2+z^2$
A_{1u}	1	1	1	1	1	-1	-1	-1	-1	-1	
A_{2g}	1	-1	1	-1	1	1	-1	1	-1	1	
A_{2u}	1	-1	1	-1	1	-1	1	-1	1	-1	
E_g	2	0	2	0	-1	2	0	2	0	-1	z^2, x^2-y^2
E_u	2	0	2	0	-1	-2	0	-2	0	1	
T_{1g}	3	1	-1	-1	0	3	1	-1	-1	0	
T_{1u}	3	1	-1	-1	0	-3	-1	1	1	0	x,y,z
T_{2g}	3	-1	-1	1	0	3	-1	-1	1	0	xz, yz, xy
T_{2u}	3	-1	-1	1	0	-3	1	1	-1	0	

Figure 9.12 The character table for the group O_h

In this area of chemistry the distinction between the first-row metals and the heavier metals is not a very useful one because similar compounds are very often formed by all three metals in a triad, for example $M(C_6H_6)_2$ (M = Cr, Mo, W) and $M(CO)_5$ (M = Fe, Ru, Os). Even when this is not so, the general chemistry is usually similar and comparisons of the behaviour of the three metals are very instructive — more so than in the area of simple compounds and classical complexes dealt with in preceding chapters.

The chemistry of the 4d and 5d metals will, however, be emphasised in what follows by choosing, where appropriate, their compounds as specific examples to illustrate general trends and by discussion of those aspects of the π-complex chemistry of 4d and 5d metals that differ most from that of the 3d metals. There are some areas where fairly considerable differences occur: the 4d and 5d metals, compared with the 3d, (a) form direct metal–metal bonds more readily, both with transition-metal ions and with metals such as Hg(II), (b) form more stable   bonds to carbon and particularly to hydrogen, and (c) show a tendency for their cyclo-pentadienyl compounds to occur in higher formal oxidation states.

In order to confine this discussion within reasonable bounds of space, it is largely restricted to comparatively simple compounds, in the sense that those containing more than two or three different ligands are mentioned as little as possible. For example, the structure of the compound $Ru_6C(CO)_{14}\{C_6H_3(CH_3)_3\}$, though of great interest to the connoisseur, would form a less basic and typical vehicle for illustration of the properties of metal-arene complexes than would $Mo(C_6H_6)_2$.

9.5 The Binary Carbonyls and Their Derivatives

With the exceptions of nickel and iron, cabon monoxide does not react easily with transition metals and the carbonyls are usually prepared by treating a compound (salt or complex) of the metal in a normal oxidation state with carbon monoxide in the presence of a reducing agent. The reducing agents employed cover a wide range. They include alkali metals in strongly solvating ethers such as tetrahydrofuran or diglyme, aluminium alkyls, and on occasion excess CO acts as its own reducing agent. Pressures of about 10–20 MPa (100-200 atm) and temperatures of about $100°$ are typical, so that an autoclave system is necessary for much preparative work in this area.

The formulae, preparative routes and structures of all the well-characterised metal carbonyls are summarised at the end of this chapter.

9.5.1 Reactions of the Metal Carbonyls
Although a number of the reactions of the carbonyls yield products of such a complex and surprising nature that a single-crystal X-ray diffraction study is the only conclusive way of determining their structure, the majority of their reactions are readily classified and are predictable and explicable.

9.5.2 Substitution Reactions
In these, carbon monoxide is displaced by an alternative ligand, usually a π-bonding group such as PPh_3 $AsPh_3$, $Ph_2PCH_2CH_2PPh_2$, $PhNC$, C_6H_6, $C_5H_5^-$ or norbornadiene. The three-electron donor NO is also reactive, but as might be expected, does not normally give one-for-one substitutions. In the case of phosphines or arsines the empty 3d or 4d orbitals act as the π acceptor; in the case of olefins the $\pi*$ orbital performs this role, and the bonding in complexes of aromatic systems involves the molecular orbitals of the carbon rings.

Triphenylphosphine is a ligand whose carbonyl-substitution reactions have been extensively investigated. Some examples follow.

$$Ni(CO)_4 \xrightarrow[20°]{PPh_3} Ni(PPh_3)(CO)_3 + CO \xrightarrow[80°]{PPh_3} Ni(PPh_3)_2(CO)_2 + CO$$

There is evidence that this reaction has an S_N1 dissociative mechanism. No more than two PPh_3 groups can be introduced in this way, probably because the remaining two CO groups are more extensively π bonded to the nickel atom. This increased π bonding arises from the reduced π bonding and increased σ bonding capabilities of the PPh_3 groups compared with the CO groups they replace. This view is supported by the considerable electric dipole moment of $Ni(PPh_3)_2(CO)_2$ (3.82 D), which demonstrates that a large electron drift occurs. That this drift takes place partly at least into the $\pi*$ orbitals of the CO groups is shown by the considerable decrease in the CO force constant, the CO stretching frequency being reduced from 2128 cm^{-1} in $Ni(CO)_4$ to 2010 cm^{-1} in $Ni(CO)_2(PPh_3)_2$.

Exactly similar arguments apply to the reaction of PPh_3 with other metal carbonyls. Thus, when two of the CO groups of $Mo(CO)_6$ are directly substituted by PPh_3, the CO infrared-active frequency is reduced from 1985 cm^{-1} in $Mo(CO)_6$ to 1900 cm^{-1} in *trans*-$Mo(CO)_4(PPh_3)_2$, the corresponding CO force constants being 1.65 and 1.54 $kN\,m^{-1}$† (1 $kN\,m^{-1}$ = 10 millidynes/Å).

---

† See F.A. Cotton and C.S. Kraihanzel, *J. Am. chem. Soc.*, 84 (1962), 4432 for a fuller discussion; the general conclusions of this paper (though modified in detail by later work) remain of significance.

Chelate phosphines such as o-$C_6H_4(PEt_2)_2$ or $Ph_2PCH_2CH_2PPh_2$ react similarly and more vigorously. Thus three- or fourfold substitution is possible in $Ni(CO)_4$

$$Ni(CO)_4 + (tri\text{-}P) \longrightarrow [Ni(CO)(tri\text{-}P)] + 3CO$$

$$tri\text{-}P = PhP(o\text{-}C_6H_4PEt_2)_2$$

$$Ni(CO)_4 + 2o\text{-}C_6H_4(PEt_2)_2 \longrightarrow Ni\{o\text{-}C_6H_4(PEt_2)_2\}_2 + 4CO$$

The resulting complexes are quite stable thermally but are readily oxidised by air. The generality of *cis* substitution in disubstituted carbonyls of the type $M(CO)_4(PPh_3)_2$ may be noted. This may be attributed to the better π bonding of carbonyls in the *cis* configuration, which has been authenticated structurally for $[Mo(CO)_4\{(PPh_2)_2CH_2\}]$ (see figure 9.13), where the diphosphine constrains a *cis* geometry. The CO groups *trans* to P are notably more closely bound, which is consistent with the above argument. Here the mutually *trans* carbonyls are competing for π density with each other. The dipole moment of this type of compound (6.15–6.7 D) is as expected for a *cis* configuration.

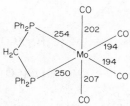

Figure 9.13 The structure of $Mo(CO)_4\{(PPh_2)_2CH_2\}$ showing the effect of competition for π-electron density on the bond lengths

Phosphorus trihalides readily displace CO from carbonyls. Thus $Ni(PCl_3)_4$ or $Ni(PF_3)_4$ can be obtained directly from $Ni(CO)_4$ and the appropriate trihalide. It is very likely that this reactivity is connected with the increased ability of the phosphorus 3d orbitals to accept electrons consequent on the electron-withdrawing effect of the electronegative halogen atoms.

In those cases where phosphine-substituted compounds cannot be made directly from carbonyls for reasons of unfavourable reaction kinetics, alternative methods may be used such as vigorous reduction of higher oxidation-state compounds in the presence of excess phosphine.

$$MoCl_5 + o\text{-}C_6H_4(PEt_2)_2 \xrightarrow{\;Na^+C_{10}H_{10}^-\,;\, THF\;} Mo\{o\text{-}C_6H_4(PEt_2)_2\}_3$$

Arsines react similarly to phosphines but rather less vigorously and the products are often slightly less stable. Phosphines do not always react with metal carbonyls to yield simple substitution products only. Thus when PPh_3 reacts with $Os_3(CO)_{12}$, there are nine products. Three of them are orthodox substitution products $Os_3(CO)_{12-n}(PPh_3)_n$ ($n = 1, 2, 3$), but the other six include examples of Os–$P(Ph_2)$–Os bridging, Ph.Ph formation, Os–H formation, pairs of Os atoms bridged by *ortho*-C_6H_4 or C_6H_5 groups, and the P—(o-C_6H_4)—Os grouping

Arsines react similarly to phosphines but rather less vigorously and the products are often slightly less stable.

The cyclopentadienyl anion and benzene both act as donors of six electrons and normally displace three CO groups. The six electrons in question are nominally those originally occupying the three bonding molecular orbitals of the aromatic system. The ring antibonding orbitals are capable of acting as π acceptors in the same way as the π^* orbitals of CO. Compared with $C_5H_5^-$, C_6H_6 behaves as a weaker ligand. Its complexes are almost always formed with metals in low oxidation states (suggesting that metal-to-ligand π bonding is predominantly necessary for stabilisation) and are usually unstable to air or heat. Metal cyclopentadienyls, however, are often formed by metals in 'normal' oxidation states and are often very stable, suggesting that a ligand-to-metal electron drift predominates. Some examples follow of substitutions by aromatic systems.

$$M(CO)_6 \xrightarrow[170^\circ]{P_2R_4} (CO)_5M.PR_2.PR_2.M(CO)_5 \xrightarrow{240^\circ} (CO)_4M \underset{PR_2}{\overset{PR_2}{\diagdown\diagup}} M(CO)_4$$

where M = Cr, Mo, W; R = Me, Et, Ph

$$Mo(CO)_6 + 1,3,5\text{-}C_6H_3Me_3 \longrightarrow Mo(C_9H_{12})(CO)_3 + 3CO$$

$$Mo(C_9H_{12})(CO)_3 + terpy \longrightarrow Mo(terpy)(CO)_3 + C_9H_{12}$$

$$Mo(CO)_6 + C_5H_5^- \longrightarrow [Mo(C_5H_5)(CO)_3]^- + 3CO$$

$$2Mo(CO)_6 + 2C_5H_6 \longrightarrow (C_5H_5)(CO)_3Mo\text{—}Mo(CO)_3(C_5H_5) + 6CO + H_2$$

$$[Nb(CO)_6]^- + C_5H_5^- + HgCl_2 \longrightarrow Nb(CO)_4(C_5H_5) + 2CO + Hg + 2Cl^-$$

$$Re_2(CO)_{10} + 2C_5H_6 \longrightarrow 2Re(CO)_3(C_5H_5) + 4CO + H_2$$

The product in the fourth reaction is a reduced version of the dimeric product of the fifth reaction, one extra electron taking the place of the metal–metal bond to maintain the eighteen-electron rule. In the sixth reaction, mercuric ion functions as an oxidising agent, converting Nb(–I) to Nb(I).

Halide ions can function similarly as substituting agents

$$M(CO)_6 + X^- \longrightarrow [MX(CO)_5]^- + CO$$

where M = Cr, Mo, W; X = Cl, Br, I.

$$Rh_3(CO)_{12} \xrightarrow[(X = Cl, Br, I)]{X_2} cis\text{-}[Rh(CO)_4X_2]$$

Carbonyl halide anions may in some cases be prepared from metal halides

$$RuCl_3.3H_2O \xrightarrow[HCO_2H]{HX} Ru(CO)X_5{}^{2-}, Ru(CO)_2X_4{}^{2-}, Ru(CO)_3X_3{}^-$$

$$RuCl_3.3H_2O \xrightarrow[CO]{HCl} [Rh(CO)_2Cl_2]^-$$

9.5.3 Formation of Anions and Hydrides

These two classes of compound will be considered together because there are frequent examples of the reaction

$$MH(CO)_n \rightleftharpoons[\text{water}] H_3O^+[M(CO)_n]^-$$

For instance in the reactions

$$H^+ + [FeH(CO)_4]^- \rightleftharpoons FeH_2(CO)_4$$

$$H^+ + [Re(CO)_5]^- \rightleftharpoons ReH(CO)_5$$

the proton is taken up by the anion without alteration to the equivalent atomic number of eighteen electrons. There is, however, usually a change of stereochemistry, and a pair of electrons, which in the anion was engaged in π bonding, becomes a σ-bonding pair in the hydride (see figure 9.14).

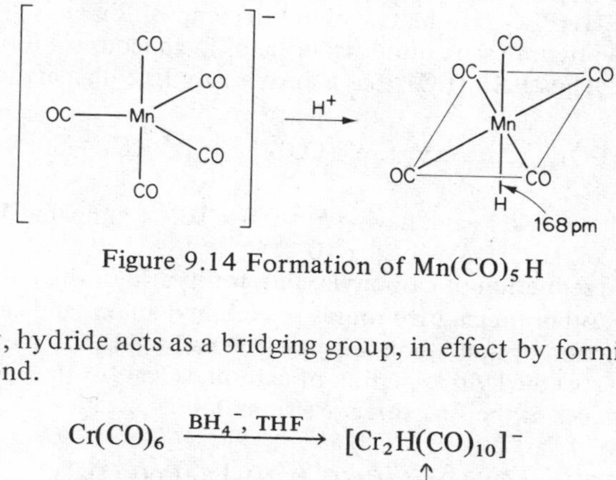

Figure 9.14 Formation of $Mn(CO)_5H$

Occasionally, hydride acts as a bridging group, in effect by forming a protonated metal–metal bond.

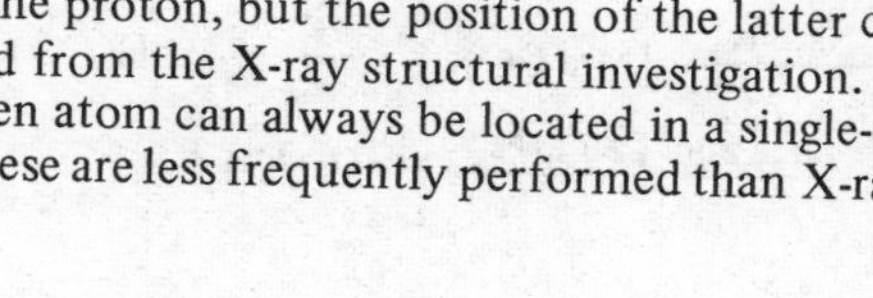

$$Cr(CO)_6 \xrightarrow{BH_4{}^-, THF} [Cr_2H(CO)_{10}]^-$$

$$Cr(CO)_6 \xrightarrow{Na/Hg, THF} [Cr_2(CO)_{10}]^{2-}$$

with H^+ connecting the two.

The structure of the anion $[Cr_2H(CO)_{10}]^-$ includes a Cr–Cr bond (figure 9.15), presumably containing the proton, but the position of the latter could not, in this instance, be determined from the X-ray structural investigation. In metal hydrides, however, the hydrogen atom can always be located in a single-crystal neutron-diffraction study, but these are less frequently performed than X-ray studies. However,

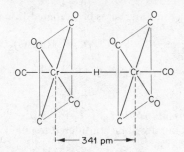

Figure 9.15 The structure of the $[Cr_2H(CO)_{10}]^-$ ion

the *presence* of the metal–hydride grouping can always be proved by the n.m.r. sig-. nal characteristically at very high field and by the M–H infrared absorption, characteristically around 1900 cm^{-1} (as are CO frequencies) but lowered in the approximate ratio $1:1/\sqrt{2}$ by deuteration.

Metal carbonyl hydrides are a rather unstable class of complex, being not only easily oxidised so that handling in an inert atmosphere is usually necessary, but often being thermally unstable also. $OsH_2(CO)_4$ is exceptional in being moderately stable, decomposing at $100°$; $CoH(CO)_4$ decomposes at its melting point of $-26°$. Substitution of some or all of the CO groups by PR_3, or $C_5H_5^-$ often leads to greatly increased stability; the reason for this is uncertain. Thus *trans*-$OsH_2\{o\text{-}C_6H_4(PEt_2)_2\}_2$ melts without decomposition at $293°$.

Some carbonyl hydrides are quite strong acids in aqueous solution; in contrast others are weak. Thus $FeH_2(CO)_4$ has a fairly ready first dissociation

$$FeH_2(CO)_4 \rightleftharpoons [FeH(CO)_4]^- + H^+ \quad K_1 = 3.6 \times 10^{-5}$$

but $MnH(CO)_5$ is a weaker acid, having $K = 0.8 \times 10^{-7}$ (compare CH_3COOH, $K = 1.76 \times 10^{-5} : C_6H_5OH, K = 1.28 \times 10^{-10}$).

In addition to reduction of carbonyls, and, for hydrides, the protonation of carbonyl anions, other preparative routes to carbonyl anions and carbonyl hydrides include (a) the action of non-π-bonding bases on carbonyls, (b) the reduction of carbonyl halides, (c) the hydrogenation of carbonyls and (d) hydrogen-abstraction reactions. Examples of the first three reactions follow.

(a) $Co_2(CO)_8 + C_5H_5N \rightarrow [Co(C_5H_5N)_6]^{2+}[Co(CO)_4]_2^-$

The driving force for this oxidative disproportionation is doubtless the instability of a zero-valent pyridine carbonyl complex of cobalt; pyridine is a poor π acceptor, and so is more stable in combination with Co(II).

(b) $MnI(CO)_5 \xrightarrow{Na, NH_3} [Mn(CO)_5]^- \xrightarrow{H^+} MnH(CO)_5$

(c) $Mn_2(CO)_{10} \xrightarrow{H_2, 200°, 20 \text{ MPa}} MnH(CO)_5$

Examples of class (d) are

$$[OsBr_6]^{2-} \xrightarrow{\text{PPh}_3, \text{HOCH}_2\text{CH}_2\text{OMe}} OsHBr(CO)(PPh_3)_3$$

$$[Ru_2Cl_3(PEt_2Ph)_6]Cl \xrightarrow{\text{KOH, EtOH}} RuHCl(CO)(PEt_2Ph)_3$$

$$cis\text{-}PtCl_2(PPh_3)_2 \xrightarrow{\text{KOH, EtOH}} trans\text{-}PtHCl(PPh_3)_2$$

In this last type of synthesis, both the CO and the H, which make their appearance rather like rabbits out of a conjuror's hat, come from the $-CH_2OH$ group of the solvent. The platinum metals (Ru, Os, Rh, Ir, Pd, Pt) are particularly prone to this type of reaction, and the products are usually formally related to very stable conventional complexes in oxidation states II or III, such as $RhCl_3(PR_3)_3$ or $PtCl_2(PR_3)_2$, which do not necessarily involve an eighteen-electron configuration, the square planar sixteen-electron arrangement of Pt(II) or Ir(I) being an alternative.

Hydrides of polynuclear carbonyls have been studied in detail recently.

$$Re_2(CO)_{10} \xrightarrow{\text{NaBH}_4} H_6Re_4(CO)_{12}{}^{2-}$$

$$Re_2(CO)_{10} \xrightarrow[120°]{\text{H}_2, \text{decalin}} H_3Re_3(CO)_{12}, H_4Re_4(CO)_{16}$$

$$Ru_3(CO)_{12} \xrightarrow[90°]{\text{H}_2, \text{octane}} H_4Ru_4(CO)_{12}$$

$$OsO_4 \xrightarrow[\text{H}_2]{\text{CO}} H_2Os(CO)_4 + H_2Os_2(CO)_8$$

$$M_3(CO)_{12} \xrightarrow{\text{H}_2\text{SO}_4} [HM_3(CO)_{12}]^+$$

where M = Ru, Os.

$$Ru_3(CO)_{12} \xrightarrow[\text{Mn(CO)}_5{}^-]{\text{THF}} H_2Ru_6(CO)_{18}, H_4Ru_4(CO)_{12}$$

Some uncertainty surrounds their structures. $Re_4H_6(CO)_{12}{}^{2-}$ contains a tetrahedron of rhenium atoms, each bound to three carbonyls. The hydrogen atoms were not located in the X-ray study, but may bridge the edges of the tetrahedron (see figure 9.16). $H_2Ru_6(CO)_{18}$ has a nearly octahedral array of metal atoms, each again bound to three carbonyl groups. Again X-ray study did not locate the hydrogen atoms (they may triply bridge opposite *faces* of the octahedron).

In many cases anionic polynuclear carbonyls have interesting structures

or
$$\begin{array}{c} Fe(CO)_5 \\ \\ Fe_3(CO)_{12} \end{array} \xrightarrow[\text{heat}]{\text{pyridine}} [Fe(py)_6]^{2+}[Fe_4(CO)_{13}]^{2-}$$

The structure of the anion is based on a $Fe_3(CO)_9$ unit (3 bridging COs) with a $Fe(CO)_3$ unit occupying an apical position, the iron atoms forming a tetrahedral array. The remaining CO molecule bridges the Fe_3 base of the tetrahedron.

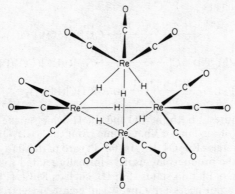

Figure 9.16 The structure of $[H_6Re_4(CO)_{12}]^{2-}$ showing the probable positions of the hydrogens (edge-bridging)

Reaction of rhodium carbonyl chloride with carbon monoxide provides a variety of carbonyls

$$Rh_2(CO)_4Cl_2 \xrightarrow[H_2O]{CO} Rh_4(CO)_{12} \xrightarrow[OH^-]{CO} [Rh_{12}(CO)_{30}]^{2-} \xrightarrow[OH^-]{CO} [Rh_7(CO)_{16}]^{3-}$$

$$\xrightarrow[OH^-]{CO} [Rh_6(CO)_{14}]^{4-} \xrightarrow[Na]{CO} [Rh(CO)_4]^-$$

The anion $[Rh_{12}(CO)_{30}]^{2-}$ contains two octahedral units connected by a metal–metal bond, while $[Rh_7(CO)_{16}]^{3-}$ has an extremely unusual structure (mentioned here only to illustrate the complexities arising in this field) involving a monocapped octahedron of rhodium atoms, and seven terminal, six doubly bridging and three face-bridging carbonyls.

Reaction of $Re_2(CO)_{10}$ with borohydride affords $[Re_4(CO)_{16}]^{2-}$, which has a *planar* Re core; each Re is bound to four terminal carbonyls. The absence of bridging carbonyls in this case is quite typical of the behaviour of 5d metals.

Although 'extra' electrons, in the form either of an anionic charge or of bridging hydrogen atoms, commonly confer stability on polynuclear ('cluster') carbonyls, they are not always necessary. Thus pyrolysis of $Os_3(CO)_{12}$ in a sealed tube at 195–200°C for 12 hours gives several polynuclear carbonyls $Os_x(CO)_y$, where $x,y = 4, 13; 5, 16; 6, 18; 7, 21; 8, 23$, respectively. $Os_6(CO)_{18}$ has been characterised by X-ray studies and its structure is shown in figure 9.17.

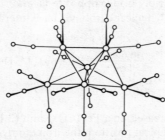

Figure 9.17 The structure of $Os_6(CO)_{18}$ (after R. Mason, K.M. Thomas and D.M.P. Mingos, *J. Am. chem. Soc.*, **95** (1973), 3803)

9.5.4 Heteronuclear Carbonyls

In recent years some attention has been given to heteronuclear carbonyls, the preparation of which is interesting for a number of reasons. They are not confined to transition metals, but some are known in which elements such as zinc or mercury are included (although not being bonded to CO).

$$Fe(CO)_5 \ + \ Re_2(CO)_{10} \ \xrightarrow{h\nu} \ Re_2 Fe(CO)_{14} \qquad\qquad (i)$$

$$Mn_2(CO)_{10} \ + \ Ru_3(CO)_{12} \ \xrightarrow{205°} \ (CO)_5 MnRu(CO)_4 Mn(CO)_5 \qquad (ii)$$

These reactions employ homolysis of metal–metal bonds (either by thermal or radiative processes) to generate reactive intermediates. $MM'_2(CO)_{12}$ $(M = Ru,$ $M' =$ Os: $M = Fe, Os, M' = Ru$), $MM'_2(CO)_{14}$ $(M = Os, M' = Mn, Re)$ and $MnRe(CO)_{10}$ may be prepared similarly.

$MnRe(CO)_{10}$ may be prepared in better yield by

$$NaMn(CO)_5 \ + \ Re(CO)_5 Br \ \rightarrow \ MnRe(CO)_{10} \ + \ NaBr$$

This is thought to have a similar structure to the homometallic analogues. It is of interest that reaction of $NaRe(CO)_5$ with $Mn(CO)_5 Br$ yields mainly $Mn_2(CO)_{10}$ and $Re_2(CO)_{10}$.

Examples involving nontransition metals are

$$Me_2 M + Re(CO)_5 H \rightarrow M\{Re(CO)_5\}_2$$

where $M = Zn, Cd, Hg$, and

$$Hg + Re(CO)_5 H \rightarrow Hg\{Re(CO)_5\}_2$$

Anionic heteronuclear carbonyls are also formed; for example

$$Ni(CO)_4 + [W_2(CO)_{10}]^{2-} \ \xrightarrow{THF} \ [W_2 Ni_3(CO)_{16}]^{2-}$$

This involves a trigonal bipyramidal $W_2 Ni_3$ unit.

9.5.5 Oxidation Reactions

It is often difficult to define the oxidation state of the metal ion in π-bonded complexes. Thus in the isoelectronic sequence

$$NaMn(CO)_5 \ \xrightarrow{H^+} \ MnH(CO)_5 \ \xrightarrow{HBr} \ MnBr(CO)_5$$

the metal is formally Mn(−I) in $[Mn(CO)_5]^-$ and formally Mn(I) in $MnBr(CO)_5$ but neither H^+ nor HBr are usually regarded as oxidising agents, though it might be said that HBr is acting in this way here. Among a series of isoelectronic covalent complexes, the idea of oxidation state loses its usefulness, just as it would not be useful to refer to NF_3 and NH_3 as containing N(III) and N(−III) respectively. However, we can regard reactions of carbonyls with, for example, halogens as oxidations and most metal carbonyls and their derivatives will react in this way.

Metal–metal bonds are readily split by halogens to give carbonyl halides.

$$Re_2(CO)_{10} \xrightarrow{\quad I_2 \quad} 2ReI(CO)_5$$

$$Fe(CO)_5 \xrightarrow{\quad I_2 \quad} FeI_2(CO)_4 + CO$$

However, some carbonyl halides are best prepared by the converse reaction of CO with a metal halide; for example

Stannic chloride can add to iron carbonyl in an oxidative way to give a metal–metal bond

$$Fe(CO)_5 + SnCl_4 \rightarrow FeCl(SnCl_3)(CO)_4 + CO$$

Methyl mercuric chloride also behaves analogously

$$[Ta(CO)_6]^- + CH_3HgCl \rightarrow CH_3Hg.Ta(CO)_6 + Cl^-$$

giving a seven-co-ordinate product. The post-transition metals quite frequently form such metal–metal bonds with transition metals. A further example, this time involving gold, is

$$Os_3(CO)_{12} + Ph_3PAuCl \rightarrow Cl.Os(CO)_4.Os(CO)_4.Os(CO)_4AuPPh_3$$

9.6 Metal Cyclopentadienyl and Arene Systems

$Fe(C_5H_5)_2$, di-π-cyclopentadienyliron, ferrocene, whose discovery was one of the most important events in inorganic chemistry, is an orange crystalline substance, m.p. 174°, soluble in benzene, insoluble in water, volatile in steam and stable at 500°. Ruthenocene and osmocene are similar. Other compounds $M(C_5H_5)_2$ are formed by nearly all the d transition metals, but are much less stable, both to air and to heat. The most general method of preparation is

$$MCl_2 + 2NaC_5H_5 \rightarrow M(C_5H_5)_2 + 2NaCl$$

Cyclopentadiene, C_5H_6, is a weak acid

It is always stored as dicyclopentadiene, its Diels–Alder dimer, from which it is made as required by heating the dimer, and to which it reverts on standing at room temperature. The acidic nature is a consequence of the stability of the symmetrical planar aromatic anion $C_5H_5^-$, which is in turn an effect of the complete occupation by its six π electrons of the three bonding molecular orbitals of the five carbon p_z orbitals arranged as a regular pentagon (see figure 9.18).

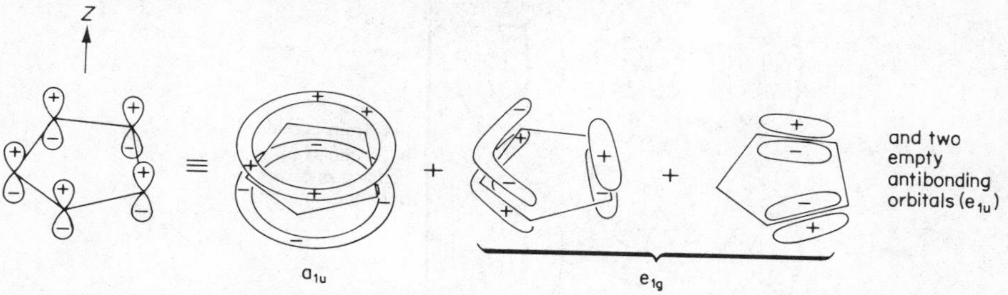

Figure 9.18 Molecular orbitals of the $C_5H_5^-$ anion

In nonhydroxylic solvents, such as ethers, the formation of the anion is ready and complete

$$C_5H_6 + Na \xrightarrow{THF} Na^+C_5H_5^- + \tfrac{1}{2}H_2$$

It is, however, instantly hydrolysed by traces of water

$$C_5H_5^- + H_2O \rightarrow C_5H_6 + OH^-$$

9.6.1 Bonding in Ferrocene

Most divalent metals will form a di-π-cyclopentadienyl; thus $Ca(C_5H_5)_2$ has a structure similar to that of ferrocene but is instantly hydrolysed, presumably owing to lack of effective covalent metal–ring bonding. Ferrocene is not hydrolysed in this ready way and the extra stability of this compound and its analogues can be accounted for according to the bonding scheme outlined below, where it can be seen that the metal s and d_{z^2}, and d_{xz}, d_{yz} orbitals are of compatible symmetry for bonding with the aromatic ring orbitals while the d_{xy} and $d_{x^2-y^2}$ remain largely nonbonding (but are stabilised to some extent by interaction with the unoccupied antibonding ring orbitals; see figure 9.19). One bonding, one nonbonding and one antibonding orbital are formed from two metal orbitals (s and d_{z^2}) and the a_{1g} orbital made up from the combination of both cyclopentadienyl anions. The metal ion thus has room for six essentially nonbonding electrons but should have its remaining orbitals unoccupied to receive the bonding electron pairs from the $C_5H_5^-$ groups. Thus a d^6 ion like Fe^{2+} will form a stable compound with three strong covalent bonds each spanning the iron ion and both $C_5H_5^-$ groups, and will have its remaining three orbitals ($d_{x^2-y^2}$, d_{xy} and a linear combination of s and d_{z^2}) occupied with nonbonding electrons.

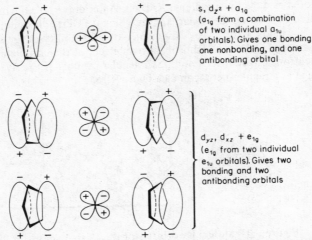

s, d_{z^2} + a_{1g}
(a_{1g} from a combination
of two individual a_{1u}
orbitals). Gives one bonding
one nonbonding, and one
antibonding orbital

d_{yz}, d_{xz} + e_{1g}
(e_{1g} from two individual
e_{1u} orbitals). Gives two
bonding and two
antibonding orbitals

Figure 9.19 Orbital interactions in a bis(π-cyclopentadienyl) complex

In addition we note that only six of the twelve π electrons of the two $C_5H_5^-$ rings have been used in bonding, so that six nonbonding electrons remain in the ring orbitals. The total is thus eighteen electrons and ferrocene itself obeys the eighteen-electron (equivalent atomic number) rule; cobaltocene (19 electrons) is easily oxidised to the very stable cobalticene, $[Co(C_5H_5)_2]^+$.

The exact extent to which the individual orbitals are stabilised or destabilised is not yet established but there is agreement that the E_{1g} bonding orbitals are stable and important. The whole problem is difficult to tackle quantitatively since no less than 19 separate orbitals are involved, all of which except two E_{2u} ligand antibonding orbitals can in principle undergo metal–ligand interaction.

9.6.2 Hydrides and other Derivatives

There is a marked tendency for the dicyclopentadienyls to bond with hydrogen, particularly in the case of the complexes of the second- and third-row transition metals. The isoelectronic series (which obeys the eighteen-electron rule) $TaH_3(C_5H_5)_2$, $WH_2(C_5H_5)_2$, $ReH(C_5H_5)_2$, $[OsH(C_5H_5)_2]^+$ is of considerable interest. Synthesis is by action of NaC_5H_5 on the chloride under reductive conditions

$$TaCl_5 + NaC_5H_5 + NaBH_4 \xrightarrow{THF} TaH_3(C_5H_5)_2$$

Thermal decomposition of MCp_2H_3 (M = Nb, Ta) gives $M(C_5H_5)_2$. The structure of these last compounds is not as simple as the formula indicates. They are in fact dimers, as has been established crystallographically for $\{Nb(C_5H_5)(C_5H_4)H\}_2$ (see figure 9.20). Each metal atom is bound by π orbitals to two C_5 rings, by a σ orbital to one C_5 ring and a σ orbital to a hydrogen; there is probably some metal–metal bonding too (Nb–Nb 310.5 pm). The osmium compound $Os(C_5H_5)_2$ is protonated

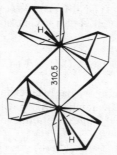

Figure 9.20 The structure of the 'niobocene' dimer (after L.J. Guggenberger and F.N. Tebbe, *J. Am. chem. Soc.*, **93** (1971), 5924)

in strong acid. The hydrides are quite stable; thus $ReH(C_5H_5)_2$ decomposes only at 250°. Typical very high-field metal hydride n.m.r. absorption is shown, for example at τ 23 for $ReH(C_5H_5)_2$. The tungsten and rhenium compounds can be protonated, giving the polyhydridic cations $[WH_3(C_5H_5)_2]^+$ and $[ReH_2(C_5H_5)_2]^+$. X-ray studies of the molybdenum compound $MoH_2(C_5H_5)_2$ have shown that the rings are not parallel, and it is likely that in the protonated compound the extra hydrogen atom is accommodated as shown in figure 9.21 because (a) theory indicates that a

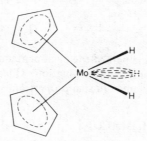

Figure 9.21 The probable position of the incoming proton in $[MoH_3(\pi\text{-}C_5H_5)_2]^+$ is indicated by H

lone pair extends in this direction in $MoH_2(C_5H_5)_2$ and (b) n.m.r. studies show that the H ligands in $[MoH_3(C_5H_5)_2]^+$ form an A_2B system (that is, the environment of one is slightly different from that of the other two, which have identical environments). $MoH_2(C_5H_5)_2$ may be readily converted into the corresponding chloride.

$$MoH_2(C_5H_5)_2 \xrightarrow[\text{boil}]{CHCl_3} MoCl_2(C_5H_5)_2$$

This last compound is an example of the cyclopentadienyls formed by 4d and 5d metals in their higher oxidation states (Ti and V also behave in this way). Other examples are $MCl_2(C_5H_5)_2$, M = Ti, Zr, Hf; V; W and $MCl_3(C_5H_5)_2$, M = Nb, Ta. They can be made by the action of a reactive cyclopentadienyl on the appropriate metal halide, avoiding an excess of the organometallic reagent, which has a reducing character. They are of the general type $M(C_5H_5)_2X_n$, where X may be a halide ion or a σ cyclopentadienyl or other σ-alkyl or σ-aryl group, and are usually rather stable to heat or hydrolysis; for example

$$MoI_4 + 4TlC_5H_5 \rightarrow Mo(\pi\text{-}C_5H_5)_2\ (\sigma\text{-}C_5H_5)_2 + 4TlI$$

In the case of titanium the chemistry has been examined in some detail and derivatives $Ti(\pi\text{-}C_5H_5)_2X_2$ are known, where X = F, Cl, OAc, Me, Ph and σ-C_5H_5.
Examples of the behaviour of the molybdenum cyclopentadienyls follow.

$$MoH_2(C_5H_5)_2 \xrightarrow{Mo(CO)_6} (C_5H_5)_2MoH_2Mo(CO)_5$$

In this latter example, the lone pair of the cyclopentadienyl complex is active in bonding, as in the earlier example of protonation. One notable example of a nitrosyl complex, which exhibits some unusual properties, is $(C_5H_5)_3Mo(NO)$. It may be prepared as follows

$$Mo(CO)_6 \xrightarrow{C_5H_5Na} [(\pi\text{-}C_5H_5)Mo(CO)_3]^- \xrightarrow[H_2O]{NO} (\pi\text{-}C_5H_5)Mo(CO)_2(NO) \qquad \text{(i)}$$

$$(\pi\text{-}C_5H_5)Mo(CO)_2NO \xrightarrow{I_2} \{(\pi\text{-}C_5H_5)Mo(NO)I_2\}_2 \qquad \text{(ii)}$$

$$\{(\pi\text{-}C_5H_5)Mo(NO)I_2\}_2 \xrightarrow[\text{(4 moles)}]{TlC_5H_5} (C_5H_5)_3Mo(NO) \qquad \text{(iii)}$$

The structure determined by X-ray analysis (figure 9.22) shows that one ring is σ bound (*monohapto*, h^1); the other two are 'tilted'. Taking NO to be a three-electron donor, application of the eighteen-electron rule would require the rings to approximate to four-electron donors. The ^1H n.m.r. spectrum is also interesting. At room temperature, there is only one narrow line ($\tau \approx 4.3$) for all fifteen protons, but at $-52°$, an AA'BB'X spectrum due to a h^1 ring is observed. Thus at room temperature the C_5H_5 rings are rapidly exchanging their structural relationships ('fluxion').

Crystallographic study on $(C_5H_5)_2Mo(NO)Me$ likewise shows the skew type of cyclopentadienyl co-ordination discussed above.

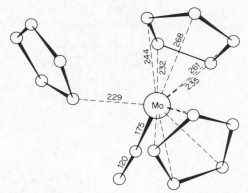

Figure 9.22 The structure of $Mo(C_5H_5)_3NO$ (after F.A. Cotton, *J. Am. chem. Soc.*, **91** (1969), 2523

9.7 Other Transition-metal π Complexes

The cyclopentadienyls are the most stable compounds formed between metals and unsaturated systems, but there are many other systems that form π-complexes. Complexes of two-atom systems (olefins and acetylenes) have already been mentioned and to complete the series, π-allyl, π-cyclobutadiene, π-benzene, π-tropylium (cycloheptatrienyl) and π-cyclo-octatetraene complexes are all known[†]. A few representative examples of these complexes follow, beginning with carborane complexes.

9.7.1 Transition-metal Carborane Complexes

Analogous to the π-cyclopentadienyl complexes are those of carboranes. Carborane, $C_2B_{10}H_{12}$ (1, 2-dicarbaclosodecaborane), which has an icosahedral structure, is conveniently prepared thus

$$B_{10}H_{14} + C_2H_2 \xrightarrow[\text{base}]{\text{Lewis}} C_2B_{10}H_{12} + 2H_2$$

Degradation with strong base removes a BH fragment to give the carbollide ion $[C_2B_9H_{11}]^{2-}$ (see figure 9.23).

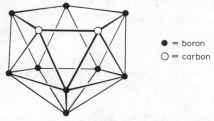

$\bullet$ = boron
$\bigcirc$ = carbon

Figure 9.23 The carbollide ion $[C_2B_9H_{11}]^{2-}$

[†] Certain of these compounds have already been mentioned in chapter 7.

The open face of the structure resembles the $C_5H_5^-$ ion in its binding characteristics, and forms a large number of complexes, some of which are shown below

$$[Fe(C_5H_5)(1,2\text{-}C_2B_9H_{11})] \xleftarrow[\ (2)\ \ O_2\ ]{(1)\ FeCl_2,\ C_5H_5^-} 1,2\text{-}C_2B_9H_{11}{}^{2-} \xrightarrow{[(C_6H_5)_4C_4PdCl_2]_2} \{\pi\text{-}[(C_6H_5)_4C_4]Pd(1,2\text{-}C_2B_9H_{11})\}$$

$$CoCl_2 \searrow \qquad\qquad \downarrow Re(CO)_5Br \qquad\qquad \swarrow {(1)\ AuCl_3 \atop (2)\ H_2O_2}$$

$$[Co(1,2\text{-}C_2B_9H_{11})_2]^- \qquad\qquad [(1,2\text{-}C_2B_9H_{11})Re(CO)_3]^- \qquad\qquad [Au(1,2\text{-}C_2B_9H_{11})_2]^-$$

X-ray study of the iron complex shows it to be π-bonded (see figure 9.24). The cobalt complex may be brominated to yield a hexabromo-derivative in which three

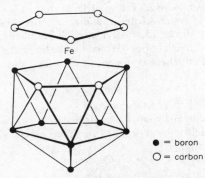

Fe

● = boron
○ = carbon

Figure 9.24 The carbollide complex $[Fe(\pi\text{-}C_5H_5)(\pi\text{-}1,2\text{-}C_2B_9H_{11})]$

bromine atoms have replaced three hydrogens bound to boron in each 'carbollide' unit, while X-ray study on the rhenium compound shows that the carbollide unit is again symmetrically π-bonded. The gold compound has been shown by X-ray studies to have a 'slipped' configuration (see figure 9.25). This has been described in terms of π-allyl type bonding.

An extensive series of these compounds is related by oxidation or reduction

$$[Au(1,2\text{-}C_2B_9H_{11})_2]^- \xrightarrow[H_2O_2]{Na/Hg} [Au(1,2\text{-}C_2B_9H_{11})_2]^{2-}$$

$$Ni(C_2B_9H_{11})_2 \xleftarrow{Fe^{3+}} [Ni(1,2\text{-}C_2B_9H_{11})_2]^- \xrightarrow{Na/Hg} [Ni(1,2\text{-}C_2B_9H_{11})_2]^{2-}$$

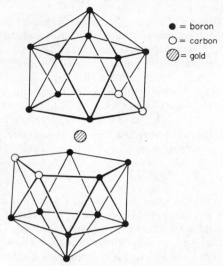

● = boron
○ = carbon
◉ = gold

Figure 9.25 The 'slipped' configuration adopted by $[Au(1,2\text{-}C_2B_9H_{11})_2]^-$ and several other carbollide complexes

A number of unusual formal oxidation states has been obtained, for example Co(IV), Ni(IV), Cu(III) and Au(II).

Many complexes have been prepared with more complex carbollide-type ligands, which fall outside the scope of this discussion.

9.7.2 π-Allyl Complexes

These are particularly stable in the case of the platinum metals, especially palladium. They can be obtained from the allyl Grignard reagent $CH_2=CH-CH_2MgBr$ and a metal halide. Several complexes are known of the type $M(C_3H_5)_n$; thus $n = 4$, M = Zr, Nb, Mo; $n = 3$, M = V, Cr, Fe, Co; $n = 2$, M = Ni, Pd, Pt, Mo. $M(C_3H_5)_2$ (M = Cr, Mo) are dimers (Cr–Cr = 197 pm; Mo–Mo = 218 pm). Each metal atom is bound to two bridging allyls and one 'chelating' π allyl.

The bridged dimeric complex $Pd_2Cl_2(C_3H_5)_2$ undergoes an interesting reaction with NaC_5H_5 giving $Pd(\pi\text{-}C_3H_5)(\pi\text{-}C_5H_5)$, the palladium atom being co-ordinated by both a three-membered system and a five-membered system. The allyl group may also bond in a one-electron, σ-bonded way, as in $(CO)_5MnCH_2 \cdot CH=CH_2$, and in an unsymmetrical way ideally described as

$$
M \longleftarrow
\begin{array}{c}
CH_2 \\
\| \\
CH \\
\diagup \\
CH_2
\end{array}
$$

as in $PdCl\{CH_2C(CH_3)CH_2\}PPh_3$.

9.7.3 Cyclobutadiene Complexes

These are relatively far fewer and less stable than cyclopentadienyl complexes but seem to be particularly stable when combined with iron and as the tetraphenyl derivative C_4Ph_4. Their instability reflects the instability of cyclobutadiene (the synthesis of which has, however, been reported). The bridged complex $(C_4Ph_4)BrPdBr_2PdBr(C_4Ph_4)$, obtained from $[PdCl_4]^{2-}$ and $PhC \vdots CPh$, contains the π-C_4Ph_4 group. It forms a convenient source of this group, giving respectively on treatment with $Fe(CO)_5$, $Co_2(CO)_8$ and $Co(C_5H_5)_2$ the compounds $Fe(CO)_3(C_4Ph_4)$, $CoBr(CO)_2(C_4Ph_4)$ and $Co(C_5H_5)(C_4Ph_4)$. It can be seen from these formulations that this group acts as a four-electron donor, just as does a non-cyclic diolefin such as 1,3-butadiene or norbornadiene.

9.7.4 Arene Complexes

These have actually been known since the 1920s (but were not understood then) when Hein isolated, for example, $Cr(C_6H_6)(C_6H_5 . C_6H_5)$. They have only more recently become well established. The C_6H_6 ring has π orbitals of similar symmetry to the $C_5H_5^-$ ring, together with an extra antibonding orbital, which cannot bond with the metal orbitals. Thus the bonding is formally the same as in $C_5H_5^-$ complexes. Since the benzene ring is uncharged and stable — that is, it has a high first ionisation potential — it is not likely to be such a good donor as $C_5H_5^-$, which is probably a reason why it forms less stable complexes. Thus $Cr(C_6H_6)_2$ is instantly oxidised by air, in great contrast to $Fe(C_5H_5)_2$, which is effectively isoelectronic. The benzene system forms complexes more readily when electron-repelling methyl groups are substituents, as in mesitylene $(1,3,5,C_6H_3Me_3)$ or hexamethylbenzene. Thus the isoelectronic species $W(C_6H_6)_2$, $[Re(C_6H_6)_2]^+$, $[Os(C_6H_3Me_3)_2]^{2+}$ and $[Ir(C_6H_3Me_3)_2]^{3+}$ are known. The benzene system retains its sixfold symmetry and acts as a six-electron donor. The bis-arene complexes may be obtained in a reductive manner; for example

$$MoCl_5 + Al + AlCl_3 + C_6H_6 \longrightarrow [Mo(C_6H_6)_2]^+$$

$$\downarrow OH^-$$

$$[Mo(C_6H_6)_2] + MoO_4^{2-}$$

Co-condensation of molybdenum vapour with benzene at 77 K also yields $[Mo(C_6H_6)_2]$. 'Piano-stool' complexes such as $Mo(CO)_3(C_6H_3Me_3)$ may be obtained by direct substitution of carbonyls; for example

$$Mo(CO)_6 + C_6H_3Me_3 \xrightarrow{u.v.}$$

Complexes of substituted Dewar-benzenes, such as hexamethylbicyclo [2,2,0] hexa-2,5-diene(hexamethyldewarbenzene), HMDB, are related to benzene complexes. $PdCl_4^{2-}$ reacts with HMDB in methanol to form $PdCl_2(C_{12}H_{18})$, thought to have the structure shown in figure 9.26. With $RhCl_3$ in methanol, however, $RhCl_2(C_5Me_5)$ is formed; it may be dimeric.

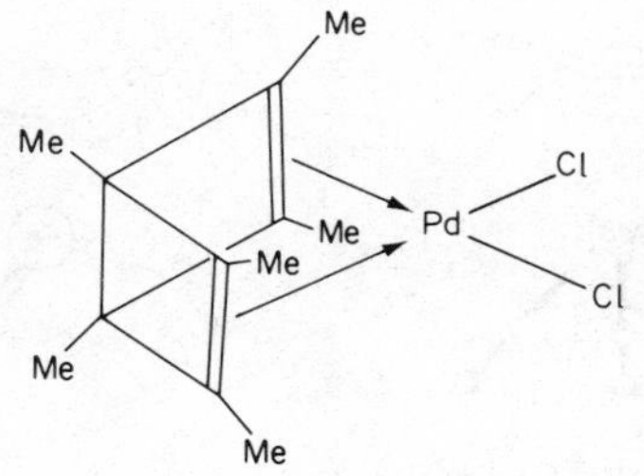

Figure 9.26 The probable structure of the hexamethyldewarbenzene complex $PdCl_2(C_6Me_6)$

9.7.5 Tropylium Complexes

The tropylium cation $C_7H_7^+$ is iso-π-electronic with $C_5H_5^-$ and C_6H_6 and its lower-lying orbitals have similar symmetries. There are rather few known examples of its occurrence as a ligand but in these examples it is fairly stable, perhaps surprisingly so when it is remembered that the donor properties of cations are obviously poor. Its complexes may be obtained by hydride abstraction from cycloheptatriene complexes; for example

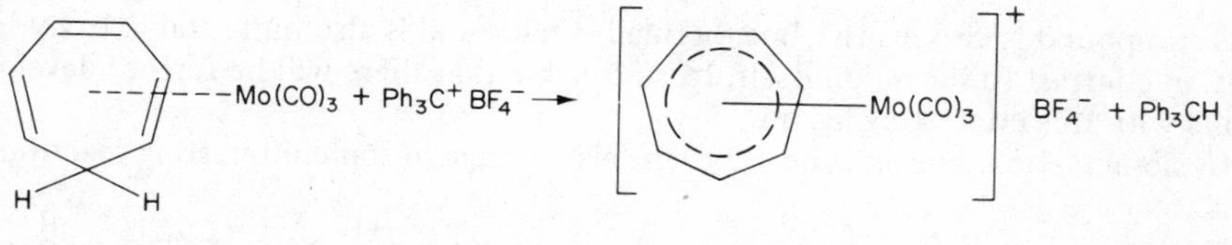

One example of a complex involving a C_7H_7 ring is $(\pi\text{-}C_7H_7)Mo(CO)_2$ $(\sigma\text{-}C_6F_5)$, whose structure has been established by X-ray studies (see figure 9.27).

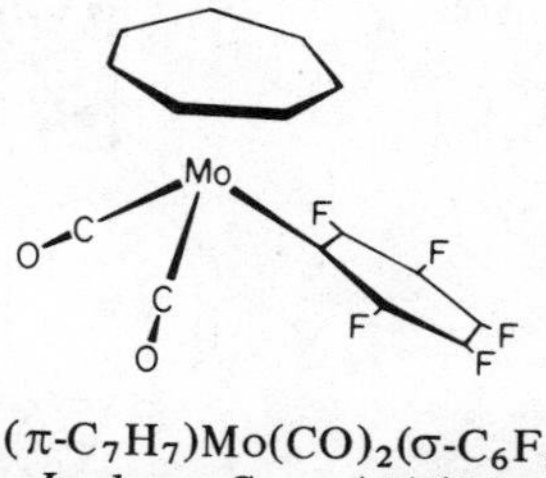

Figure 9.27 The structure of $(\pi\text{-}C_7H_7)Mo(CO)_2(\sigma\text{-}C_6F_5)$ (after M.R. Churchill and T.A. O'Brien. *J. chem. Soc., (A)* (1969), 1110)

9.7.6 Cyclo-octatetraene Complexes

Cyclo-octatetraene itself is not planar, but it forms a doubly charged anion $C_8H_8^{2-}$, which theory predicts should be planar. The uncharged ligand is known to use either two or three of its double bonds for co-ordination, as indicated by X-ray structure determinations of $Fe(CO)_3(C_8H_8)$ and $Mo(CO)_3(C_8H_8)$ (see figure 9.28).

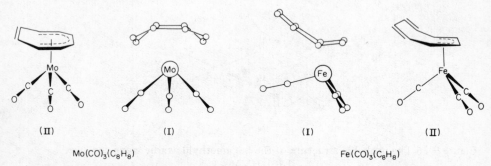

(II)   (I)   (I)   (II)

$Mo(CO)_3(C_8H_8)$ $Fe(CO)_3(C_8H_8)$

Figure 9.28 Structures of $Mo(CO)_3(C_8H_8)$ and $Fe(CO)_3(C_8H_8)$ in the solid state (I) and idealised solution structures (deducted from ^1H n.m.r.) (II)

The iron complex was one of the first recognised examples of fluxional compounds. It has only one ^1H n.m.r. signal, a singlet, the protons being made dynamically equivalent by a rapid rotation of the ring. However, the planar $C_8H_8^{2-}$ system is known, as in uranocene $U(C_8H_8)_2$, which is made as follows

$$UCl_4 + 2K_2C_8H_8 \rightarrow U(C_8H_8)_2 + 4KCl$$

This compound is air sensitive but thermally stable and is also quite stable to hydrolysis, in contrast to the anion itself. It is probable that there will be further developments with this interesting ligand.

Cyclo-octa-1,5-diene and the 1,3-diene also engage in some interesting reactions

Osmium, however, behaves differently

$$\text{Os}_3(\text{CO})_{12} \quad h\nu \qquad \text{Ph}_3\text{C}^+$$

Os(CO)$_3$ Os(CO)$_3$

$$\downarrow \text{BH}_4^-$$

Os(CO)$_3$ Ph$_3$C$^+$ Os(CO)$_3$

9.8 Transition-metal Alkyls and Aryls

A number of compounds containing σ-bonded alkyls and aryls have been known for some time, for example [(PtMe$_3$Cl)$_4$], but generally they are rare for transition metals in 'normal' oxidation states, when π-bonding ligands (such as PR$_3$, CO, π-C$_5$H$_5$) are not present. In recent years, however, a number of stable compounds that do not contain π-bonding ligands have been isolated (some of these compounds, involving metals in 'normal' oxidation states, are mentioned in the appropriate chapter).

There is ample evidence that the transition-metal–carbon bond is not weak, and that the metal–carbon distance is not significantly influenced by other ligands (whether π or σ bonding). A number of platinum(II) and (IV) compounds containing σ-bonded alkyl groups have been examined by X-ray analysis, and the Pt–C distance is always found to be ≈ 210 pm. Some Cr(III), Rh(III) and Co(III) alkyls also have considerable stability. Limited thermodynamic data accord with this view.

Instability is the rule in many cases, however: the Sn–C and Ti–C force constants in Me$_4$Sn and Me$_4$Ti are virtually identical (≈ 225 N m^{-1}), but although the former is extremely stable the latter decomposes readily in ether at -40°C.

Early attempts to rationalise the stability of certain σ-bonded alkyls were based *not* on the postulate of a weak σ bond but rather on the possibility of decomposition occurring via other processes involving pathways with a low activation energy. It was suggested that promotion from a non-bonding d orbital to antibonding metal–carbon orbitals is the first step in metal–carbon bond fission. It should therefore follow that many of the complexes concerned should be photosensitive, since the energies concerned correspond to that of u.v. radiation; they are not.

Recently, it has been observed that use of ligands that are sterically hindered and cannot eliminate alkene according to the reaction

$$\text{M} \longrightarrow \text{H} \;+\; \text{CH}_2 \!=\! R_2$$

can lead to thermally stable compounds. In particular $-CH_2SiMe_3$ has been shown to form compounds such as $V(CH_2SiMe_3)_4$, $Cr(CH_2SiMe_3)_4$ and $M_2(CH_2SiMe_3)_6$ (M = Mo, W). Other groups that can play a similar role are benzyl, methyl and phenyl, in compounds such as $M(CH_2Ph)_4$ (M = Ti, Zr, Hf), WMe_6, $Au(C_6F_5)_3$ and $Ti(C_6H_5)_4$.

Of course, there are other decomposition pathways, such as attack on the solvent. Adduct formation (for example, with phosphines) can inhibit this process by 'blocking' vacant sites in an otherwise co-ordinately unsaturated molecule. It has also been pointed out that the electronic configuration of the metal can influence the activation energy for dissociation of the metal–carbon bond; that is, if the direct product of the ground- and excited-state wave functions has the same symmetry as a vibrational stretching mode, then homolytic fission is facilitated. It has therefore been suggested that t_{2g}^1, t_{2g}^3 and t_{2g}^6 configurations (and obviously t_{2g}^0) are likely to yield the most stable octahedral complexes, as is indeed observed.

9.9 The Binary Carbonyls. Synthesis and Structure

9.9.1 The d^5 Metals

A carbonyl, $V(CO)_6$, is formed by V, but Nb and Ta give only the anion $[M(CO)_6]^-$.

Preparation[†] VCl_3, Na, CO; 160°, 20 MPa, diglyme; the resulting $[V(CO)_6]^-$ is then treated with acid.

Structure[‡] Probably octahedral.

Appearance Blue-black solid, decomposes at 70°.

9.9.2 The d^6 Metals

One type of carbonyl, $M(CO)_6$, is formed by Cr, Mo and W.

Preparation $CrCl_3$, Na, CO; 0°, 20 MPa, diglyme. $MoCl_5$, $AlEt_3$, CO. WCl_6, $AlEt_3$, CO.

Structure Octahedral (X-ray).

Appearance Colourless volatile solids, decompose ≈ 200°.

9.93 The d^7 Metals

One type of carbonyl, $M_2(CO)_{10}$, is formed by Mn, Tc and Re.

Preparation $MnCl_2$, Na, Ph_2CO, CO; 100°, 10 MPa, THF. Tc_2O_7, CO; 250°, 20 MPa. Re_2O_7, CO; 250°, 20 MPa.

[†] Preparative methods are listed in the form: reactants; temperature, pressure, solvent. Only one method is given for each carbonyl. In most cases, several alternatives are known.

[‡] The methods by which the structure was determined or indicated are given in parentheses.

Structure Two octahedra joined in a staggered configuration by a metal–metal bond (X-ray).

Appearance Orange (Mn) or colourless (Tc, Re) solids.

9.9.4 The d^8 Metals

$M(CO)_5$ is formed by Fe, Ru and Os; $M_2(CO)_9$ by Fe and Os; $M_3(CO)_{12}$ by Fe, Ru and Os; $Os_4(CO)_{13}$, $Os_5(CO)_{16}$, $Os_6(CO)_{18}$, $Os_7(CO)_{21}$ and $Os_8(CO)_{23}$ have also been reported.

Preparation Fe, CO; 200°, 20 MPa $\rightarrow$ Fe(CO)$_5$

$\quad$ Fe(CO)$_5$, light $\rightarrow$ Fe$_2$(CO)$_9$

$\quad$ Fe(CO)$_5$, OH$^-$, MnO$_2$; MeOH/H$_2$O $\rightarrow$ Fe$_3$(CO)$_{12}$

$\quad$ Ru(acac)$_3$, CO, H$_2$; heptane $\rightarrow$ Ru(CO)$_5$

$\quad$ RuCl$_3$.3H$_2$O, CO; methanol $\rightarrow$ Ru$_3$(CO)$_{12}$

$\quad$ OsO$_4$, CO; 100°, 5 MPa $\rightarrow$ Os(CO)$_5$

$\quad$ Os(CO)$_5$, light; $-40°$ $\rightarrow$ Os$_2$(CO)$_9$

$\quad$ OsO$_4$, CO; methane $\rightarrow$ Os$_3$(CO)$_{12}$

$\quad$ Os$_3$(CO)$_{12}$ $\xrightarrow{\text{pyrolise}}$ Os$_x$(CO)$_y$ $(x, y = 4, 13; 5, 16; 6, 18; 7, 21; 8, 23)$

Structure Fe(CO)$_5$ is a trigonal bipyramid (X-ray). Fe–C $\approx$ 182 pm, Ru(CO)$_5$ and Os(CO)$_5$ are probably similar. Fe$_2$(CO)$_9$ comprises two Fe(CO)$_6$ pseudo-octahedra, three vicinal CO groups being common (X-ray). Os$_2$(CO)$_9$ is of unknown structure (but it is thought that only one CO bridges). M$_3$(CO)$_{12}$ are based on a triangular arrangement of metal atoms; the Fe compound alone contains bridging CO.

Appearance Fe(CO)$_5$; colourless liquid; Fe$_2$(CO)$_9$: yellow solid, strangely involatile, decomposes 100°; Fe$_3$(CO)$_{12}$: volatile black solid. Ru(CO)$_5$: colourless liquid; Ru$_3$(CO)$_{12}$: yellow solid. Os(CO)$_5$: colourless liquid; Os$_2$(CO)$_9$: orange-yellow crystals; Os$_3$(CO)$_{12}$: yellow solid; Os$_6$(CO)$_{18}$: brown solid.

9.9.5 The d^9 Metals

Three types of carbonyl are formed, $M_2(CO)_8$ (Co, Rh†); $M_4(CO)_{12}$ (Co, Rh, Ir) and $M_6(CO)_{16}$ (Co, Rh).

Preparation Co(CO)$_3$, CO, H$_2$; 150°, 35 MPa $\rightarrow$ Co$_2$(CO)$_8$. Co$_2$(CO)$_8$; 50° $\rightarrow$ Co$_4$(CO)$_{12}$. Co$_2$(CO)$_8$; (i) 60°, EtOH; (ii) Fe^{3+} $\rightarrow$ Co$_6$(CO)$_{16}$. Rh$_4$(CO)$_{12}$, CO, pressure $\rightarrow$ Rh$_2$(CO)$_8$. Rh$_2$Cl$_2$(CO)$_4$, CO, OH$^-$; hexane $\rightarrow$ Rh$_4$(CO)$_{12}$. RhCl$_3$.3H$_2$O, CO, MeOH; 60°, 5 MPa $\rightarrow$ Rh$_6$(CO)$_{16}$. IrCl$_3$.3H$_2$O, CO, MeOH; 60°, 5 MPa $\rightarrow$ Ir$_4$(CO)$_{12}$.

$\dagger$ Rh$_2$(CO)$_8$, however, is only stable at high pressures of CO, rapidly converting to Rh$_4$(CO)$_{12}$ otherwise.

Structure $Co_2(CO)_8$ has each metal atom pseudo-octahedrally co-ordinated to three terminal CO, two bridging CO and one metal (X-ray). $M_4(CO)_{12}$ are based on a tetrahedral array of metal atoms with three bridging CO (Co, Rh) or all terminal CO (Ir) (X-ray). $Rh_6(CO)_{16}$ has an octahedral array of Rh atoms. (X-ray); $Co_6(CO)_{16}$ is probably similar.

Appearance All solids: $Co_2(CO)_8$ orange, $Co_4(CO)_{12}$ and $Co_6(CO)_{16}$ black; $Rh_4(CO)_{12}$ red, $Rh_6(CO)_{16}$ black; $Ir_4(CO)_{12}$ yellow.

9.9.6 The d^{10} Metals
$Ni(CO)_4$ only.

Preparation Ni, CO.

Structure Tetrahedral (X-ray), Ni$-$C = 184 pm.

Appearance Colourless liquid, b.p. 42°.

The question of whether or not a polynuclear carbonyl will contain bridging groups is very open. It seems that there is very little difference in energy between bridged and nonbridged structures; for example, there is i.r. evidence for nonbridged $(CO)_4Co-Co(CO)_4$ in solution (see figure 9.29).

There is now firm evidence for the existence of several ordinarily unstable metal carbonyls at low temperatures when stabilised in a solid rare gas matrix. Carbonyls, often of uncertain stoichiometry, of Al, Sn, Ta, Pd, Pt, Cu, Ag, Nd, Yb and U have been reported. The metal is heated to, for example, 1500°C *in vacuo* and the resulting monatomic vapour is condensed at 20 K together with a large excess of Kr containing a small proportion of CO. The characteristic carbonyl infrared stretching frequencies are used to identify the species present as far as possible; ^{13}C isotopic substitution has been used to aid this. In the palladium-carbon monoxide system, there is evidence for PdCO, $Pd(CO)_2$ (linear), $Pd(CO)_3$ (trigonal planar) and $Pd(CO)_4$ (tetrahedral). The latter is stable to 80 K. The Pd$-$CO force constant is only 82 N m^{-1} (180 for $Ni(CO)_4$).

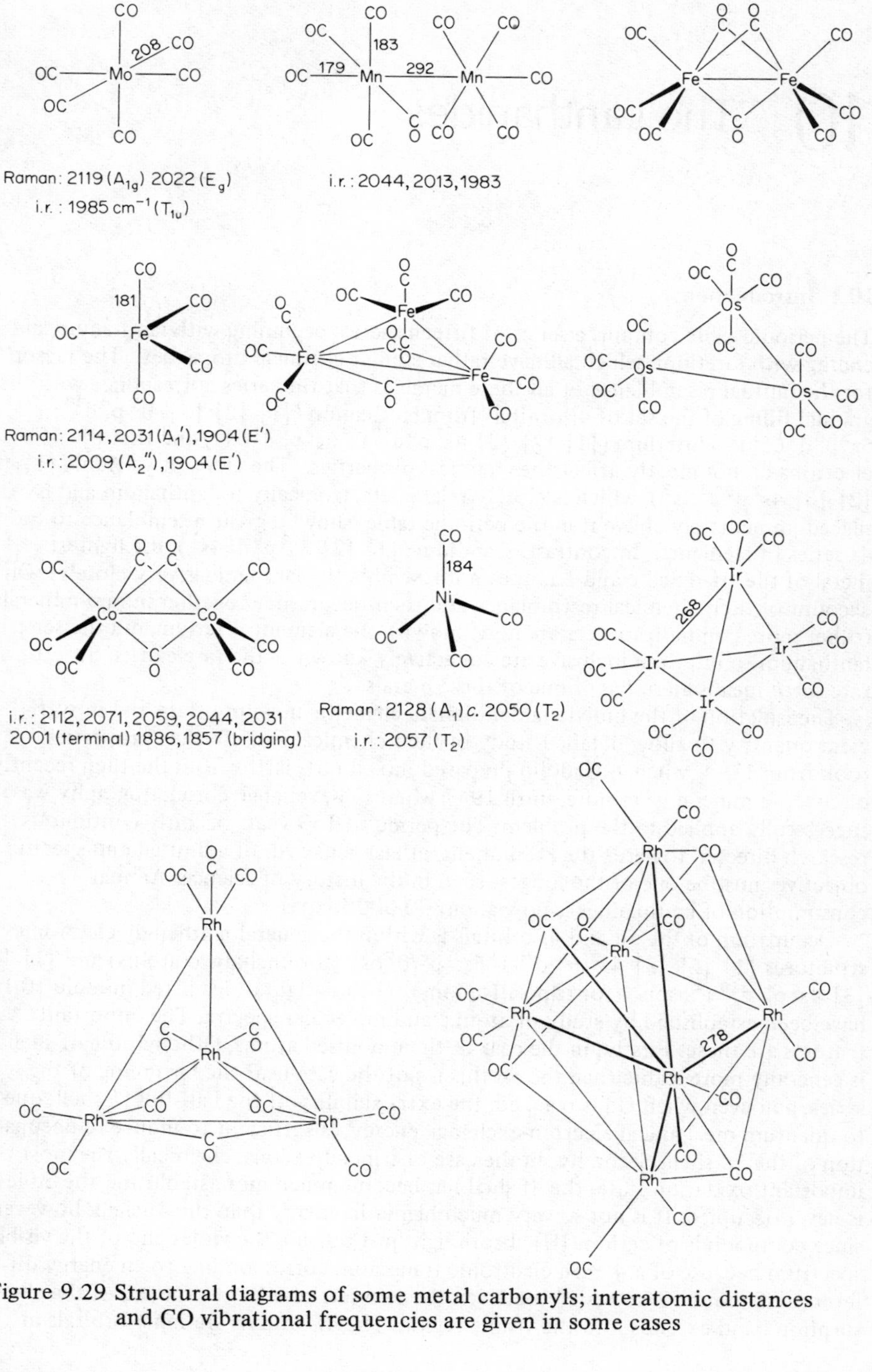

Raman: 2119 (A$_{1g}$) 2022 (E$_g$)
i.r. : 1985 cm^{-1} (T$_{1u}$)

i.r. : 2044, 2013, 1983

Raman: 2114, 2031 (A$_1'$), 1904 (E')
i.r. : 2009 (A$_2''$), 1904 (E')

i.r.: 2112, 2071, 2059, 2044, 2031
2001 (terminal) 1886, 1857 (bridging)

Raman: 2128 (A$_1$) c. 2050 (T$_2$)
i.r. : 2057 (T$_2$)

Figure 9.29 Structural diagrams of some metal carbonyls; interatomic distances and CO vibrational frequencies are given in some cases

10 The Lanthanides

10.1 Introduction

The periodic table contains a series of fifteen metals beginning with lanthanum and ending with lutetium, which all have rather similar chemical properties. The reason for the mutual resemblance of all these metals is that the series corresponds with the gradual filling of the set of 4f orbitals from lanthanum ([1] [2] [3] $4s^2p^6d^{10}$, $5s^2p^6d^16s^2$) to lutetium ([1] [2] [3] $4s^2p^6d^{10}f^{14}5s^2p^6d^16s^2$) and that the 4f electrons do not greatly affect the chemical properties. The element yttrium ([1] [2] [3] $4s^2p^6d^15s^2$), which is closely related electronically to lanthanum and is placed immediately above it in the periodic table, shows a great resemblance to the 4f series of elements. In contrast, scandium ([1] [2] $3s^2p^6d^14s^2$), the lightest metal of the triad Sc, Y and La, does not resemble the lanthanides very closely. On account of their chemical resemblance and their occurrence together in rare minerals (other more common sources are now known), the element ytterium and the series lanthanum to lutetium inclusive are collectively known as the rare earths. Strictly, rare *earth* means an *oxide* of one of these metals.

The isolation of the individual lanthanide elements in a pure state and in sufficient quantity to allow detailed study of their chemical and physical properties, took from 1794, when J. Gadolin prepared mixed rare earths from the then recently discovered mineral gadolinite, until 1947 when ion-exchange chromatography was successfully applied to the problem. The period of 153 years of fairly continuous research directed towards the attainment, at last achieved, of a limited and specific objective must be one of the longest such in the history of science. Annual consumption of lanthanides is now about 30 000 tonnes.

Occupation of the 5d and 4f orbitals is within the general lanthanide electronic structures [1] [2] [3] $4s^2p^6d^{10}f^x5s^2p^6d^y6s^2$ (for uncharged atoms) and [1] [2] [3] $4s^2p^6d^{10}f^x5s^2p^6$ (for tripositive ions). The configurations listed in table 10.1 have been established by study of atomic and molecular spectra. The important features are these. Firstly, in the case of the unionised atoms, although the 4f shell is generally more stable than the 5d this is not the case near the beginning of the series, and even when Gd is reached, the extra stability of the half-full f^7 shell, due to quantum-mechanical electron-exchange energy, is sufficient to induce re-occupation of the 5d shell. Secondly, in the case of tripositive ions, chemically the most important oxidation state, the 4f shell has become much more stable and the 5d level is never occupied. It is not so very much higher in energy than the 4f shell, however, since compounds of cerium(III) absorb light just outside the violet end of the visible spectrum because of a 4f→5d electronic transition, corresponding to an energy difference of about 50 000 cm^{-1} between the two sets of orbitals; the tail of the absorption band extends into the visible region. The 4f orbitals are inner orbitals in

Table 10.1: Outer Electronic Configurations of Lanthanide Atoms and the M^{3+} Ions

Element	Symbol	M	M^{3+}	Element	Symbol	M	M^{3+}
lanthanum	La	$d^1 f^0$	f^0	terbium	Tb	$d^0 f^9$	f^8
cerium	Ce	$d^1 f^1$	f^1	dysprosium	Dy	$d^0 f^{10}$	f^9
praseodymium	Pr	$d^0 f^3$	f^2	holmium	Ho	$d^0 f^{11}$	f^{10}
neodymium	Nd	$d^0 f^4$	f^3	erbium	Er	$d^0 f^{12}$	f^{11}
promethium[†]	Pm	$d^0 f^5$	f^4	thulium	Tm	$d^0 f^{13}$	f^{12}
samarium	Sm	$d^0 f^6$	f^5	ytterbium	Yb	$d^0 f^{14}$	f^{13}
europium	Eu	$d^0 f^7$	f^6	lutetium	Lu	$d^1 f^{14}$	f^{14}
gadolinium	Gd	$d^1 f^7$	f^7				

[†] Radioactive: ^{147}Pm (β^-, 2.5 years) is most useful isotope

the sense that the space that their electrons occupy lies mainly inside that occupied by the outermost shell of the 5s and 5p electrons; paradoxically, although enclosed well within the latter shell, their electrons are less stable than the 5s and 5p and have lower ionisation potentials. The less stable 5d orbitals extend nearer to the $5s^2 5p^6$ closed shell. Evidence for this situation is provided by the electronic absorption spectra of compounds of the tripositive lanthanide ions, where a change of the element or group bonded to the lanthanide does not substantially alter the energy of electronic transitions within the well-shielded 4f shell, although 4f→5d transitions are much more strongly perturbed.

We have mentioned already that the tripositive oxidation state of the lanthanides is of considerable importance. The reason for this is simply that these elements and yttrium all happen to form their most stable compounds when in the tripositive state. The word 'happen' is used intentionally; it is an almost accidental consequence of their ionisation potentials and ionic radii that the lanthanides are predominantly tripositive. That this need not necessarily have turned out in this way is shown by the considerable stability of particular nontripositive lanthanide species, notably Ce^{4+} and Eu^{2+}, and by the very variable oxidation states adopted by the lighter actinides. Let us consider the stability of cerium difluoride. If it is unstable, it will disproportionate into the trifluoride and metallic cerium

$$3CeF_2(s) \rightarrow 2CeF_3(s) + Ce(s)$$

Whether or not this reaction takes place will depend (neglecting entropy changes) on the sign of ΔH, the enthalpy change. We can break the process down into its component processes, which when summed give the original equation

$3CeF_2(s) \rightarrow 3Ce^{2+}(g) + 6F^-(g)$ (lattice energy of CeF_2)

$Ce^{2+}(g) + 2e \rightarrow Ce(g)$ (first and second ionisation potentials of Ce)

$Ce(g) \rightarrow Ce(s)$ (heat of sublimation of Ce)

$2Ce^{2+}(g) \rightarrow 2Ce^{3+}(g) + 2e$ (third ionisation potential of Ce)

$2Ce^{3+}(g) + 6F^-(g) \rightarrow 2CeF_3(s)$ (lattice energy of CeF_3)

Thus applying Hess's law we see that the sign of ΔH depends on the respective magnitudes of the lattice energies of CeF_2 and CeF_3, on all three ionisation potentials of cerium and on its heat of sublimation. In principle it is possible to determine ΔH in this way, making certain valid assumptions, for example that bonding is substantially ionic and that CeF_2 (actually unknown except when stabilised in a host lattice as a dilute solid solution) would assume the fluorite structure. The clear point remains, however, that only the precise quantitative relationship of all these quantities determines the stable oxidation state adopted by a lanthanide in the solid state.

Exactly the same type of consideration applies to the question of the stability in water solution of differing oxidation states. Consider the reduction of water by Yb^{2+} ions

$$Yb^{2+}(aq) + H^+(aq) \rightarrow Yb^{3+}(aq) + \tfrac{1}{2}H_2(g)$$

Here the process $H^+ + e \rightarrow \tfrac{1}{2}H_2$ requires a free-energy change (expressed in terms of standard electrode potential) of about 0.6 V. Whether or not this free energy is available depends on the magnitude of the third ionisation potential of ytterbium and the difference in free energies of hydration of Yb^{2+} and Yb^{3+}. If we simply wish to compare the stability of two adjacent lanthanide divalent species, such as Tm^{2+} and Yb^{2+}, we note that hydration energies, which are determined mainly by the ionic radius of the species, will be very similar for any two adjacent lanthanides. Thus relative stability in aqueous solution (or for that matter in the solid state) depends largely on the third ionisation potential of the two lanthanides in question. If we were considering the relative stability in water of two adjacent tetravalent species such as Ce^{4+} and Pr^{4+} we would compare the processes

$$M^{4+}(aq) + H_2O \rightarrow M^{3+}(aq) + \tfrac{1}{2}O_2(g) + 2H^+(aq)$$

depending relatively on the fourth ionisation potentials of the lanthanides being considered.

In this way the oxidation states of the lanthanides, whether as simple compounds such as fluorides or oxides, or as hydrated ions, may be rationalised. In particular there exists a correspondence between the stability of the dipositive and tetrapositive species and the number of electrons in the 4f shell. This is because there exists a quantum-mechanical exchange stabilisation of electrons in a particular set of orbitals, which is proportional to $n!$, where n is the number of electrons with parallel spins. Hence a half-full shell will be particularly stable for two reasons. Firstly, it will have an unusually high ionisation potential, I. Secondly, such a configuration will be easily attained: I_3 (20.6 eV) for Gd will have no loss-of-exchange component since Gd^{2+} is $f^8 = f^7$ (all spins parallel) $+ f^1$. I_3 for other elements will have a component larger than this. Similarly, the removal of its single f electron from Ce^{3+} (f^1) will be an unusually easy process. Consideration will show that the benefit will extend to a lesser extent to tetrapositive states f^1 and f^8 and dipositive states f^6 and f^{13}. Thus if we accept that in general lanthanides tend to be tripositive we might expect that the species Eu^{2+} (f^7), Yb^{2+} (f^{14}), Sm^{2+} (f^6), Tm^{2+} (f^{13}), Ce^{4+} (f^0), Tb^{4+} (f^7), Pr^{4+} (f^1) and Dy^{4+} (f^8) should be anomalously stable. *In general* this is found to be true. Thus Eu^{2+} (I_3 = 24.9 eV) is stable in water, Yb^{2+} and

Sm^{2+} reduce water to hydrogen; while Tm^{2+} is stable only as the solid TmI_2. Of the tetrapositive ions, Ce^{4+} (the well-known volumetric reagent) is the only one to be stable in water; Tb^{4+} forms TbO_2 and TbF_4; Pr^{4+} forms PrO_2 and PrF_4; Dy^{4+} forms only a complex fluoride.

10.2 The Metals

The lanthanide metals and yttrium are quite soft and are silvery in colour. The lighter members of the series are rapidly tarnished in air but the heavier members are stable. They are all attacked by water and acids. Three crystal structures are adopted, all based on the typical metallic close packing of spheres. Certain of the metals have unusual properties. Thus gadolinium has a Curie point of 26°, below which it is ferromagnetic and above which it is paramagnetic. In the ferromagnetic state it has a magnetic moment at saturation (263.5 c.g.s. units) greater than that of iron (221 c.g.s. units). Lanthanum is a superconductor below 4.9 K (hexagonal form) or 5.9 K (cubic form). The isotopes of samarium have exceptional neutron capture capabilities.

The isolation of the metals in a pure state is a matter of some difficulty because of their electropositive character and high melting points. These elements are so high in the electrochemical series that electrolysis of aqueous solutions is unavailable as a means of isolation. However, electrolysis of the aqueous chloride using a mercury cathode does yield the metal as an amalgam from which most but not all of the mercury can be removed by distillation; this method is now superseded by reduction of anhydrous halides or oxides by electrolytic or metallothermic means. There has been some interest in the possible metallurgical and nuclear applications of the rare-earth metals and methods have been described for their production on the 1–10 kg scale. Metallothermic reduction in a tantalum crucible of the lanthanide fluoride with calcium at 1450–1700° is usually preferred.

Those metals with a considerable tendency towards a dipositive oxidation state, namely samarium, europium and ytterbium, cannot be obtained by this method owing to formation of the stable difluoride. They have, however, been obtained by exploitation of the greater volatility of these three metals compared with lanthanum (b.p.s: La, 3469°; Sm, 1900°; Eu, 1349°; Yb, 1427°). The equilibrium

$$2La + (Sm, Eu, Yb)_2O_3 \rightleftharpoons La_2O_3 + 2(Sm, Eu, Yb)$$

is displaced to the right by removal of the more volatile metal to a condenser. The high melting points and high reactivity of the lanthanides leads to considerable difficulty in their electrolytic preparation. This method is used chiefly to obtain the mixture of lanthanum (25 per cent), cerium (50 per cent) and neodymium (20 per cent) known as mischmetall. This alloy is used in metallurgy, since 1–2 per cent confers creep resistance on magnesium alloys. Lighter flints are made from mischmetall further alloyed with 50 per cent or less iron. Mischmetall is made by electrolysis at 900° in iron pots of fused light lanthanide chlorides mixed with alkali chlorides.

Although methods such as vacuum melting, scavenging or zone-refining are sometimes useful, the most effective method of purification is vacuum distillation. The metal may contain elements such as calcium, oxygen and tantalum (from the crucible) as impurity, and these are eliminated by distillation, which proceeds at temperatures between $1400°$ and $2000°$ and a pressure of 1 mPa (10^{-5} mm Hg), yielding bright metallic crystals. The process is applicable to all the rare-earth metals and also scandium. By this process impurity levels are reduced to about 100 p.p.m.

10.3 Simple Compounds of Rare-earth Metals

10.3.1 Introduction

The simple compounds of the lanthanides all resemble the corresponding compounds of yttrium. In general the latter show just the properties that would be expected from the position of yttrium in the periodic table. It may well be asked, therefore, if there is any need for a close study of the individual lanthanide compounds. There are, however, several features of unusual interest in lanthanide compounds which repay attention.

The very fact that there are fifteen metals which have such a close mutual resemblance means that we can study fairly closely the influence of ionic radius (the main variable factor) on co-ordination geometry and crystal structure. The possible existence and stability of compounds in the dipositive and tetrapositive oxidation states is, of course, dependent on the identity of the lanthanide under consideration and these oxidation states provide an interesting comparison not only of one lanthanide with another but of lanthanides with metals of Groups IIA and IVA, for example Ba or Hf. Another field of study concerns the physical properties of the compounds. Even though two compounds may be chemically alike, they may have interesting and important variations in, for example, magnetic susceptibility, visible and ultraviolet absorption and emission, temperature dependence of electrical resistance, or electron spin resonance parameters. These properties are, of course, just those that depend on the f-electrons whose presence, if not too obvious from a chemical point of view, is made evident by their physical effects.

The crystal structures of the simple compounds such as oxides and halides show high co-ordination numbers. This is an effect of the large ionic radii of the tervalent ions (see figure 10.1). The overall trend is a decrease in ionic radius as atomic number increases. The changes in radius are largest at the lighter end of the series where they are about 2 pm, decreasing to about 1 pm at the heavier end. The cause of this effect (known as the lanthanide contraction, but also observable for the d transition elements and for the actinides) is the poor shielding quality of the f electrons. The additional attractive force experienced by the filled 5p outer shell, as each successive proton is added to the nucleus to complete the lanthanide series, is not quite neutralised by the repulsive effect of the corresponding additional f electron and so the outer shell contracts.

It may be noticed from figure 10.1 that the whole curve for La^{3+} – Lu^{3+} is actually made up of two half-curves intersecting at Gd^{3+}, whose radius and that of its neighbours is anomalously slightly high. This discontinuity, together with discontinuities in chemical properties (mainly stability constants, which, however, are not necessarily a consequence of the ionic radius anomaly) is sometimes referred to as the 'gadolinium break'. An exactly similar effect is observed in the case of, for example, the dipositive 3d transition metals, with Mn^{2+} having an anomalously large radius. There is every reason to think that the causes of the two phenomena are

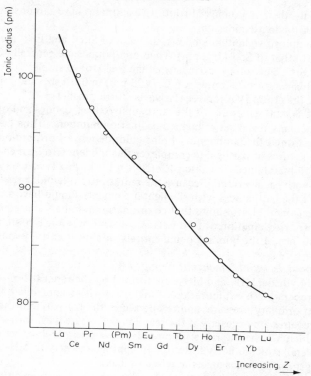

Figure 10.1 Ionic radii of tervalent lanthanide (M^{3+}) ions

similar, that is, a slight preferential orientation of the d or f electrons in directions between the anions, leading to decreased shielding between the metal cation and the anion with a resultant shorter bond length. This type of stabilisation is not available to Mn^{2+} (d^5) or Gd^{3+} (f^7) since they both have a spherically symmetrical half-filled electronic shell. Thus it would be more correct to consider that the radii of La^{3+}, Gd^{3+} and Lu^{3+} are normal, while the other radii are anomalously small.

Before finally leaving the subject of ionic radii, we may consider the size relationships between lanthanide ions and other metallic cations. Firstly, yttrium (radius of Y^{3+}, 98 pm) has a radius similar to those of the lanthanides. In particular its radius lies between those of Ho^{3+} and Er^{3+}, which elements it consequently resembles in its chemical behaviour more closely than it resembles the other lanthanides. Thus the natural increase of size of La^{3+} over Y^{3+}, owing to its having one extra complete electronic shell, is eliminated by the lanthanide contraction when we move across the series to Ho^{3+}. Scandium, the lightest element in Group IIIA, with a radius for Sc^{3+} of only 68 pm, has chemical properties quite distinct from those of the lanthanides and yttrium. It is less basic and is a stronger complexing agent.

The actinide ions in their tripositive states are only slightly larger than corresponding lanthanide ions and if we except the consequences of their very different redox potentials, the compounds of tripositive actinides have rather similar properties to those of the lanthanides.

Among the dipositive lanthanide ions, the radii of Sm^{2+} and Eu^{2+} are almost exactly equal to that of Sr^{2+} (112 pm). Their compounds are generally isomorphous with those of Sr^{2+}. The tetrapositive lanthanide ions (for example, Ce^{4+}, 92 pm) show a resemblance to the actinide ion U^{4+} (93 pm) in both size and properties. Zr^{4+} (79 pm), the group IVA representative, is a little smaller.

Among the crystal structures of the lanthanide oxides, halides and other simple salts, there are many instances of high co-ordination numbers. Thus $LaCl_3$ has a co-ordination around the lanthanum of nine chloride ions. Furthermore, among the crystal structures of lanthanide complexes this tendency to high co-ordination numbers is well maintained, the anion $[Ce(NO_3)_6]^{3-}$ having twelve oxygen atoms co-ordinated with each cerium. Because the radius-ratio rule indicates that most compounds of the lanthanides with the lighter elements should have a co-ordination number greater than six, it follows that co-ordination numbers exceeding six are not evidence for participation of f orbitals in covalent bonding but are the most likely consequence of the bonding being entirely or almost entirely ionic.

10.3.2 Compounds with Elements of Groups I–V

Two types of hydride, MH_2 and MH_3, are formed by the rare earth metals. The method of preparation is equilibration of the metal with a controlled pressure of hydrogen. At ordinary pressures and temperatures the trihydride is the more stable. Quite low pressures are necessary if the reaction is to cease with formation of the dihydride, which has the fluorite structure. However, very high pressures are necessary if the overall stoichiometry is desired to approach closely to MH_3; somewhat lower hydrogen/metal ratios are otherwise obtained.

The electronic structures of the hydrides are best regarded as mainly ionic in nature. Thus the compounds MH_3 will essentially have the structure $M^{3+}3H^-$, while MH_2 will essentially consist of $M^{3+}2H^-.e^-$, where the third electron is delocalised in a conduction band formed by overlap of the metal orbitals. The electrical conductance properties of the hydrides support a semimetallic structure for the strongly conducting dihydrides but a salt like character in the case of the feebly conducting trihydrides. It is noteworthy that those metals with a fairly strong tendency towards divalence, Eu and Yb, give the dihydride either solely (Eu) or unless 6 MPa pressure is applied, when $YbH_{2.5}$ is obtained.

The borides are of interest more from the point of view of the chemistry of boron than from that of the rare earths. They may be obtained by direct union of the elements *in vacuo* at temperatures up to 2000° or by other methods such as reduction of lanthanide oxides at 1500–1800° by boron carbide or by boron alone. For yttrium, phases YB_2, YB_4, YB_6 and YB_{12} exist. YB_6 is a very stable compound, being unattacked by hot acids or alkalis. It contains the usual octahedra of boron atoms known in other metallic hexaborides together with Y^{3+} ions; it shows metallic conductivity. The lower borides are not so stable, reverting to the hexaboride on heating *in vacuo,* and being hydrolysed by water to give boron hydrides.

There are three principal types of rare earth carbide: M_3C, M_2C_3 and MC_2. They may be prepared directly from the elements, which are heated together in a tantalum crucible. The phase M_3C is more stable in the case of the heavier rare earths; its structure contains isolated carbon atoms. The carbides M_3C are hydrolysed by water to methane and hydrogen, the appearance of methane being in accord with the isolated situation of the carbon atoms within the crystal structure.

The sesquicarbides M_2C_3, which are known for the elements La–Ho and yttrium, contain paired carbon atoms at an olefinic distance apart. Hydrolysis with water gives a mixture of acetylene, ethylene, ethane and hydrogen. The acetylides MC_2, which react with water to give mainly acetylene, have the orthodox CaC_2 structure.

The carbides present an interesting problem when their electronic structure is considered. It is likely that the acetylides MC_2 are essentially $M^{3+}C_2^{3-}$, where each C_2^{2-} ion carries an additional electron in an antibonding orbital, this electron being stabilised by some delocalisation caused by interaction with the metal orbitals. The metal ions have been shown by neutron diffraction to be in the tripositive state. YbC_2, however, has structural parameters and hydrolysis properties consistent with the formulation $Yb^{2+}C_2^{2-}$, which would be in accord with the considerable tendency of ytterbium to adopt the dipositive oxidation state. The constitution of the sesquicarbides M_2C_3 is probably similar in principle, but each C_2 unit would have to support four negative charges, which would necessitate considerable stabilisation by delocalisation and interaction with metal ions. These views are supported by the metallic electrical conductivity of the carbides MC_2 and M_2C_3.

The carbides M_3C are best considered as essentially metallic structures in which C atoms interact with the metallic conduction bands, thus giving completely delocalised orbitals which hold the electrons released by the M^{3+} ions.

A number of silicides are formed by yttrium and the lanthanides: M_5Si_3, M_5Si_4, MSi, M_3Si_5 and MSi_2; any individual lanthanide does not form all five compounds. They may be made by direct union or by reduction of an oxide by heating to 1500°. They are solids of high melting point, and are attacked by aqueous hydrochloric acid.

The nitrides are straightforward. They have the stoichiometry MN and have the sodium chloride structure. They may be made by direct combination at 1000°. The nitrides are very stable thermally, YN melting at 2570°, but they are hydrolysed to ammonia by water.

Compounds MX (X = P, As, Sb or Bi) have been prepared from the metals La, Ce, Pr, Nd and Sm by the direct action of the nonmetal at about 1000° in a closed tube. They have the sodium chloride structure and MP, MAs and MSb are hydrolysed to the nonmetal hydride. A phase study of the system La–Sb showed the existence of the further compounds La_2Sb, La_3Sb_2 and $LaSb_2$.

10.3.3 The Oxides

The sesquioxides M_2O_3 can all be obtained and are the most stable oxide except for cerium, praseodymium and terbium, whose stable oxides contain the metal either wholly (Ce) or partly (Pr, Tb) in the +4 oxidation state. The monoxide of europium is also known.

Ignition of the hydroxide, carbonate, sulphate or other suitable salt will give the sesquioxide, but the reaction must be performed in an atmosphere of, for example, hydrogen if the sesquioxides of cerium, praseodymium or terbium are required. The latter three compounds may also be obtained by reduction of the higher oxides in hydrogen; CeO_2 is reduced only at 2000° and 15 MPa pressure but Pr_6O_{11} and Tb_4O_7 are reduced at 900° and atmospheric pressure.

There is some tendency towards formation of hydroxide and carbonate in moist air, especially in the cases of the lighter lanthanides. These typically basic oxides dissolve in mineral acids with an ease dependent on their temperature of ignition and on their position in the series; thus comparable samples of lanthanum oxide and lutetium oxide take a few seconds and several hours, respectively, to dissolve in cold 70 per cent perchloric acid. The colours of the oxides are similar to those of the hydrated ions, except that neodymium oxide appears distinctly more blue than the mauve hydrated ion.

The sesquioxides show three different crystal structures, termed rare earth A, rare earth B and rare earth C. Interconversion between these structures is slow and the transition temperatures rather high ($\approx 1000°$). The A form is rather complex, each lanthanide ion having seven-co-ordination in the form of an irregular octahedron with an additional oxide ion above one face. The C form is simpler (α-Mn_2O_3 structure), being directly derived from a slightly distorted fluorite structure from which one quarter of the oxide ions have been removed in such a way as to leave half the metal ions with six-co-ordination to oxide ions situated on the corners of a cube from which two of them lying on a face-diagonal have been removed. In the case of the other half, two oxide ions lying on the body-diagonal of the cube are missing. The B form is complex, having three nonequivalent types of lanthanide ion, one with octahedral six-co-ordination and two with seven-co-ordination derived from a trigonal prism with an extra oxide ion beyond one rectangular face.

Higher oxides   Cerium metal or ceric hydroxide, if heated in oxygen, give ceric oxide, which is a pale yellow powder, inert towards hydrochloric or nitric acids but converted into ceric sulphate by concentrated sulphuric acid. It is inert towards aqueous or fused alkalis but is attacked by reducing agents such as stannous chloride, forming cerous salts. Ceric oxide has the fluorite structure, but several phases exist in the region of composition intermediate between Ce_2O_3 and CeO_2. Thus there is X-ray powder photograph and other evidence for the successive phases of ideal composition CeO_2, $Ce_{32}O_{58}$, $Ce_{32}O_{57}$, $Ce_{18}O_{31}$, C-Ce_2O_3 and A-Ce_2O_3.

The action of heat on a suitable praseodymium salt gives a black powder of approximate composition Pr_6O_{11}. This may be converted into PrO_2 by oxygen at 10 MPa pressure and 500°. As in the case of cerium, a sequence of phases exists between PrO_2 and Pr_2O_3. Praseodymium dioxide is a strong oxidising agent, converting Mn^{2+} into MnO_4^- and Ce^{3+} into Ce^{4+}.

The action of atomic oxygen on terbium sesquioxide at 450° is necessary in order to obtain the dioxide: the dark brown product obtained by heating the sesquioxide in air is approximately Tb_4O_7. An investigation of the system has shown the intermediate phase $Tb_{32}O_{55}$.

Lower oxides   X-ray powder photography of the products of treatment of the sesquioxide with excess metal at high temperature implies the following compounds. LaO, CeO, NdO, SmO, EuO, YbO and possibly Yo. They have the NaCl structure. SmO, however, has been shown to be $SmO_{1/2}N_{1/2}$ and there is doubt about several other 'lower' oxides. EuO and Eu_3O_4 appear to be genuine phases, however.

10.3.4 Sulphides, Selenides and Tellurides
Much work has been carried out on the lanthanide chalcogenides on account of their

semiconducting properties but there are experimental difficulties and structural information is not available in a number of instances. The stoichiometries MS, M_3S_4, M_2S_3 and MS_2 observed for the sulphides also appear with the selenides and tellurides. However, nonstoichiometry is rampant in this area of chemistry, arising from random vacancies within the normal crystal structures. The monosulphides MS may be obtained by direct union at 1000°. They have the sodium chloride structure but the rare earth ion is tripositive and the compounds show a pseudo-metallic conductance. However, EuS, SmS and YbS have dipositive cations, thus showing the behaviour expected of these metals. Indeed, EuS cannot be made in the usual way and must be obtained by the action of H_2S gas on heated $EuCl_3$

$$2EuCl_3 + 3H_2S \rightarrow 2EuS + 6HCl + S$$

The monoselenides and monotellurides are rather similar.

The sesquisulphides M_2S_3 may be prepared by the action of H_2S gas on the heated chloride

$$3H_2S + 2MCl_3 \rightarrow M_2S_3 + 6HCl$$

Eu_2S_3 is made by thermal dissociation of EuS_2. The sulphides M_2S_3 may exist in four different crystalline modifications; one of these is based on the Th_3P_4 structure with cation vacancies. The sesquiselenides are similar; the sesquitellurides are made by direct combination. In contrast to the lower chalcogenides, these compounds may be regarded as normal salts with no delocalised electrons.

10.3.5 The Hydroxides

The radius of the OH^-ion is similar to that of the F^- ion and it is thus not surprising that the same crystal structure (the structure of tysonite, LaF_3) is adopted in the case of the lighter lanthanide fluorides and the hydroxides. The latter may be obtained by raising the pH of a solution of a rare earth salt, the hydroxide being precipitated, collected and dried to give the anhydrous hydroxide $M(OH)_3$. At higher temperatures (200°), LnO(OH) is formed. In the case of cerium(III), air must be excluded owing to ready formation of Ce(IV).

Fractional precipitation of the hydroxide has often been used as a method for the separation of the lanthanides as a group from other cations which are either considerably less or considerably more basic. The method was also much used as a means of mutual separation of the lanthanides before the advent of column chromatography.

10.3.6 Fluorides

The anhydrous trifluorides may be made by dehydration of the hemihydrate or by heating the sesquioxide in an atmosphere of hydrogen fluoride at 700° for eight hours. The hemihydrate $MF_3 \cdot \frac{1}{2}H_2O$ is easily obtained by the action of aqueous hydrofluoric acid on a hot solution of the lanthanide or yttrium nitrate; the fluorides are the only rare earth metal halides to be insoluble in water. The hemihydrate loses its water at about 300° *in vacuo* or at about 600° in a stream of hydrogen fluoride gas. The melting points are quite high (La, 1493°). In contrast to the other halides, the fluorides are not deliquescent.

The lighter lanthanide fluorides adopt the tysonite structure in which each metal ion is nine-co-ordinate with fluorides at distances of 242–264 pm.

10.3.7 The Chlorides, Bromides and Iodides

The anhydrous chlorides cannot be obtained simply by heating a hydrated chloride; this produces an oxychloride. The hydrates may, however, be successfully dehydrated in a stream of hydrogen chloride, preferably at reduced pressure. Thus treatment of the hydrated chlorides at temperatures rising to 400° over 36 hours at 6.66 kPa (50 mm Hg) yields a pure product, while brief heating at temperatures rising to 100° at atmospheric pressure in a stream of hydrogen chloride is sufficient to give anhydrous chlorides of the light lanthanides.

Alternative methods are the action of heat on a mixture of excess ammonium chloride and either the oxide or the hydrated chloride. Since the lanthanide chlorides are slightly volatile, a very pure product may be obtained from chloride contaminated with oxychloride by distillation at about 900°C and a pressure of 100 μPa (1 micrometre of Hg).

The anhydrous chlorides are very deliquescent and must be handled in a dry atmosphere. As in the cases of the oxides and fluorides, the chlorides adopt more than one crystal structure, depending on the size of the cation. All are layer structures. Thus La, Ce, Pr, Nd, Sm, Eu and Gd adopt the nine-co-ordinated uranium trichloride structure, the co-ordination polyhedron being the trigonal prism with three further anions normal to each vertical face. $TbCl_3$ has the eight-co-ordinate $PuBr_3$ structure, while the smaller ions Dy, Ho, Er, Tm, Yb and Lu adopt the six-co-ordinate layer structure of ferric chloride. The tribromides, in contrast to the chlorides, may be prepared by heating the hydrated bromide from 70–170° *in vacuo*. The iodides are best made by direct combination of the elements (other methods for both series are of course available). The structures are of the same types as for the trichlorides: La, Ce, Pr have the UCl_3 structure; Nd–Eu, the $PuBr_3$ and Gd–Lu the $FeCl_3$. For the iodides, La–Nd have the $PuBr_3$ and Sm–Lu the $FeCl_3$.

10.3.8 Lower Halides

A number of these are known. The difluorides of Sm, Eu and Yb can be made by calcium reduction of the trifluoride. They have the fluorite structure. The difluorides of all the lanthanides may be obtained in dilute solid solution in fluorite by reduction of solid solutions of the trivalent ions by calcium vapour. These systems form a convenient basis for studying the spectra of the divalent lanthanide ions.

The dichlorides of Nd, Sm, Eu, Dy and Yb have been characterised by reduction of the trichloride with the appropriate lanthanide; some other phases have also been detected in these systems. The dibromides of Sm, Eu, Tm and Yb and the di-iodides of La, Ce, Pr, Nd and Gd are obtained by similar means, but di-iodides may be obtained by thermal decomposition of the tri-iodides of Sm, Eu, Yb. The majority of these compounds MCl_2, MBr_2 and MI_2, particularly the Sm, Eu and Yb compounds are genuine M^{2+} salts. Some of them, notably the La, Ce and Pr di-iodides are best formulated $M^{3+}(X^-)_2$.e because they are dark-coloured and are good electrical conductors. The extra electron would be delocalised. All the dihalides except those of europium are decomposed by water. They have been comparatively little investigated chemically.

10.3.9 Tetrafluorides

These compounds are formed by Ce, Pr and Tb but by no other lanthanide. They are isostructural, having the UF_4 structure with square antiprismatic co-ordination. They are prepared by fluorination of the dioxide (Ce), fluorination of the trifluoride (Tb) or by the action of liquid HF on the complex Na_2PrF_6, itself prepared by fluorination of a NaF/PrF_3 mixture. TbF_4 (and presumably PrF_4 also) is relatively unstable, decomposing to TbOF on heating to 400° in air.

10.4 Electronic Spectra and Magnetic Properties of the Lanthanides

Many of the lanthanide ions are coloured and show pastel shades of pink, mauve, green, or cream which are rather pleasing to the eye. We now proceed to a consideration of the electronic states of the lanthanides from which these colours ultimately arise. Many of the ions are strongly paramagnetic and the nature of the lowest-energy electronic states will determine the magnetic properties of each metal ion.

The tripositive lanthanide ions follow the Russell–Saunders coupling scheme rather closely. According to this approximation, we first consider the strong coupling of the electron-spin angular momenta of the individual f electrons in a particular lanthanide ion. This can be done according to a classical vector-addition procedure such that the individual electron-spin angular momentum vectors of value $[s(s+1)]^{\frac{1}{2}}h/2\pi$ (where s = spin quantum number = ½) are coupled either 'parallel' or 'anti-parallel', according as m_s (the spin magnetic quantum number) takes either of its two allowed values $+s$ or $-s$. This results in a value of the resolved component of the angular momentum parallel to the z axis of $m_s h/2\pi$ units for each electron. Figure 10.2 may make this clearer. The angular momentum vectors are shown in

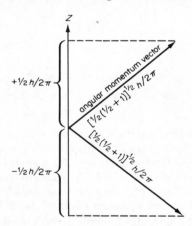

Figure 10.2 Angular momentum vectors for one electron

their two allowed orientations relative to the z axis, leading to resolved components of magnitude $m_s h/2\pi$ where m_s takes values +½ or −½.

Since each individual electron has its resolved component m_s (± ½) the total resolved component in a multi-electron atom with n electrons will be the sum of these n values of $+\frac{1}{2}$ or $-\frac{1}{2}$ and can clearly take any value from $+n/2$ to $-n/2$ in integral steps. This value is the *resultant spin quantum number* for the multi-electron atom; it is denoted by S. It is merely the total resolved components of the spin angular momenta along the z axis and therefore the actual total spin angular momentum of the atom will be $[S(S+1)]^{1/2}h/2\pi$. An atom having $S = 0$ is said to be in a singlet state, $S = \frac{1}{2}$ a doublet, $S = 1$ a triplet and so on through quartet, quintet, sextet, etc. These designations arise because an atom whose electron spins are coupled together to give, say, $S = 3/2$ may, on introduction of a new magnetic field (perhaps arising from the electronic orbital momenta) have S so orientated that the component of the total spin angular momentum $[S(S+1)]^{1/2}h/2\pi$ is resolved parallel to the new field to give values of M_s ranging from $+S$ to $-S$ in integral steps; that is, for $S = 3/2$, the four values $3/2, 1/2, -1/2$, or $-3/2$ are possible, giving four resolved components and hence $S = 3/2$ is called a quartet state. Figure 10.3 may clarify this; note the resemblance to the diagram for a single electron spin.

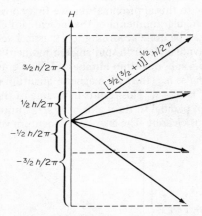

Figure 10.3 Angular momentum vectors for three electrons

We next consider the orbital angular momentum, designated by the quantum number l, of a single electron. This quantity is classically associated with the orbital motion of the electron considered as a particle. Its magnitude is characteristic of the type of electron; thus f electrons have $l = 3$. An f electron having $l = 3$ may have m_l, the component of l resolved along the z axis, equal to any of the values 3, 2, 1, 0, -1, -2 or -3. These seven possible values of m_l correspond with the seven possible f orbitals. Like the spin angular momentum, the total orbital angular momentum is $[l(l-1)]^{1/2}h/2\pi$ and the resolved part is $m_l h/2\pi$.

For two f electrons, the two individual values of m_l may be added vectorially to give M_L, which is the resolved part of the total orbital angular momentum, and which in this simple case could clearly take values 6, 5, 4, 3, 2, 1 or 0. This set of values of M_L means that $L = 6$, where L is the total orbital angular momentum quantum number and the atom concerned is said to be in an I state (for $L = 0, 1, 2, 3, 4, 5, 6, 7$, etc., the state symbols are S, P, D, F, G, H, I, K, etc.).

It can be shown that two f electrons can give rise not only to an I state but to all the states up to I, namely S, P, D, F, G and H, as well.

In addition to the strong coupling between the ls to give L and between the ss to give S, there is a weaker coupling (the spin–orbit coupling) between S and L to give their various possible vector sums. The resulting quantum number J could take values, for an example of two f electrons where $S = 1$ and $L = 5$[†], of $5 + 1 = 6$, $5 + 0 = 5$ or $5 - 1 = 4$. Since $S = 1$ is a triplet state and $L = 5$ is an H state, we could write these three possible states as ${}^3H_6$, ${}^3H_5$ and ${}^3H_4$, because the general notation is ${}^{2s+1}L_J$.

To obtain an idea of the meaning of these symbols in terms of energy, an alteration of S or L involves an energy difference of the order of 10 000 cm^{-1}, but an alteration of J (S and L remaining constant) involves energies of only about 1000 cm^{-1}.

The ground state of the ion complies with the following three rules (Hund's rules): (a) S is as high as possible; (b) L is as high as possible consistent with (a); (c) J is as high as possible for a shell more than half filled and as low as possible for a shell less than half filled. Thus the ground state of, say, Eu^{3+} (f^6) will have $S = 3$, $L = 3 + 2 + 1 + 0 + (-1) + (-2) = 3$ (no two electrons with the same m_s can have the same m_l — Pauli principle) and $J = L - S = 0$. The state will therefore be ${}^7F_0$.

We can now in principle understand the term scheme for any lanthanide; as examples the term schemes for Pr^{3+} (f^2) and Eu^{3+} (f^6) are shown in figure 10.4. The relative positions of the various energy levels have been determined experimentally by spectroscopy; they cannot be calculated *a priori*.

10.4.1 Absorption and Fluorescence Spectra

A lanthanide ion, for example an aqueous solution of Pr^{3+} as its chloride, absorbs light quanta of energy suitable for inducing excitation from its ground state, in this case ${}^3H_4$, to an excited state, perhaps ${}^3P_0$, ${}^3P_1$ or ${}^3P_2$. These three transitions, ${}^3H_4 \rightarrow {}^3P_0$, ${}^3H_4 \rightarrow {}^3P_1$ and ${}^3H_4 \rightarrow {}^3P_2$ can be observed as fairly sharp, weak bands at about 20 750, 21 100 and 22 300 cm^{-1}, respectively, which, together with the ${}^3H_4 \rightarrow {}^1D_2$ transition, give the Pr^{3+} ion its characteristic green colour (see figure 10.5). Extinction coefficients (ϵ) are in the region 1–10[‡]. These very low values arise from the fact that not only are these f $\rightarrow$ f transitions, and thus Laporte-forbidden (as are the d $\rightarrow$ d bands of d transition complexes), but that the f electrons are only very feebly perturbed by the ligands. Thus effects such as take place in the ion $[CoCl_4]^{2-}$ ($\epsilon = 600$ at 690 nm), where the tetrahedral crystal field introduces some p character into the transition, giving fairly intense absorption, do not occur in the lanthanides. Vibronic transitions are also not very intense for similar reasons.

In order to understand f–f spectra a little more thoroughly, it is necessary, however to take the ligands into account. In the case of lanthanide complexes, the crystal-field splitting is of the order of only 10–100 cm^{-1} and we can thus think of it as

---

[†] S and L can, of course, take other values in this case; we just take $S = 1$ and $L = 5$ as an example.

[‡] Compare with the d transition metals where $\epsilon \approx 100$. Fully allowed transitions in organic dyes or as charge-transfer bands in complexes have $\epsilon \approx 10\,000$.

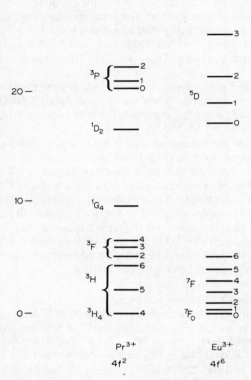

Figure 10.4 Term schemes for two lanthanide tripositive ions

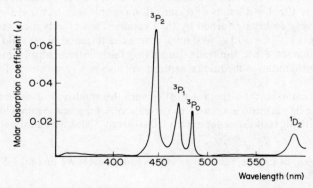

Figure 10.5 Electronic absorption peaks in the spectrum of aqueous $PrCl_3$ (after T. Moeller and J.C. Brantley, *Anal. Chem.*, 22 (1950), 433)

a perturbation of the situation obtaining *after* consideration of the stronger spin-orbital coupling between L and S, which has given particular J values. In other words, J is a meaningful quantum number and the effect of the ligand field depends only on (a) the symmetry of the field, (b) its strength and (c) the value of J.

This situation differs from that of the d transition metals, which have *weak* spin-orbital coupling and *strong* crystal-field splittings; in this case we apply the crystal-field perturbation to a situation characterised by the quantum number L; for example, any D state ($L = 2$) is first split into two components by an octahedral crystal field, or into three components by a tetragonal field; only then should we consider (it is not usually necessary to do so) the weak spin-orbital coupling.

It turns out that the splitting of a lanthanide-ion term is similar for integral J values to the splitting of a d transition-metal term having a numerically equal value of L. Thus the $^7F_2$ term of Eu^{3+} is also split by an octahedral field into two components, described as E_{1g} and T_{2g} (the g subscripts occur because an angular momentum representation is even to inversion).

If we now consider the $^7F_0 \rightarrow {}^5D_1$ transition of Eu^{3+} in a typical complex having very low symmetry, the ground $^7F_0$ state cannot split because $J = 0$ but the excited state $^5D_1$ with $J = 1$ will split into three ($= 2J+1$) levels and the absorption band will appear as a closely spaced triple peak. For a transition from a state having say $J = 3$ to one having $J = 5$ there could be up to $(2 \times 3 + 1) \times (2 \times 5 + 1) = 77$ separate peaks spaced over a total of say 200 cm^{-1}. It would be impossible to resolve these in solution using any instrument but an attempt might be made in the solid state at low temperatures.

Not all of the components of the transitions will in general be active since not all will comply with the selection rules. Not only ordinary electric-dipole transitions, but also magnetic-dipole transitions are observed. These latter are very weak compared with electric-dipole transitions of moderate intensity, but compared with the weak lanthanide electric-dipole transitions, they are observable.[†]

A useful application of a lanthanide fluorescent transition occurs in colour television, where the red phosphor is commonly YVO_4 doped with about 3 per cent Eu^{3+} ions. The $^5D_0 \rightarrow {}^7F_2$ transitions of the excited Eu^{3+} ions in the coating of the tube provide a brilliant red light of a quality superior to that of other phosphors. Another application is in the neodymium-doped laser, which uses the $^4F_{3/2} \rightarrow {}^4I_{9/2}$ transition at $10\,000$ cm^{-1}. An interesting small-scale application is the $^4I_{9/2} \rightarrow {}^4G_{5/2}$ transition of Nd^{3+}, which absorbs sharply at the wavelength of the sodium D line and hence is used for glassblowers' goggles to reduce the glare from the hot glass. A more abstract application is the deduction of the point-group symmetry of lanthanide (usually Eu^{3+}) complexes by detailed examination, along the lines outlined above, of the crystal-field splittings of the fluorescence spectrum of the complex. In this way, for example, a D_2 square antiprismatic configuration has been assigned to the $[Eu(PhCOCHCOPh)_4]^-$ anion, a slightly distorted D_{4d} square antiprism to $[Eu(pyO)_8]^{3+}$ and a slightly distorted D_3 structure to $[Eu(terpy)_3]^{3+}$.

[†] The selection rules are as follows. The direct product of the representations to which the initial and final crystal-field-split levels belong must contain a representation transforming (a) as x, y, z for electric dipole or (b) as $R_{x,y,z}$ for magnetic-dipole transitions.

The last two results were confirmed by X-ray diffraction. The so-called 'hypersensitive transitions' are a limited number of transitions in the electronic spectra of the lanthanides, which are characterised by considerable variations in intensity in different environments. Nd^{3+} exhibits this phenomenon; thus in aqueous solution, the shape and oscillator strength of some bands are quite different from those of, say, Nd^{3+} in 10M HCl. This is well exemplified by the $^4I_{9/2} \rightarrow {}^2H_{9/2}$, $^4F_{5/2}$ and $^4I_{9/2} \rightarrow {}^4G_{5/2}$, $^2G_{7/2}$ transitions. The effect can be ascribed to a lowering of symmetry; it is a general rule that the less symmetric the molecule, the more intense the electronic transitions will be.

10.4.2 Magnetic Properties

Associated with the quantum number J is a magnetic moment of magnitude $g[J(J+1)]^{1/2}$ B.M. The magnetogyric ratio, g, (the 'g factor') is given by

$$g = \{S(S+1) - L(L+1) + 3J(J+1)\}/2J(J+1)$$

This situation is not dissimilar to the 'spin-only' approximation for d transition metals, where owing to the quenching of the orbital angular momentum by the crystal field we often need only consider the spin angular momentum S and thus $\mu = g[S(S+1)]^{1/2}$ B.M. In this case, however, g has the free-electron value of almost exactly 2. Thus the bulk magnetic moments of lanthanide compounds at ordinary temperatures can be calculated very simply, since the crystal-field splittings are too small to lead to any observable deviations. However, the crystal field does lead to splitting into closely spaced energy levels, as outlined above, each of which has associated with it particular values of g_1, g_2 and g_3 (these are the three mutually perpendicular components of the g tensor, whose averaged, isotropic, value is g). Thus the thermal populations of the crystal-field-split states and the orientation of the molecular axis must be taken into account in low-temperature and in single-crystal studies, respectively. However, for bulk measurements at room temperature the expression $\mu = g[J(J+1)]^{1/2}$ gives an accurate correspondence with the observed value.

The foregoing assumes that only one $^{2S+1}L_J$ state is populated and this is usually true; thus for Ce^{3+} the $^2F_{5/2}$ state is 1000 cm^{-1} above the $^2F_{3/2}$ ground state and is nearly unpopulated at room temperature ($kT \approx 200$ cm^{-1}). However, in the case of Eu^{3+} and to a lesser extent Sm^{3+}, it is not true. Figure 10.4 shows that the separation between the $^7F_0$ ground state (which is diamagnetic since $J = 0$) and the first excited state $^7F_1$ (paramagnetic, $J = 1$) is about 300 cm^{-1}. $^7F_1$ is accordingly populated to an extent predicted by the Boltzmann ratio $e^{-300/kT}$, and Eu^{3+} compounds are paramagnetic with $\mu = 3.4$ B.M. The effect of Hund's third rule, which states that $J = L-S$ for a shell that is less than half-filled but is $L+S$ for the configurations f^8 to f^{13}, is noteworthy. The lanthanides in the first half of the series thus have much smaller magnetic moments than those in the latter half.

Some interesting results have been obtained by single-crystal studies on the magnetic properties of the tri-iodo hexakis(antipyrine) lanthanides, which possess D_{3d} local symmetry. They may be interpreted in terms of a point-charge crystal-field model, and it appears that the orbital reduction factor k is very close to unity, commensurate with essentially 'ionic' bonding in lanthanide complexes.

Typical room-temperature magnetic moments for tripositive lanthanide ions are given in table 10.2.

Table 10.2 Magnetic Moments of the Tripositive
Lanthanide ions, at Room Temperatures

	f^n	Ground term	Theory (B.M.)	Observed (B.M.)
La	0	$^1S_0$	0	0
Ce	1	$^2F_{5/2}$	2.54	2.46
Pr	2	$^3H_4$	3.62	3.48–3.60
Nd	3	$^4I_{9/2}$	3.68	3.44–3.65
Pm	4	$^5I_4$	2.83	–
Sm	5	$^6H_{5/2}$	1.55–1.65	1.54–1.65
Eu	6	$^7F_0$	3.40–3.51	3.32–3.54
Gd	7	$^8S_{7/2}$	7.94	7.9–8.0
Tb	8	$^7F_6$	9.72	9.69–9.81
Dy	9	$^6H_{15/2}$	10.63	10.4–10.6
Ho	10	$^5I_8$	10.60	10.4–10.7
Er	11	$^4I_{15/2}$	9.59	9.4–9.5
Tm	12	$^3H_6$	7.57	7.5
Yb	13	$^2F_{7/2}$	4.54	4.3–4.5
Lu	14	$^1S_0$	0	0

10.5 Co-ordination Compounds of the Lanthanides

The co-ordination chemistry of these elements shows very substantial differences
from the more familiar co-ordination chemistry of the d transition metals. There
are several well-known features of d transition-metal chemistry which are quite absent in that of the lanthanides, such as the preference for co-ordination numbers of
four and six and for regular co-ordination polyhedra such as the octahedron or
tetrahedron. The marked influence of crystal-field stabilisation energy on chemical
and physical properties is also absent, as is the stabilisation of complexes by π-
bonding ligands such as CO or PR_3. Thus such features as Jahn–Teller distortion,
or the distinction between kinetically inert and labile complexes are not observed
in lanthanide (or actinide) chemistry, and complexes of a similar type to, for example, $[MnCH_3(CO)_5]$, $[ReH_9]^{2-}$ or $[Pt(PPh_3)_3]$ are unknown. It should be clear,
therefore, that the co-ordination chemistry of the lanthanides has some fairly severe
restrictions. However, it also has possibilities, arising chiefly from the freedom to
adopt variable and high co-ordination numbers, with the consequent variety of co-
ordination polyhedra, which render it an interesting study. The general characteristics of lanthanide co-ordination compounds are now discussed in a little more detail.

10.5.1 The Available Oxidation States
In their co-ordination chemistry, the lanthanides adopt the oxidation states known
in their halides or oxides or other salts. The only exceptions to this rule are (a)
those cases where a complex fluoride is the only representative of an M^{4+} oxidation

state, for example Cs_3NdF_7, and (b) thoses cases where no complexes are known of an unstable M^{2+} state such as Tm^{2+}, which exists as TmI_2 but has no known complexes.

The reason for this limitation lies mainly in the largely electrostatic ('ionic') nature of bonds formed by lanthanides and in the small crystal-field splitting of the 4f orbitals. The crystal-field splitting of individual components of a given multiplet term is commonly of the order of 100 cm^{-1} and this quantity of energy is quite insufficient to stabilise an otherwise unstable oxidation state by means of an unusually high crystal-field stabilisation energy for some particular combination of f-electron configuration and molecular shape. The small extent of the crystal-field splitting is to be expected from the well-shielded nature of the 4f orbitals, which means that they are not much affected by the ligands, either in an electrostatic (crystal field) manner or by covalent σ or π interactions (molecular-orbital viewpoint). Thus there are no effects similar to the stabilisation of the Co(III) ammines or of Cr(0) hexacarbonyl.

10.5.2 The Principal Ligands

With a limited number of interesting exceptions, the lanthanides form complexes only with the halogens, oxygen or nitrogen as donor atoms. Thus lanthanide salts do not complex with such well-known ligands as CO, PPh_3 or SMe_2. Instead, typical ligands are H_2O, $CH_3COCHCOCH_3^-$, Ph_3PO, Me_2SO, EDTA, 2,2'-bipyridyl and anions such as Cl^-, Br^-, NCS^-, NO_3^-, SO_4^{2-}, CH_3COO^-.

This type of behaviour, namely a greater affinity for O and N rather than the heavier elements S and P, is characteristic of metal ions belonging to group A in the Chatt–Ahrland classification and is further shown by the lanthanides in their greater association constants for F^- than those for Cl^- in aqueous solution.

$$La^{3+}(aq) + F^-(aq) \rightleftharpoons LaF^{2+}(aq); \log K = 3.56$$

$$La^{3+}(aq) + Cl^-(aq) \rightleftharpoons LaCl^{2+}(aq); \log K = -0.11$$

Water as a ligand occupies a special position with the lanthanides because the hydration energy of the lanthanides is considerable; for example

$$La^{3+}(g) + nH_2O \rightarrow La^{3+}(aq) \qquad \Delta H = -4622 \text{ kJ } (-1104 \text{ kcal})$$

Also the complexes are labile, that is to say, ligand substitution reactions take place rapidly. This means that many lanthanide complexes are substantially dissociated in aqueous solutions into the hydrated ion and free ligands and cannot therefore be prepared in an aqueous medium. This particularly applies to complexes of uncharged monodentate ligands; such complexes, when prepared by other means, are, however, usually quite stable thermally.

The small size of the water molecule, combined with its high affinity for lanthanide ions and the high co-ordination numbers of the latter, frequently leads to the inclusion of water molecules in the co-ordination sphere of the metal ion under ordinary preparative conditions. Thus the acetylacetonate $[La(acac)_3(H_2O)_2]$ may be obtained; similar hydration often produces an uncertainty in assignment of co-ordination number not encountered in d transition-metal chemistry with consequent uncertainty in, for example, interpretation of the electronic-absorption spectrum.

10.5.3 Co-ordination Number and Co-ordination Polyhedra

Typical co-ordination numbers are 7, 8, 9 and 10 as shown by available X-ray diffraction data. It is especially noteworthy that in lanthanide chemistry co-ordination numbers lower than six are extremely rare, and even six is unusual. This makes a big contrast with d transition-metal chemistry and is an effect of the generally larger ion (La, 106 pm; Lu, 85 pm) and absence of directional covalent bonding. Thus, the eight-co-ordinate acetylacetonate $[La(acac)_3(H_2O)_2]$ polymerises on dehydration, rather than be reduced to six-co-ordination. However, five, four and even three-co-ordination are known in a few cases.

The very high co-ordination number of twelve is attained only by chelate anions having their donor atoms close together, such as NO_3^-. Co-ordination polyhedra of the lower numbers seven, eight, nine and ten are usually irregular owing to the different steric requirements of the various ligands present, but where all the ligands are identical as in $[Nd(H_2O)_9]^{3+}$ or $[Ce(NO_3)_6]^{3-}$ a regular figure is adopted in those cases where one exists, that is for co-ordination numbers eight, nine and twelve. These features are again what would be expected from a bonding situation that is largely electrostatic in nature.

10.5.4 Complexes of Oxygen-based Ligands

The co-ordination number of lanthanide ions in aqueous solution is uncertain, but appears to be about nine; crystalline hydrated salts have been studied by X-ray diffraction. Thus the isomorphous series of lanthanide bromates $M(BrO_3)_3.9H_2O$ contains $[M(H_2O)_9]^{3+}$ complex cations with the face-centred trigonal prismatic arrangement. The hydrated chloride $GdCl_3.6H_2O$ has an eight-co-ordinate polyhedron $[GdCl_2(H_2O)_6]^+$ (as does the Nd analogue) with both chloride and water co-ordinated. The hydrated nitrate $Pr(NO_3)_3.6H_2O$ also has a complex ion containing co-ordinated anions, in this instance with co-ordination number ten, namely $[Pr(NO_3)_3(H_2O)_4]$. The sulphate $La_2(SO_4)_3.9H_2O$ is interesting in that it contains two different sorts of La^{3+} ion; one of these is co-ordinated to twelve oxygen atoms belonging to SO_4^{2-} ions while the other has the face-centred trigonal prismatic co-ordination, six water molecules forming the prism and three oxygen atoms from SO_4^{2-} ions entering the face-central positions. Some of these hydrate polyhedra are depicted in figure 10.6.

It is of interest that in $Ce_2Mg_3(NO_3)_{12}.24H_2O$ the usual $[Mg(H_2O)_6]^{2+}$ octahedral complex accounts for that part of the water which is co-ordinated (the remaining water molecules are retained by hydrogen bonds), while the cerium co-ordinates with anions only, forming twelve co-ordinate $[Ce(NO_3)_6]^{3-}$ whose symmetry is icosahedral. $Ce(NO_3)_5^{2-}$ ions have also been characterised crystallographically. They contain an array of five bidentate nitrate groups around the metal; the complex is thus ten-co-ordinate (the five nitrogen atoms are arranged as a trigonal bipyramid).

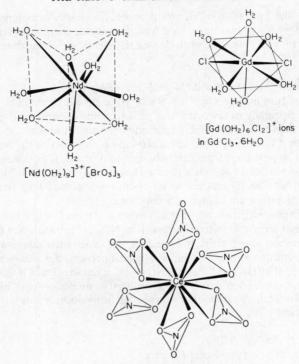

$[Nd(OH_2)_9]^{3+}[BrO_3]_3$

$[Gd(OH_2)_6Cl_2]^+$ ions
in $GdCl_3 \cdot 6H_2O$

$[Ce(NO_3)_6]^{3-}$ in $Ce_2Mg_3(NO_3)_{12} \cdot 24H_2O$

Figure 10.6 Co-ordination polyhedra of three simple lanthanide complexes

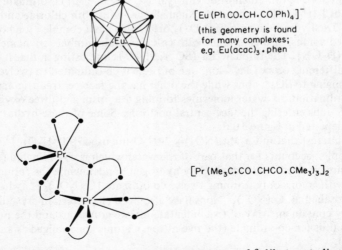

$[Eu(PhCO \cdot CH \cdot COPh)_4]^-$
(this geometry is found
for many complexes;
e.g. $Eu(acac)_3 \cdot phen$

$[Pr(Me_3C \cdot CO \cdot CHCO \cdot CMe_3)_3]_2$

Figure 10.7 Structures of lanthanide complexes of β-diketonate ligands

Lanthanides form many complexes with β-diketones of the types
$[M(RCOCHCOR)_3(H_2O)_n]$, $[M(RCOCHCOR)_4]^-$ and $[M(RCOCHCOR)_3L]$,
where M is a lanthanide, R is typically alkyl, C_6H_5 or perfluoroalkyl and L is an
uncharged ligand such as pyridine. Co-ordination numbers are seven or eight and
some representative examples are shown in figure 10.7. Preparation is typically in
buffered aqueous solution from a soluble lanthanide salt and the β-diketone. The
chelates $[M(RCOCHCOR)_3]$ have attracted much study as n.m.r. shift reagents (see
section 10.7). The volatility of these complexes (commonly sublimation is used to
effect purification) varies. Generally the most volatile and thermally stable chelates
are those in which the lanthanide is well shielded by bulky alkyl groups R (for
example $-CMe_3$), which tend to reduce solvation and any trend towards
polymerisation.

The oxides of amines, phosphines and arsines combine readily with lanthanide
salts to give rather stable complexes. This stability is unusual for formally uncharged
monodentate ligands; probably the negative charge arising from the canonical form
$R_3\overset{+}{P}-\overset{-}{O}$ is an important factor. Among amine oxides, pyridine N-oxide forms the
cations $[M(C_5H_5NO)_8]^{3+}$ (see figure 10.8), while triphenylphosphine oxide and
triphenylarsine oxide yield a variety of products, of which the complex nitrates
shown in figure 10.9 are typical. The ready extraction of lanthanide salts into tri-*n*-
butylphosphate, $OP(OBu)_3$, is dependent on the formation of complexes with the
phosphoryl oxygen in a fashion similar to the co-ordination with $OPPh_3$.

Some generalisations are possible for complexes of general formula ML_nX where
M = a lanthanide, L = an uncharged monodentate ligand and X = an anion. Where
X has only weak co-ordinating tendencies, for example ClO_4^-, $n = 6$–9. If X co-
ordinates more strongly so that it displaces L, n is usually 3 or 4. For example,
when L = $OP(NMe_2)_3$ and X = Cl, $n = 3$ giving an octahedral complex (as the meri-
dianal isomer), while when X = ClO_4, $n = 6$. Again, when X = NO_3^- and L = Me_2SO
the complex $La(NO_3)_3(Me_2SO)_4$ is formed but with ClO_4^- as anion,
$La(Me_2SO)_8(ClO_4)_3$ results.

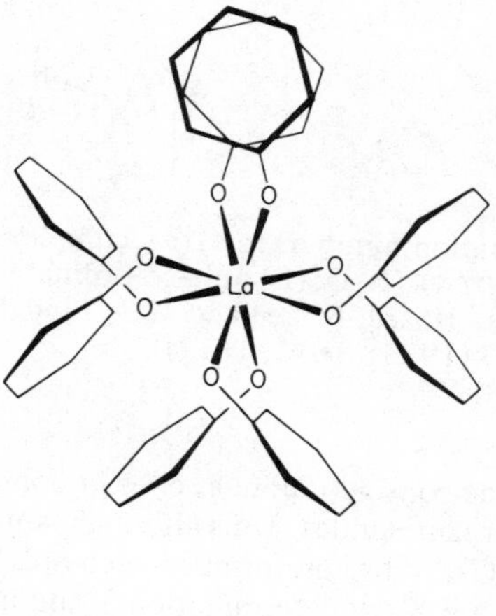

Figure 10.8 The $La(C_5H_5NO)_8{}^{3+}$ ion; C_5H_5NO is pyridine oxide ⬡N→O (after
A.R. Al-Karaghouli and J.S. Wood, *Chem. Commun.*, (1972), 516)

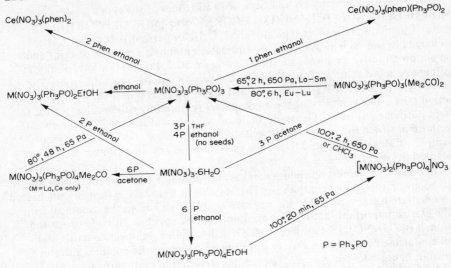

Figure 10.9 Reactions of lathanide nitrates with triphenylphosphines oxide (from D.R. Cousins and F.A. Hart, *J. inorg. nucl. chem.*, **27** (1967), 1745)

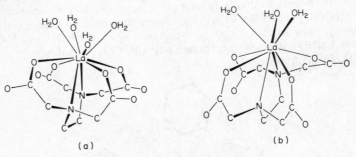

Figure 10.10 High co-ordination numbers in EDTA complexes: (a) ten-co-ordinate La(OH$_2$)$_4$HEDTA in La(HEDTA).7H$_2$O; (b) nine-co-ordinate [La(OH$_2$)$_3$EDTA]$^-$ in KLa(EDTA).8H$_2$O (after J.L. Hoard, B. Lee and M.D. Lind, *J. Am. chem. Soc.*, **87** (1965), 1611, 1612)

Because of the lanthanide contraction, in a series of complexes ML$_n$X$_3$ n is sometimes greater for the lighter lanthanides. Thus although lanthanum forms ten-co-ordinated La(NO$_3$)$_3$(Me$_2$SO)$_4$, lutetium forms nine-co-ordinated Lu(NO$_3$)$_3$(Me$_2$SO)$_3$ (both are X-ray determinations), and in the series M{OS(CH$_2$)$_4$}$_n$(ClO$_4$)$_3$, n = 8 when M = La – Gd, 7½ for Tb – Er and 7 for Tm-Lu.

10.5.5 *Complexes of Nitrogen-based Ligands*

Negative charge seems to have a favourable effect on co-ordinating ability in the case of the oxygen-based ligands, stable complexes being formed only by anionic ligands ($CH_3COCHCOCH_3$) or dipolar ligands ($Ph_3\overset{+}{P}O$). Because typical nitrogen-based ligands such as pyridine or ethylenediamine are uncharged, a strict comparison of the relative affinity of oxygen and nitrogen for lanthanide ions is difficult. Amines do, however, form stable complexes with lanthanide ions. These complexes are always quite stable thermally and in many cases, especially where the amine is polydentate, are stable in water solution also.

The best-defined complexes are those of polydentate aromatic amines and may be prepared directly from the amine and the lanthanide salt in a polar organic solvent such as ethanol or acetone. Co-ordination numbers are typically eight, nine and ten. Typical ligands are 2,2'-bipyridyl, 2,2'6',2''-terpyridyl and the tetradentate Schiff base $C_5H_4N.CH:NCH_2CH_2N:CH.C_5H_4N$. The crystal structure of [La(bipy)$_2$(NO$_3$)$_3$], shown in figure 10.11, again demonstrates the low symmetry (in this case C_2) and high co-ordination number typical of lanthanide complexes.

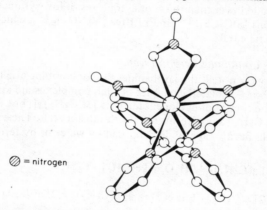

= nitrogen

Figure 10.11 The structure of La(bipy)$_2$ (NO$_3$)$_3$ (after A.R. Al-Karaghouli and J.S. Wood, *J. Am. chem. Soc.,* **90** (1968), 6548)

However, the lanthanide complexes [M terpy$_3$](ClO$_4$)$_3$ have rather symmetrical (D$_3$ propeller-shaped cations of unusual geometry with nine co-ordinated nitrogen atoms, the highest number known for any metal complex. Eight nitrogen atoms are (presumably) co-ordinated in the lanthanide complexes ML$_4$(ClO$_4$)$_3$ (L = 1,10-phenanthroline or 1,2-diaminoethane).

Pr(terpy)Cl$_3$.5H$_2$O is an example of low symmetry and of possible uncertainty in structure assignment when water molecules are present. X-ray measurement, however, shows it to contain [Pr(terpy)Cl(OH$_2$)$_5$]$^{2+}$ cations with a monocapped square antiprismatic structure.

The very bulky ligand $-N(SiMe_3)_2$ is effective in lowering the co-ordination number to three, the lowest known for the lanthanides

$$3LiN(SiMe_3)_2 + LaCl_3 \rightarrow La\{N(SiMe_3)_2\}_3 + 3LiCl$$

The symmetry of the molecules is only C_3, a curious trigonal pyramidal structure being adopted in the solid state (X-ray); in solution the zero dipole moment indicates planarity. Adducts $M\{N(SiMe_3)_2\}_3OPPh_3$ are formed with $OPPh_3$ which are presumably four-co-ordinated. Though thermally stable and volatile, these compounds are rapidly hydrolysed to give $M(OH)_3$ and free amine.

10.5.6 Halide Complexes

Lanthanide ions in a +3 oxidation state have a comparatively small affinity for halide ions, but do form mixed halides $M^IM^{III}X_4$. Thus $KLaF_4$ has the fluorite structure with a statistical distribution of eight-co-ordinated K^+ and La^{3+} ions, while $NaLaF_4$ contains six-co-ordinated Na^+ ions and nine-co-ordinated La^{3+} ions.

In nonaqueous solvents, however, the lanthanide–halogen bond is sufficiently strong to resist dissociation and complexes $(Ph_3PH)_3[MX_6]$ (X = Cl, Br and I) are obtainable in the solid state. The iodide is very unstable. The complex ions appear from their electronic spectra to be six-co-ordinate and octahedral, an unusual geometry for lanthanides. The absorption bands have extinction coefficients that are an order of magnitude lower than those of complexes of low symmetry. The complex thiocyanate ions $[M(NCS)_6]^{3-}$ and nitrites $[M(NO_2)_6]^{3-}$, which are N-bonded, are also octahedral.

10.5.7 Compounds Containing Carbon Ligands

Although, as previously noted, the lanthanides do not combine at all readily with donors such as S, P or C, isolated examples of such complexes are known. Thus there is a well-defined series of cyclopentadienyls $[M(C_5H_5)_3]$, but this is not unexpected since the $C_5H_5^-$ anion forms salts with most metals. These complexes, which are quite volatile, are rapidly decomposed by water or by ferrous chloride.

$$La(C_5H_5)_3 + 3H_2O \rightarrow La(OH)_3 + C_5H_6$$

$$2La(C_5H_5)_3 + 3FeCl_2 \rightarrow 3Fe(C_5H_5)_2 + 2LaCl_3$$

They form unexpected adducts $M(C_5H_5)_3CNC_6H_{11}$ with cyclohexylisocyanide; isocyanides are, of course, isoelectronic with CO but no carbonyl adducts are known. The structure of $Sm(C_5H_5)_3$ is not simple. The C_5H_5 groups are *penta-hapto* but some are within bonding distance of neighbouring Sm atoms, thus forming chains, of which the structure contains two distinct types.

Compounds $Ln(C_9H_7)_3$ are also formed with indenyl

by reaction of $LnCl_3$ with sodium indenyl. N.M.R. evidence was interpreted in terms of σ bonding but a crystallographic study shows that they also are *penta-hapto*

$$LnCl_3 + 3NaC_9H_7 \xrightarrow{\text{THF}} Ln(\pi\text{-}C_9H_7)_3 + 3NaCl$$

Lanthanides form stable (although not to air and moisture) complexes with cyclo-octatetraene, $Ln(C_8H_8)Cl.2THF$; crystallographic study of the cerium compound shows it to be dimeric, with two chlorine bridges. Each cerium is thus bound 'π' to a C_8H_8 ring, two oxygen atoms (THF) and two chlorines. The bonding to the C_8H_8 ring seems to be more electrostatic than in the case of the actinides (reaction with UCl_4 gives uranocene in 89 per cent yield). In the more straightforward compound $[K\ diglyme_2]^+[Ce(C_8H_8)_2]^-$, the anion consists of a staggered D_{8d} sandwich.

Studies on phenyls of the lanthanides indicate them to be polymeric; they are highly unstable to air and water. The products are of two types

$$YCl_3 + 3LiPh \xrightarrow{\text{THF}} YPh_3 + 3LiCl$$

$$LaCl_3 + 4LiPh \xrightarrow{\text{THF}} LiLaPh_4 + 3LiCl$$

The use of the 2,6-dimethylphenyl group leads to a thermally stable (but, of course, readily hydrolysed) compound in which the lanthanide is co-ordinated tetrahedrally to four σ-aryl ligands, with $Lu-C = 242 - 250$ pm (see figure 10.12).

$$LuCl_3 + 4LiC_6H_3Me_2 \xrightarrow{\text{THF}} [Li(THF)_4][Lu(C_6H_3Me_2)_4] + 3LiCl$$

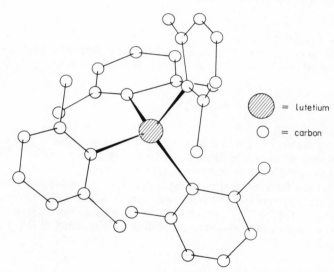

      ⬚ = lutetium

      ○ = carbon

Figure 10.12 The structure of the ion $[Lu(2,6\text{-dimethylphenyl})_4]^-$ (after S.A. Cotton, F.A. Hart, M.B. Hursthouse and A.J. Welch, *Chem. Commun.* (1972), 1225)

Relatively stable compounds are also obtained by employing the groups $-CH_2CMe_3$ or $-CH_2SiMe_3$ (=L). In this case the products are $ML_3(THF)_2$. All three of these groups are bulky and also lack a β-hydrogen (see page 183). It is not clear which of these properties is the more necessary.

Some divalent lanthanide organimetallics are known; thus reduction of $Sm(\pi\text{-}C_5H_5)_3$ with potassium naphthalenide affords $Sm(C_5H_5)_2$. Europium forms a methylacetylide

$$Eu \xrightarrow[-78°]{NH_3/MeCCH} Eu(MeC{\equiv}C)_2$$

This brown solid is pyrophoric in air and reacts with water to form $Eu(OH)_2$ and methylacetylene. The analogous reaction with ytterbium affords a mixture of $Yb(MeC{\equiv}C)_2$ and $Yb(NH_2)_2$. Reaction of europium with acetylene gives $Eu(C{\equiv}CH)_2$; on heating to 90° *in vacuo* EuC_2 forms. Europium and ytterbium react with alkyl and aryl iodides in THF to give solutions that undergo standard Grignard reactions. Europium and ytterbium also form cyclo-octatetraene complexes $M(C_8H_8)$.

Tetravalent cerium forms a number of organo-derivatives, notably with poly*hapto* anionic ligands of delocalised charge. Examples are $Ce(C_5H_5)_3Cl$ and $Ce(indenyl)_2Cl_2$. These molecules form alkyl and aryl derivatives such as $Ce(C_5H_5)_3Ph$ or $Ce(indenyl)_2Me_2$ on treatment with Grignard or lithium reagents. The additional carbon ligands are presumably σ bonded. Increased covalent character and hence possibly increased stability would be expected in the Ce(IV)–C bond as opposed to Ce(III)–C because of the greater inductive effect of the tetrapositive ion.

10.5.8 Other Ligand Atoms
A stable series of complexes of the dimethyldithiocarbamate anion exists in which the bonding is believed to be bidentate through the sulphur atoms

The complexes are of the types $La(DMDTC)_3$ and $[La(DMDTC)_4]^-$. There are very few well-established lanthanide–sulphur complexes; Ph_3PS and $MeSCH_2CH_2SMe$, for example, apparently fail to react with lanthanide salts.

A final example features what might be a metal–metal-bonded compound involving lanthanides; a mercury–erbium mixture reacts with $Hg\{Co(CO)_4\}_2$ in THF to form a compound $Er\{Co(CO)_4\}_3.4THF$. Further investigation of such systems is of interest; in this instance the erbium to cobalt linkage is possibly of the type Er–OC–Co.

10.5.9 Cerium(IV) Complexes
Cerium is the only lanthanide to have any substantial co-ordination chemistry in the tetravalent state, where it rather resembles zirconium or hafnium. There is some evidence for the hydrated ion in very strong $HClO_4$ solution; in other acids co-ordination of anions probably occurs. The best-known complex is possibly the double nitrate $(NH_4)_2Ce(NO_3)_6$, which is obtained from nitric acid solution; the $[Ce(NO_3)_6]^{2-}$ ion is twelve-co-ordinate, the twelve oxygen atoms forming an icosahedral array about the cerium. This complex reacts with Ph_3PO to form the orange $[Ce(NO_3)_4(Ph_3PO)_2]$, which is ten-co-ordinated (see figure 10.13). CeO_2 reacts with

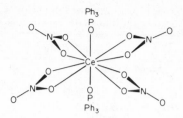

Figure 10.13 The ten-co-ordinate $Ce(NO_3)_4(Ph_3PO)_2$

HCl in nonaqueous solvents (for example, dioxan, pyridine) to yield salts such as $(pyH)_2[CeCl_6]$. A considerable number of fluoro-complexes exist, for example

$$CeF_4 \xrightarrow[H_2O]{NH_4F} (NH_4)_4CeF_8, (NH_4)_2CeF_6, (NH_4)CeF_5$$

$$CeO_2 \xrightarrow[MCl]{F_2} Na_2CeF_6, Cs_3CeF_7$$

Alkoxides, rather similar to the Zr and Hf analogues, are known; thus

$$CeCl_6{}^{2-} \xrightarrow[NH_3]{ROH} Ce(OR)_4$$

Most of these are polymeric, although the isopropoxide sublimes *in vacuo*.

Some lanthanide β-diketonate complexes, notably $Ce(Me_3C.CO.CH.CO.CMe_3)_4$, show promise as nontoxic antiknock agents in motor fuels.

10.6 Stability Constants of Lanthanide Ions in Water Solution

The stability constants $K_1, K_2, \ldots$ where

$$K_1 = \frac{[(ML_{aq})^{(3-n)^+}]}{[(M_{aq})^{3+}][(L_{aq})^{n-}]}$$

(M is a lanthanide and L is an uncharged or anionic ligand) have been determined for a very large number of different ligands, many of them anionic chelates. The values of K are in many cases rather similar to those expected for a poorly co-ordinating d transition metal ion such as Mn^{2+}. Some typical values are given in table 10.3 where they are compared with corresponding data for other metal ions. Some salient features are as follows.

The stability of the complex species is usually greater for La^{3+} than for Ca^{2+}, as would be expected on the basis of the comparable ionic radii (106 and 99 pm respectively) and the greater charge of the La^{3+} ion. The Mn^{2+} ion gives, as just

Table 10.3 Aqueous stability constants ($\log K_1$) for La^{3+}
and other metal ions with various ligands

	La^{3+}	Ca^{2+}	Mn^{2+}	Cu^{2+}	Fe^{3+}
F^-	2.7	$\langle 1.0$	(K_1 very small)	0.70	5.2
Cl^-	−0.12	(K_1 very small)	0.0	0.11	0.46
NO_3^-	−0.26	0.28	–	–	−0.50
$HONCMeCMeNO^-$	6.6	–	–	11.9(50% dioxan)	–
$CH_3COCHCOCH_3^-$	5.1	–	4.2	8.3	11.4
$EDTA^{4-}$	15.3	10.7	13.8	17.7	25.1
dipyridyl	(K_1 small)	–	2.5	6.4	–
H_2O	0.23(Nd^{3+} in MeOH)	–	–	−0.2 (in EtOH)	–.

mentioned, complexes of roughly equal stability, both the charge and ionic radius (80 pm) being smaller. The Cu^{2+} ion is crystal-field stabilised and complexes of those ligands with a Δ value higher than that produced by water are correspondingly stabilised. The Fe^{3+} ion, with an equal charge and much smaller radius (≈ 60 pm) gives comparatively very stable complexes. The only unusual feature is the apparent low stability of lanthanide amine complexes. There are few numerical data available, but $\log K_1$ for the tetradentate amine $C_5H_4N.CH:NCH_2CH_2N:CH.C_5H_4N$, is approximately 2, which is certainly a low value for a ligand of this type.

The stability in aqueous solution of complexes of the polydentate amino-acid ligands such as EDTA, NTA and DTPA has been rather closely studied. The cause of the very high values of K for the EDTA complex is the very favourable entropy change on formation of the sexadentate complex; the heat change is only marginally favourable.

The graph of K against atomic number of the lanthanide ion is of considerable interest, since at first sight it suggests the presence of a crystal-field stabilisation effect similar to that observed for the d transition metals in a plot of, for example, lattice energies against atomic number. Thus for EDTA K shows a small positive deviation from the smooth increase expected as a consequence of decreasing ionic radius, except at Gd^{3+} the unstabilised f^7 ion. This general behaviour is also shown by other similar ligands such as NTA or DTPA.

However, more detailed consideration shows that the facile crystal-field interpretation is unlikely to be entirely correct. It is known from lanthanide spectra that the splitting of a particular term due to the crystal field is about 100 cm^{-1}. Thus the stabilisation of an f^8 Tb^{3+} complex relative to a similar f^7 Gd^{3+} complex in water solution would be of this relatively insignificant magnitude (1.2 kJ$\equiv$0.3 kcal).

If ΔG for complex formation is resolved into its components ΔH and ΔS, it is found in the case of EDTA that the inflection at Gd does not mainly arise from

ΔH, as it should for a crystal-field effect, nor does ΔH inflect at Gd but at Tb instead. Hence it is most likely that the inflection of the K value at Gd is caused by a change in structure, probably of the aquo-cation (not of the complex because the effect occurs for several different ligands), involving an alteration in this region of the lanthanide series in the number of co-ordinated water molecules. Thus at some point around Gd–Tb there may be a decrease in co-ordination number with decrease of ionic radius, with consequent decrease of free energy of hydration and increase of ease of formation of the complex species. This would be a general phenomenon for all ligands, but it is clear that in the case of any particular ligand, the structure of the complex ion also might undergo a change of co-ordination number at some point in the series (as observed in the solid-state structures of some series of lanthanide complexes), giving a superposition of changes in ΔH and ΔS and a complicated graph of K against atomic number. The postulated change in hydration number of the aquo-cation is strongly supported by measurement of the heats of solution of the series of isostructural hydrated bromates and ethylsulphates; a discontinuity is observed before the middle of the series. Furthermore, if the heats of formation of well-defined lanthanide complex species, such as $[M(2,6\text{-dipicolinate})_3]^{3-}$, are measured starting from the isostructural series of hydrated bromates, the resulting plot of ΔH (the heat content change for this process) against atomic number is essentially smooth. There is a local maximum at gadolinium of about 2 kJ (0.5 kcal), which presumably arises from a genuine crystal-field stabilisation of species other than f^0 (La^{3+}), f^7 (Gd^{3+}) and f^{14} (Lu^{3+}).

There is also a small but definite inflection in the value of the M^{3+} ionic radius, to the extent of about + 0.5 pm, at Gd^{3+} (see figure 10.1).

10.7 Lanthanide Complexes as N.M.R. Shift Reagents

The paramagnetic nature of lanthanide complexes finds an important application in the interpretation of complex n.m.r. spectra of organic compounds and the elucidation of their structure. If an organic compound A having a nucleophilic group (—OH, =CO, —COO⁻, etc.) is in solution together with a lanthanide complex B, which has a low co-ordination number and whose ligands are not too bulky, an adduct AB will exist in solution in very rapid equilibrium with its components. Under these circumstances the magnetic moment of each lanthanide ion will provide a resultant magnetic field in the opposite direction to the applied field of the n.m.r. instrument, since this orientation of the ionic moment lies at a slightly lower energy level than other orientations (see figure 10.14). As a particular molecule of the adduct AB tumbles in solution, the field exerted by the lanthanide magnetic dipole on the nuclear dipole of a specified proton belonging to A will reverse in direction between orientation (a) in figure 10.13 and orientation (b). In (a) the lanthanide dipole will reinforce the magnetic field of the n.m.r. spectrometer but in (b) it will oppose it. In fact it can be shown that the effect on the proton nuclear moment averages to zero if the lanthanide moment is constant. In general, however, the magnetic moment of the lanthanide is not constant because the adduct is magnetically nonisotropic (symmetrical complexes such as tetrahedral or octahedral are isotropic) and the magnetic moment of the lanthanide in orienta-

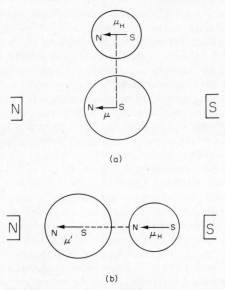

Figure 10.14 Orientations of the magnetic dipole of the lanthanide ion and the nuclear dipole of a proton in solution. (μ_H = nuclear moment of the proton; μ, μ' = moments of the lanthanide ion; rotation of the broken line represents tumbling of the complex. The larger circle represents the lanthanide ion and the smaller circle the proton)

tion (a) of AB is thus not equal to that for orientation (b). Hence the proton experiences a resultant field and its resonance position is shifted. This effect is known as a pseudo-contact shift.[†] The magnitude of the shifts in a specific adduct can be shown to be given by

$$\Delta H/H = \frac{K(3 \cos^2 \theta - 1)}{r^3}$$

where r is the lanthanide–proton distance, θ is the angular distance of the proton from the axis of the adduct and K is a constant. Figure 10.15 shows the operation

[†] Paramagnetic shifts in n.m.r. spectra may also be produced by two further mechanisms. Firstly, there is a bulk paramagnetic effect, which produces a high-field shift in all n.m.r. spectra of paramagnetic solutions; as all protons in a given solution are shifted equally this effect can be disregarded, but may, if desired, be used as a method for the determination of magnetic moments. Secondly, where there is some covalent bonding between a paramagnetic metal and a ligand, bonding orbitals extending over the metal and the proton-carrying ligand may be occupied by unpaired electrons, which will produce a magnetic field at the proton. This is the contact mechanism, and seems to operate in many instances where spectra of d transition metals show shifts. Amine complexes of lanthanides show this effect to some extent.

of this effect when the β-diketone complex $Eu(Me_3CCOCHCOCMe_3)_3$ is added to a solution of *n*-hexanol; the methylene protons, unresolved in a conventional spectrum, are now well separated and give a first-order spectrum. The operation of the geometric factor $(3 \cos^2\theta - 1)/r^3$ enables the location of the protons (or other nuclei) relative to each other to be determined from the magnitudes of their shifts. In this way, a choice may often be made between alternative structures of the organic molecule, where these cannot be differentiated by other means.

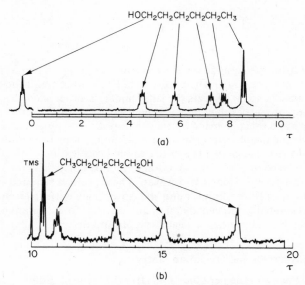

Figure 10.15 (a) 100 MHz ^1H n.m.r. spectrum of *n*-hexanol in CCl_4 containing $Eu(Me_3CCOCHCOCMe_3)_3$ (0.29 mole); (b) 100 MHz ^1H n.m.r. spectrum of *n*-pentanol (0.22 molar) in CCl_4 with $Pr(Me_3CCOCHCOCMe_3)_3$ (0.053 molar). Chemical shifts in τ units relative to Me_4Si ($\tau = 10$) (after J.K.M. Sanders and D.H. Williams, *Chem. Commun.* (1970), 422 and J. Briggs, G.H. Frost, F.A. Hart, G.P. Moss and M.L. Staniforth, *Chem. Commun.* (1970), 749)

11 The Actinides

The fifteen elements of the actinide series form a unique group in two respects. Firstly, they are, taken as a whole, chemically unlike any other transition-metal series. They resemble in some respects the lanthanides, in other ways the 4d and 5d metals, but in many properties they resemble neither. Secondly, most of them do not occur in nature and must be made by transmutation. The general chemistry of the naturally occurring members thorium and uranium has long been known, but detailed studies of these two elements and the whole investigation of the remainder of the series was delayed until the introduction of atomic bombs and nuclear power stations during and after the Second World War. These projects required the manufacture of kilogram quantities of plutonium and isotopically enriched uranium, which in turn necessitated a detailed chemical and physical investigation of the whole series of metals as far as their availability permitted.

11.1 Electronic Structures and Oxidation States

The actinides commence in the periodic table after the element radium (Ra, [1] [2] [3] [4] $5s^2 p^6 d^{10} 6s^2 p^6 7s^2$), and it is impossible to predict theoretically with complete assurance whether the extra electron contained in the next element, actinium, will reside in a 5f or in a 6d orbital. In fact it is found to be a 6d electron. The complete series, with 5f and 6d orbital occupations, is as follows.

actinium	Ac	$5f^0$	$6d^1$	berkelium	Bk	$5f^9$	
thorium	Th	$5f^0$	$6d^2$	californium	Cf	$5f^{10}$	
protactinium	Pa	$5f^2$	$6d^1$	einsteinium	Es	$5f^{11}$	
uranium	U	$5f^3$	$6d^1$	fermium	Fm	$5f^{12}$	
neptunium	Np	$5f^4$	$6d^1$	mendelevium	Md	$5f^{13}$	
plutonium	Pu	$5f^6$		nobelium	No	$5f^{14}$	
americium	Am	$5f^7$		lawrencium	Lr	$5f^{14}$	$6d^1$
curium	Cm	$5f^7$	$6d^1$				

The configurations up to curium inclusive rest on evidence from atomic spectra while the remainder are extrapolations from the behaviour of the first half of the series. Thus, just as in the d transition-metal series the energies of the d and s orbitals are similar (Ir, $5d^7 6s^2$; Pt, $5d^9 6s^1$), and in the lanthanide series the 4f and 5d are similar (La, $4f^0 5d^1$; Ce, $4f^2 5d^0$), so in the actinide series the 5f and 6d levels are of similar energy. This is particularly so in the case of the lighter actinides, after which the 5f level is the more stable. There is a larger energy gap between the 4f and 5d levels than between the 5f and 6d levels.

When the actinides form positively charged ions, the orbital having the lower principal quantum number becomes comparatively much more stable. Thus all the actinide M^{3+} ions have $5f^n 6d^0$ configurations; this behaviour is also observed in the d transition series (for example, Ti^{3+}, $3d^1$) and in the lanthanide series (for example, Ce^{3+}, $4f^1$).

The oxidation states of the actinides, as known at present, are shown in figure 11.1. They show almost as wide a range as the d transition metals. No +8 (or even higher) oxidation state is yet known, but all states from +7 to +2 inclusive are

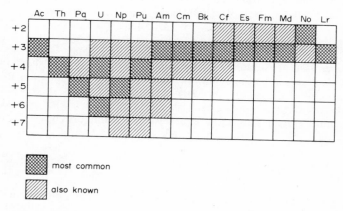

most common

also known

Figure 11.1 The incidence of oxidation states for the actinides

found. The most important difference from the d transition metals is that no oxidation states lower than +2 are formed, and there is a complete absence of stable carbonyls and a severe dearth of other π-bonded complexes. In discussing the absence of actinide carbonyls and related compounds, it should be noted that the chemistry of the heavier, electron-rich actinides, whose electronic configuration might favour π bonding, has necessarily been little investigated owing to the radioactivity and lack of availability of these metals.

In the case of the first four elements, the most stable oxidation state tends to be that in which all 5f or 6d electrons have been lost, leaving the bare core [1] [2] [3] [4] $5s^2 p^6 d^{10} 6s^2 p^6$. This behaviour is similar to that in the series La, Hf, Ta, W where the highest possible states, respectively +3, +4, +5 and +6, are the most stable. It differs from the behaviour of the lanthanides, where the +3 state is uniformly the most stable apart from Ce^{4+} which is almost equally stable with Ce^{3+}. For this reason the actinides were, during the many years when only Th and U were available, supposed to be a new d transition-metal series with the 6d shell filling rather than the 5f. This was hardly a logical deduction, however, since experimental facts such as the high stability of Th(IV) and U(VI) (as $UO_2{}^{2+}$) merely mean that the ionisation potentials of the outer electrons are comparatively low and possibly also mean that either the 5f or 6d orbitals can bond covalently to stabilise the high +6 oxidation state of uranium. No conclusions are possible from such data about whether the electrons were originally in the 5f or 6d orbitals in the unionised atom or about how the heavier elements would behave if and when they were available.

Now that all the elements are available and all have been studied chemically, even if only by tracer methods, it turns out that the elements up to uranium show essentially early d transition-metal behaviour while the elements from berkelium onwards show essentially lanthanide behaviour, the intermediate elements Np, Pu, Am and Cm effecting the transition between the two groups. This behaviour probably reflects mainly the increase of ionisation potential with atomic number and the accompanying smaller spatial extension of the 5f and 6d orbitals, thus tending to restrict the formation of high oxidation states and to restrict their stabilisation by covalent ligand-to-metal π bonding as in, for example, $MnO_4{}^-$.

11.2 Availability of the Actinides

All the isotopes of all the actinides are radioactive, the principal means of decay being α-particle (^{4}He) emission. Even the naturally occurring ^{232}Th and ^{238}U decay in this manner, with half-lives of 1.4×10^{10} and 4.5×10^9 years, respectively. These two elements have been part of the earth since its beginning. Actinium and particularly protactinium exist in radioactive equilibria in uranium ores but only to the extent of one part in 10^{10} and 10^7, respectively

$$^{235}\text{U} \xrightarrow[7 \times 10^8 \text{ years}]{\alpha} {}^{231}\text{Th} \xrightarrow[25.6 \text{ years}]{\beta} {}^{231}\text{Pa} \xrightarrow[3.3 \times 10^5 \text{ years}]{\alpha} {}^{227}\text{Ac} \xrightarrow[22 \text{ years}]{\alpha} {}^{223}\text{Fr}$$

Uranium ores also contain minute quantities of neptunium and plutonium formed by neutron-capture processes.

11.2.1 Actinium

This is obtained artificially from radium by the process

$$^{226}\text{Ra (n,}\gamma) \rightarrow {}^{227}\text{Ra} \xrightarrow[41 \text{ min}]{\beta} {}^{227}\text{Ac}$$

Purification is by means of a cation-exchange resin. It may be precipitated as the fluoride. It was, however, studied by tracer techniques many years ago, having been separated from pitchblende residues. The aqueous ion was shown to be tripositive by von Hevesy in 1927 from its rate of diffusion. The characteristic radiation was identified by Debierne in 1899. It is available in milligram quantities, but investigation of its chemistry is hampered by its β- and γ-radioactive decay products. These radiations necessitate remote-handling techniques that are unnecessary in the case of α decay alone since α particles are nonpenetrative. Of course, even for α radiation, scrupulously careful handling is required in glove boxes maintained at slightly below atmospheric pressure. This is to guard against accidental ingestion of the material, the maximum tolerable cumulative dose being of the order of 10^{-6} (^{239}Pu) to 10^{-11} g (^{242}Cm) depending on the particular isotope in question. The actinides tend to accumulate in the bones, kidneys or stomach, depending on their individual chemical properties.

11.2.2 Thorium

This element occurs as a component of monazite, a principal lanthanide ore. The ore, a phosphate, may be decomposed by strong alkali and the resulting metal hydroxides separated by solvent extraction from acid solution using standard extractants such as tributylphosphate or methylisobutylketone. Thorium (11 g per ton) is about as plentiful in the earth as lead.

11.2.3 Protactinium

We think of protactinium as associated with modern technology, and this is perfectly true in many respects. However, it was first isolated by chemical means from pitchblende residues and its atomic weight determined gravimetrically in 1934 by von Grosse. He obtained the value 230.6 ± 0.5, which compares well with the modern physically determined value 231.04. It was obtained by co-precipitation with ZrO_2 (although in the +5 oxidation state) and removal of zirconium as $ZrOCl_2$. Finally, separation from the tantalum which follows the protactinium and chemically closely resembles it was by removal of the former as a soluble peroxide.

More recently about 100 g of protactinium was recovered from residues from a uranium extraction plant and at present forms the world supply.

11.2.4 Uranium

This is extracted on the thousand-ton scale and, though accurate forecasts are difficult, is estimated to be required in the 1970s at the very high rate of about 50 000 tons per year for use in nuclear reactors. It costs about $10 per pound to produce. The ores are widely distributed, the main producers being Canada (Ontario), the U.S.A. (Colorado) and South Africa (Witwatsrand). There are about 150 known uranium minerals. Important and widely occurring ones are uraninite, UO_{2+x}, and pitchblende, U_3O_8 (pitchblende, although similar to uraninite in composition, has a hydrothermal rather than igneous origin). Carnotite, $KUO_2VO_4.1.5 H_2O$, is an important Colorado mineral. These minerals are dispersed in rock so that the overall uranium metal content of the rocky ore is only about 0.1 per cent.

A typical extraction procedure first concentrates the ore by flotation or by the rather improbable process of automatically measuring the weight and radioactivity of each lump, finding the ratio and rejecting those found wanting. The concentrate is then leached with an MnO_2–sulphuric acid mixture (the MnO_2 oxidises U(IV) to the soluble $UO_2{}^{2+}$ ion). The uranium is then removed from the filtered solution as a sulphato-anion by ion-exchange and then eluted from the resin to give a concentrated solution from which it is precipitated as the hydrated oxide. It is then purified by solvent extraction in the two-phase system tributylphosphate–hexane/nitric acid, since it forms a hexane-soluble complex $[UO_2(NO_3)_2\{OP(OBu)_3\}_2]$. The hexane solution is washed and then the uranium is removed as pure uranyl nitrate by extraction into dilute nitric acid. To obtain uranium metal, the following sequence is used

$$UO_2(NO_3)_2 \xrightarrow{450°} UO_3 \xrightarrow{H_2, 650°} UO_2 \xrightarrow{HF, 500°} UF_4 \xrightarrow{Mg, 700°} U + MgF_2$$

Impurities are in the 100 p.p.m. range. Small masses of very pure uranium may be made by decomposition of UI_4 on a hot tungsten filament.

11.2.5 Neptunium and Plutonium

These are byproducts of the operation of nuclear reactors and are consequently readily available

$^{238}U(n, 2n)$

$^{235}U(n, \gamma) \longrightarrow {}^{236}U(n, \gamma) \searrow$

$^{237}U \xrightarrow[7 \text{ days}]{\beta} {}^{237}Np \,(\alpha, \, 2.2 \times 10^6 \text{ years})$

$^{238}U(n, \gamma) \longrightarrow {}^{239}U \xrightarrow[23 \text{ mins}]{\beta} {}^{239}Np \xrightarrow[2 \text{ days}]{\beta} {}^{239}Pu(\alpha, \, 24\,000 \text{ years})$

The neptunium has no large-scale use, although some is converted into ^{238}Pu $(\alpha, 86 \text{ years})$, which is a nuclear-power source for use in space exploration. The Apollo 14 Moon mission left on the Moon an array of experiments (5 February, 1971), the power source for which was 3.8 kg (8.4 lb) of $^{238}PuO_2$ generating 1480 watts of heat, converted into 64 watts of electricity by PbTe thermoelectric elements Plutonium can be used for atomic weapons, but at present requirements are happily rather for use in nuclear reactors and are of the order of 10 tons per year.

The separation of neptunium and plutonium from the used uranium fuel rods, which also contain, as fission products, lanthanides and metals such as Zr, Nb and Ru, is chemically easy. However, in practice it is exceedingly difficult because of the intense radiation of the fission products and the large scale on which the process must be carried out. This necessitates remote handling and heavy shielding. The separation depends on principles such as the following: (a) U, Np and Pu can be oxidised to MO_2^{2+}, which can be separated off into the organic phase by solvent extraction; (b) Np and Pu can then be selectively reduced to the M^{4+} ion and removed into an aqueous phase, (c) Pu can be selectively reduced to Pu^{3+}, the Np^{4+} then being preferentially extracted into the organic phase. The relevant oxidation potentials of U, Np and Pu are discussed subsequently; the solvent-extraction procedure may use mixtures such as ether–nitric acid or hexane–tributylphosphate–nitric acid. The metals have been made by reactions such as the following

$$NpF_3 \xrightarrow{Li} Np + LiF$$

$$PuF_4 \xrightarrow{Ca} Pu + CaF_2$$

Plutonium metal is noteworthy as having no less than six crystal structures.

$$\alpha(Mc) \xrightarrow{117°} \beta(Mc) \xrightarrow{185°} \gamma(Or) \xrightarrow{310°} \delta(FCC) \xrightarrow{458°} \delta'(BCTg) \xrightarrow{480°} \epsilon(BCC)$$

11.2.6 Americium, Curium, Berkelium, Californium, Einsteinium and Fermium

The first two of these elements have been made on the 100 g scale as ^{243}Am (7650 years) and ^{244}Cm (18 years), obtained by lengthy neutron irradiation of 239 Pu.

$$^{239}Pu \xrightarrow{n} {}^{240}Pu \xrightarrow{n} {}^{241}Pu \xrightarrow{n} {}^{242}Pu \xrightarrow{n} {}^{243}Pu \, (\beta, 5 \text{ h}) \rightarrow {}^{243}Am$$

65% fission   70% fission

$\downarrow n$

$^{244}Am \, (\beta, 26 \text{ min}) \rightarrow {}^{244}Cm$

^{241}Am (433 years) and ^{242}Cm (162 days) have also both been made by neutron irradiation as follows

$$^{239}Pu \xrightarrow{n} {}^{240}Pu \xrightarrow{n} {}^{241}Pu \xrightarrow[13 \text{ years}]{\beta} {}^{241}Am \xrightarrow{n} {}^{242m}Am \xrightarrow[16 \text{ h}]{\beta} {}^{242}Cm$$

The intention is to use the ^{243}Am and ^{244}Cm as neutron targets in the high-flux reactor at Oak Ridge National Laboratory, U.S.A., this process leading to production of ^{249}Bk, ^{252}Cf, ^{253}Es, ^{254}Es and ^{257}Fm. Subsequently, the ^{252}Cf (α, 2.6 years) must provide a continuous source of ^{248}Cm (α, 4.7 x 10^5 years), which is a much more desirable isotope than the ^{244}Cm (18 years) for chemical studies. It is estimated that Bk, Cf and Es will be available on the 100 mg scale but Fm on the microgram scale only. The half-lives are, however, rather short (^{249}Bk, 314d.; ^{252}Cf, 2.6yr; ^{253}Es, 20d.; ^{254}Es, 270d.; ^{257}Fm, 94d.).

In some cases, more stable isotopes of these heavier elements are known but can only be made in very small quantities. Thus ^{247}Bk (10^4 years) can be made in an accelerator but only in trace amounts owing to the nature of this technique.

It is not possible to make elements heavier than fermium by successive neutron capture in reactors, since the isotope ^{258}Fm, formed next after ^{257}Fm, undergoes complete spontaneous fission in seconds, before it can capture any more neutrons.

Another way that has been used to make elements up to ^{257}Fm and possibly beyond is to set off a small underground nuclear explosion. The tremendously high neutron flux produced converts a ^{238}U target into heavier elements, which will in general be isotopically different from those produced by reactor irradiation because neutron capture will proceed faster than β-decay or spontaneous fission. Thus such a pathway as $^xA \rightarrow {}^{x+1}A \rightarrow$ (fission products) could be replaced by $^xA \xrightarrow{n} {}^{x+1}A \xrightarrow{n} {}^{x+2}A \xrightarrow{\beta} {}^{x+2}B$.

Americium and curium trifluorides have been reduced to the metals by barium at 1200–1400°. The metals (Am, 994° and Cm, 1340°) are higher melting and less dense than neptunium and plutonium and have the double-hexagonal close-packed structure peculiar to these metals and some lighter lanthanides. Berkelium metal has been made by Li reduction of BkF_3 and adopts double-hexagonal or cubic close packing as alternative structures.

These metals are all typically tripositive in aqueous solution and their mutual separation is by ion-exchange chromatography, often on a micro scale, using complexing agents such as α-hydroxyisobutyric acid to elute the column. It is also necessary to separate them from lanthanide fission products, which are very similar chemically and even have similar ion-exchange chromatographic properties. However, a separation may be effected using a cation-exchange column and 13M aqueous ethanolic HCl because the actinides form an anionic chloride complex to a rather greater extent than the lanthanides and are hence not retained in the column so long (see figure 11.2).

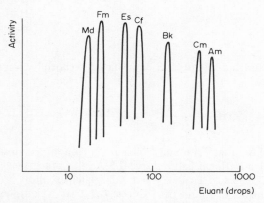

Figure 11.2 Elution of tripositive actinides from a Dowex-50 ion-exchange resin with buffered ammonium α-hydroxybutyrate (after J.J. Katz and G.T. Seaborg, *The Chemistry of the Actinide Elements,* Methuen, London (1957), p. 435)

11.2.7 Mendelevium, Nobelium and Lawrencium

The three heaviest elements, as explained above, cannot be made by successive neutron capture. However, the nuclei of light atoms such as He, B, C, N, O or Ne have been accelerated in a cyclotron or linear accelerator and used to bombard U, Np, Pu, Cm, Cf or Es. If absorbed, they obviously raise the mass number by many units and the products, after immediate loss of one or more protons or α particles will be a heavy elements. For mendelevium, the route

$$^{253}\text{Es} + {}^{4}\text{He} \rightarrow {}^{256}\text{Md} \text{ (1.5 h, E.C.)} + \text{n}$$

was successful. The scale was 10^{-12} g. Nobelium was produced by use of ^{12}C and ^{13}C nuclei with curium targets, and lawrencium by ^{10}B and ^{11}B with californium targets.

11.3 Chemical and Structural Properties of the Actinide Series

Our knowledge of the chemistry of the actinides is very unbalanced. It is probably true that the amount of purely chemical work published on uranium and its compounds exceeds that published on all other actinides combined. There are two reasons for this; firstly, uranium is one of the only two actinides (thorium being the other) that are available in any desired quantity and can be handled with minimal precautions against radioactivity; secondly, since it exhibits four different oxidation states and thorium only one, and is used as a nuclear fuel while thorium is not, it has attracted much more investigation than has thorium. In the case of the very rarely available heavy actinides, simple solution studies and tracer experiments may be all that is feasible. Not only is lack of availability a barrier to fuller investigation, but the radioactivity itself can be very troublesome. It can of course render undesirable the use of more than a few milligrams of material for chemical operations even when more is available in case a spillage might result in fairly massive contamination. Heating is also an annoyance. Thus 100 mg ^{242}Cm (162 days) puts out 12 watts of heat and this in itself would be sufficient to render solution studies difficult on that scale. Radiation-induced reactions, for example with solvent, are also a problem. Thus a solution of ^{239}Pu(VI) in water undergoes self-reduction owing to the effect of the α activity on the water. Radiation also has a disruptive effect on crystal structures, thus rendering X-ray work difficult. In some cases, however, it has been possible to carry out the X-ray diffraction experiments at temperatures near the melting point of the crystal, because the radiation-produced faults in the crystal structure are then being removed continuously by annealing.

11.3.1 *The Tripositive Actinide Halides*

These compounds form a convenient vehicle for an illustration of the similarities along the series of actinides. In general, the actinide elements are sufficiently different from one another to be best discussed separately and this is the scheme we shall adopt. However, a review of the series of actinide trihalides well illustrates the many points of resemblance, especially in structures of stoichiometrically similar compounds, along the series. In general, *properties of actinides in the same oxidation state are very closely similar, relative ease of oxidation or reduction being excepted.* The series of trivalent ions, for example, forms a uniformly similar M^{3+} series just as do the lanthanides. The general properties of the M^{3+} actinides moreover closely resemble the lanthanide M^{3+} ions. The outstanding difference between the two series is that whereas all M^{3+} lanthanides are stable, in the case of Th, Pa, U, Np, Pu and No the +3 state is unstable to a greater or lesser extent as regards oxidation or reduction.

Most of the actinides show a +3 oxidation state. The only state for actinium, it is ill defined for thorium, absent for protactinium, unstable for uranium, progressively more stable through neptunium and plutonium, and the most stable state for the rest of the series except for nobelium where +2 is more stable. The ionic radius is roughly 5 pm larger than a corresponding lanthanide (see figure 11.3), an actinide contraction being observed, which can be explained in a similar way to the lanthanide contraction.

Although a common oxidation state with actinides, the +3 state is not easy to investigate thoroughly since it is either very unstable, as with U, or when more stable than this it occurs with very radioactive elements.

Trifluorides of Ac, U, Np, Pu, Am, Cm, Bk and Cf are known. The stable trifluorides AcF_3 (white), AmF_3 (pink) and CmF_3 (white) can be made by reactions such as

$$MO_2 + 4HF \rightarrow MF_3 + \tfrac{1}{2}F_2 + 2H_2O$$

and

$$MF_3 . H_2O \rightarrow MF_3$$

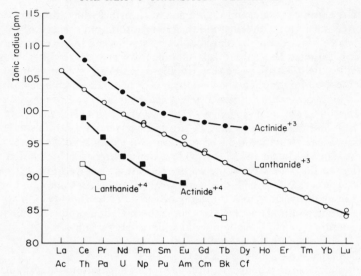

Figure 11.3 The ionic radii of lanthanide and actinide ions (after D. Brown, *Halides of the Lanthanides and Actinides,* Wiley–Interscience, London (1968))

The hydrated trifluorides are, as in the case of the lanthanides, precipitated by fluoride ion from aqueous solution. Neptunium trifluoride (purple) and plutonium trifluoride (purple), where the +3 state is of a reducing nature, require the use of a H_2/HF mixture.

$$MO_2 + \tfrac{1}{2}H_2 + 3HF \rightarrow MF_3 + 2H_2O$$

In the case of uranium, with its very reducing +3 state, the dark violet, high melting (1425°) trifluoride is only obtained by vigorous reduction.

$$UF_4 + Al \xrightarrow{\ 900°\ } UF_3 + AlF(sic)$$

Like all uranium(III) compounds it reduces water though the reaction is sluggish owing to insolubility. On strong heating it disproportionates

$$4UF_3 \xrightarrow{\ 1600°\ } 3UF_4 + U$$

The structures of all these trifluorides except CfF_3 are the same as that of LaF_3, having an irregular arrangement of nine fluoride ions around each metal with fluorine bridging to neighbouring metal ions. This high co-ordination number is as typical of actinide chemistry as it is of lanthanide chemistry. CfF_3 (and the low-temperature form of BkF_3) have the essentially eight-co-ordinated YF_3 structure.

Trichlorides are known for Ac, U, Np, Pu, Am, Cm, Bk, Cf and Es. Many methods have been used to prepare the trichlorides. Among those reported are the following

$$Ac(OH)_3 \xrightarrow[250°]{NH_4Cl, \text{ reduced pressure}} AcCl_3$$

$$Pu \xrightarrow{Cl_2, 450°} PuCl_3$$

$$AmCl_3.xH_2O \xrightarrow[\text{heat}]{NH_4Cl, \text{ reduced pressure}} AmCl_3$$

$$Es_2O_3 \xrightarrow{HCl, 500°} EsCl_3$$

These methods are suitable for those metals with a fairly stable +3 oxidation state. For uranium and neptunium, reductive methods must be used

$$U \xrightarrow{H_2, 220°} UH_3 \xrightarrow{HCl, 300°} UCl_3$$

and

$$NpCl_4 \xrightarrow{H_2, 600°} NpCl_3$$

The trichlorides are fairly high-melting solids (UCl_3, m.p. 842°), which come in assorted gay colours (Ac, white; U, olive green $\xrightarrow{450°}$ dark purple; Np, green; Pu, emerald green; Am, pink; Cm, white and Cf, green), which are more intense than the colours of the lanthanide trichlorides owing to the broader absorption bands. They are slightly volatile (UCl_3, vapour pressure 100 Pa at 1300°). They all have the same structure, which has regular face-centred prismatic D_{3h} nine-co-ordination of the metal by chloride ions. Each chloride ion bridges a given metal ion to two other metal ions. The metal ions are arranged in six-membered rings stacked vertically above one another to form tubes. This structure is also adopted by the lanthanide chlorides from La to Gd; heavier lanthanides have the six-co-ordinate $AlCl_3$ structure and it is quite probable that this or a similar structure would also be adopted by the heavier actinides in their as yet unmade trichlorides. $CfCl_3$ also adopts the eight-co-ordinate $PuBr_3$ structure as an alternative to its nine-co-ordinate form; each form is about equally stable.

Uranium trichloride is hygroscopic. It must be kept in a dry nonoxidising atmosphere. In water, it dissolves to give a purple solution, which at once gives off hydrogen to leave a green solution of U^{4+}. It gives UO_2Cl_2 on reaction with air and UCl_4 with chlorine.

Tribromides are known for Ac, U, Np, Pu, Am, Cm, Bk and Cf; as in the case of the other trihalides, the heavy-actinide halides that are unknown could all doubtless be prepared, with the possible exception of NoI_3. Many reactions have been employed to obtain the tribromides and they include the following

$$Ac_2O_3 \xrightarrow{AlBr_3, \ heat} AcBr_3$$

$$AmCl_3 \xrightarrow{NH_4Br/H_2, \ 450°} AmBr_3$$

$$U \xrightarrow{Br_2, \ 500°} UBr_3$$

$$PuBr_3.6H_2O \xrightarrow{70-170°, \ vacuum} PuBr_3$$

$$Cf_2O_3 \xrightarrow{HBr, \ 800°} CfBr_3$$

Dehydration of the hydrates, although it leads to oxychloride formation with the chlorides, gives pure bromides. Like the chlorides, the bromides are slightly volatile and can be sublimed in a vacuum above about 800°. Dark brown uranium tribromide. m.p. 755°, is very hygroscopic, forming a solid hexahydrate, which on dissolution in water evolves hydrogen. It reacts with chlorine or bromine to give the corresponding tetrahalide.

The crystal structures of the tribromides show a reduction of co-ordination number at Np from 9 to 8. Thus $AcBr_3$, UBr_3 and α-$NpBr_3$ have the same structure as the trichlorides, while β-$NpBr_3$, $PuBr_3$, $AmBr_3$, $CmBr_3$ and $BkBr_3$ have a different structure, which has essentially the same nine-co-ordinate face-centred trigonal prismatic co-ordination, but slightly distorted and with one of the face-centre halide ions removed to give eight-co-ordination. Some of the heavy-actinide bromides that have not yet been made would, if analogy with the lanthanide bromides (La–Pr, nine-co-ordinate, isostructural with $AcBr_3$; Nd–Eu, eight-co-ordinate, isostructural with $PuBr_3$ (see figure 11.4); Gd–Lu, six-co-ordinate $FeCl_3$ structure) is preserved, have the six-co-ordinate $FeCl_3$ structure.

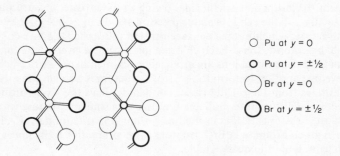

○ Pu at $y = 0$

○ Pu at $y = \pm \frac{1}{2}$

◯ Br at $y = 0$

◯ Br at $y = \pm \frac{1}{2}$

Figure 11.4 The layer structure of $PuBr_3$ (after A.F. Wells, *Structural Inorganic Chemistry,* Clarendon Press, Oxford (3rd edn, 1962)). Note that nine-co-ordination is prevented by nonbonding Br . . ., Br interactions. The notation $y = 0$ and $y = \pm \frac{1}{2}$ indicates the relative height of atoms with respect to the plane of the paper. The planes of the layers are normal to the plane of the paper

The methods for preparation of the tri-iodides are in most cases similar to those used for the other halides. The following have been used successfully

$$NpO_2 \xrightarrow{\;AlI_3,\,500°\;} NpI_3$$

$$AmCl_3 \xrightarrow{\;NH_4I/H_2,\,400°\;} AmI_3$$

$$2Pu + 3HgI_2 \rightarrow 2PuI_3 + 3Hg$$

$$2U + 3I_2 \xrightarrow{\;550°\;} 2UI_3$$

$$UH_3 + 3HI \rightarrow UI_3 + 3H_2$$

Again, the structures of the tri-iodides resemble those of the lanthanide iodides. The tri-iodides of Pa, U, Np and Pu have the eight-co-ordinate $PuBr_3$ structure while AmI_3, CmI_3, BkI_3 and CfI_3 have the six-co-ordinate $FeCl_3$ structure. The regular variations of structure type in the lanthanide and actinide halides are summarised in table 11.1. Their most important cause is clearly the variation of cation/anion radius ratio.

Table 11.1 Structures of Lanthanide and Actinide Trihalides

L	A	LF_3	AF_3	LCl_3	ACl_3	LBr_3	ABr_3	LI_3	AI_3
La	Ac	9†	9	9'‡	9'	9'	9'	8‖	
Ce	Th	9		9'		9'		8	
Pr	Pa	9		9'		9'		8	8
Nd	U	9	9	9'	9'	8	9'	8	8
Pm	Np		9	9'	9'	8	9',8		8
Sm	Pu	8+1‡	9	9'	9'	8	8	6'	8
Eu	Am	8+1	9	9'	9'	8	8		6'
Gd	Cm	8+1	9	9'	9'	6'††	8	6'	6'
Tb	Bk	8+1	8+1,9	8		6'	8	6'	6'
Dy	Cf	8+1	8+1	6§	9'	6'		6'	6'
Ho	Es	8+1		6		6'		6'	
Er	Rm	8+1		6		6'		6'	
Tm	Md	8+1		6		6'		6'	
Yb	No	8+1		6		6'		6'	
Lu	Lr	8+1		6		6'		6'	

† 9 = nine-co-ordinate LaF_3 structure; ‡ 9' = UCl_3 structure; ‡ 8+1 = YF_3; ‖ 8 = $PuBr_3$; § 6 = $AlCl_3$; †† 6' = $FeCl_3$.

The trihalides of thorium and protactinium may be mentioned separately. In these elements the trivalent state is very unstable and in the case of protactinium the tri-iodide is the only reported halide in an oxidative state lower than +4. It is made by heating PaI_5 *in vacuo*. Thorium tri-iodide is made by heating the metal with ThI_4 at 600–700°. It is at once oxidised by water. There is some possibility that in these compounds the metal is wholly or partially in the +4 state with the extra electron delocalised among the iodide ions.

11.4 Electronic Spectra of the Actinides

Compounds of the actinides have absorption bands in or near the visible region of the spectrum arising from one or other of three causes. These are: (a) transitions between electronic states, involving only the 5f orbitals; (b) transitions essentially involving transfer of an electron from a 5f to a 6d orbital; (c) transitions involving transfer of an electron from a ligand orbital to a metal orbital or vice versa. These three types of absorption, namely f→f, f→d and charge transfer, also occur in lanthanide spectra. Charge-transfer bands, of course, occur in the spectra of many metal complexes with polarisable ligands such as bromide or iodide, whether the metal be of f-transitional of d-transitional type or a main-group ion of a highly polarising nature, such as Pb^{4+}. The f→d bands are ncessarily limited to lanthanides and actinides.

The actinide f→f bands are rather analogous to the d→d bands in d transition-metal spectra in that they are transitions within a particular sub-shell (therefore Laporte-forbidden) and are not very intense, having absorption coefficients of about 10–100. These absorptions are rather broader and more intense than the lanthanide f→f transitions. The reason for this arises from the mechanism for gaining intensity in f→f spectra. Apart from some magnetic-dipole transitions, which are usually weak, f→f bands mainly obtain their intensity from interaction with the oscillating electric dipole of electromagnetic radiation. These electric-dipole transitions are in principle forbidden but are allowed in the presence of an unsymmetrical ligand field, which may arise either by coupling with an unsymmetrical metal–ligand vibration so that there is a simultaneous change of electronic and vibrational energy levels (vibronic transitions) or by a permanent unsymmetrical distribution of ligands in the co-ordination sphere of the complex ion. The 5f orbitals are more capable of interaction, whether electrostatic or covalent, with the ligands than the 4f. Thus there may be obtained a fairly intense central absorption peak broadened by outlying transitions, the former due to the unsymmetrical ligand field and the latter to vibronic coupling. This effect is well seen in the spectrum of the octahedral $[UCl_6]^{2-}$ ion, where the vibronic multiplet at around 17 000 cm^{-1} lacks a central peak since this symmetrical ion needs coupling with the ν_3 metal–ligand vibration, observed at 262 cm^{-1} in the infrared spectrum, for the transition to gain intensity. However, on destroying the centre of symmetry by hydrogen bonding the purely electronic transition appears strongly. Strong vibronic structure is also shown by the UO_2^{2+} ion, where an O–U–O stretching mode is involved (see figure 11.5, 11.6).

The electronic states that give rise to the f→f and f→d spectra will now be considered. Any spectra of this general type will depend on the approximate relative magnitude of a number of parameters. In the case of the actinides these are (1) the relative orbital energies of 5f and 6d electrons, (2) the 5f interelectronic-repul-

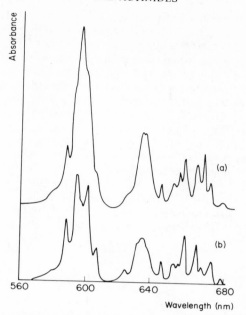

Figure 11.5 The effect of hydrogen bonding on the absorption spectrum of $[UCl_6]^{2-}$ (a) $(Et_4N)_2UCl_6$ in HCl-saturated CH_3CN; (b) $(Et_4N)_2UCl_6$ in CH_3CN (redrawn from J.L. Ryan, *Inorg. Chem.*, 3 (1964) 211)

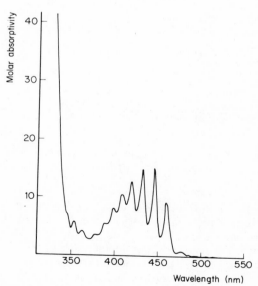

Figure 11.6 The absorption spectrum of $[UO_2(CO_2Me)_3]^-$ in CH_3CN (after J.L. Ryan and W.E. Keder, *Advances in Chemistry Series*, 71 (1967), 335)

sion energy parameters which determine the energy difference between, for example, the ^3H and ^3P levels of f^2, namely the Slater–Condon integrals F_2, F_4 and F_6, (3) the spin–orbit coupling constant ζ_{5f} and (4) the magnitude of the crystal-field splitting effect of the ligands. When the lanthanide-ion spectra were discussed in the previous chapter, it was noted that the equivalent parameters in that instance were of such magnitudes that Russell–Saunders coupling applied to a configuration purely f in character gave a satisfactory description of the electronic levels. Thus the ground state of Pr^{3+} (f^2) is $^3H_4$ and this is well separated from other levels; it is then split into up to (2 x 4 + 1) sublevels by the crystal field of the ligands. This relatively simple pattern is also observed in the actinide series in some instances but breaks down in many others. In all known actinide compounds (light actinides in low oxidation states would be an exception) the 6d levels are considerably ($\rangle$20 000 cm^{-1}) above the 5f, and have little influence on the f→f transitions themselves (see figure 11.7). In the U^{3+} aquo-ion, where the 5f-6d separation is comparatively small (it increases with atomic number and with oxidation number) there are intense allowed $5f^3 \to 5f^2 6d^1$ transitions at 25 500, 26 800 and 31 200 cm^{-1}.

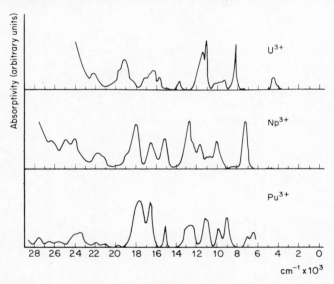

Figure 11.7 Absorption spectra of trivalent actinides in aqueous solution (after B.G. Wybourne, *Spectroscopic Properties of Rare Earths,* Interscience (1965))

However, even though the f levels may validly be considered on their own, simple Russell–Saunders coupling is in many cases inapplicable because the 5f interelectronic-repulsion parameters are only about 60 per cent of the 4f values, while the spin–orbit coupling parameters are about double those of the 4f series (ζ_{4f} 644 cm^{-1} in Ce^{3+}; ζ_{5f} = 1236 cm^{-1} in Th^{3+}). This means that the spin–orbit coupling can no longer be considered as a small perturbation and the 'intermediate' coupling system (that is, intermediate between Russell–Saunders and *jj* couplings) must usually be employed, thus rendering assignments more difficult. Both these effects are explained directly by the greater size of 5f orbitals relative to 4f; a further effect is that the crystal-field splittings of the ligands are much greater in actinide compounds than in lanthanides, since the larger orbitals are more directly affected by the ligands, and splittings of the order of 1000 cm^{-1} are produced, which lead to further mixing or overlapping of levels.

There is also observed a considerable nephelauxetic effect (expansion of the 5f electron cloud) in actinide spectra. Thus the energy separation between particular multiplets such as ^{3}H and ^{3}P for U^{4+} is considerably reduced by partial covalent bonding by ligands if the latter are fairly polarisable (for example, Br$^-$). The 5f orbitals need not necessarily themselves take part in the covalent bonding to produce the approximately 20 per cent decrease in repulsion parameters that occurs on moving from the U^{4+} aquo ion to the [UI$_6$]$^{2-}$ ion, but the probability is that they do so. The considerable shifts of bands due to the nephelauxetic effect and the very considerable crystal-field splittings, which vary from compound to compound, often lead to gross changes between spectra of different compounds of the same ion.

11.5 Actinium

Actinium, with an outer electronic configuration of 6d^{1}7s^2, is limited to the oxidation state +3. It clearly cannot have any higher state; lower states, as in the case of the other Group III metals Al, Sc, Y and La, are unstable and no examples of Ac^{2+} or Ac$^+$ are known. The very radioactive metal (^{227}Ac, $t_{1/2}$ = 22 years) is a little more electropositive than its lanthanide analogue, lanthanum. The oxidation potentials are La, 2.5 V; Ac, 2.6 V. The ionic radii are La^{3+}, 106 pm; Ac^{3+}, 110 pm. Actinium metal has m.p. 1050° and oxidises in air.

Despite their radioactivity, a number of actinium compounds have been characterised, some of them on a microgram scale using X-ray techniques. These include all the trihalides and all the oxyhalides AcOX. The trihalides all have the same crystal structures as the corresponding lanthanum compounds and may be made by similar means.

The oxyfluoride AcOF is made by heating the hydrated fluoride to 1200°. It has a cubic structure similar to fluorite. AcOCl has the eight-co-ordinate PbClF structure; it and the isostructural AcOBr are prepared by heating the trihalide with steam and ammonia at 900° and 500°, respectively.

There is only one oxide, Ac$_2$O$_3$, prepared by heating the hydroxide. It is isostructural with La$_2$O$_3$. When it is eventually investigated, the portion of the chemistry of actinium that is at present unknown will almost certainly resemble that of lanthanum just as strongly as does the presently known portion.

11.6 Thorium

This metal has a predominating +4 oxidation state, thus resembling its d transition-metal analogues Ti, Zr and Hf, but in contrast to its lanthanide analogue cerium, which has +3 rather more stable than +4. Lower thorium compounds are in fact limited to the tri-iodide and the di-iodide, while the Th^{4+} ion forms a complete range of anhydrous and hydrated salts together with an extensive range of complexes, especially with oxygen, nitrogen and halogen donor atoms.

Thorium metal, m.p. $1750°$, is obtained by Ca reduction of ThF_4 and is tarnished in air. The standard oxidation potential Th/Th^{4+}aq. is $+ 1.9$ V. The metal is slowly attacked by hot water and by dilute acids. The colourless Th^{4+}ion is slightly hydrolysed in water (compare Ac^{3+}, not hydrolysed, and Pa^{5+}, completely hydrolysed). The extent of hydrolysis, of course, depends on the pH and, since it is accompanied by the formation of hydroxy-bridged polymers, it also increases with increasing concentration. The salts isolated from aqueous solution are typically heavily hydrated, the best known being $Th(NO_3)_4 .5H_2 O$ where the Th^{4+} ion is eleven-co-ordinated (an example of this exceedingly rare co-ordination number) by four bidentate nitrate ions and three water molecules (see figure 11.8). The remain-

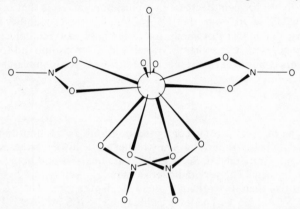

Figure 11.8 The structure of $Th(NO_3)_4.5H_2O$ (hydrogens omitted)

ing two molecules are hydrogen-bonded in the lattice but are not co-ordinated to the thorium ion. Thorium may be extracted from aqueous solution by tri-*n*-butylphosphate, a feature common to M^{3+} and M^{4+} actinide and lanthanide ions and $MO_2{}^{2+}$ actinide ions. In the case of aqueous thorium nitrate the thorium is extracted as $[Th(NO_3)_4\{ OP(OBu)_3 \}_2]$, which has sufficient bulky nonpolar groups in its molecule to be soluble in nonpolar solvents, and thus extractable in a two-phase aqueous/TBP-hydrocarbon system. Its structure is doubtless of a similar type to $[Th(NO_3)_4(OPPh_3)_2]$, which is isomorphous with the analogous ten-co-ordinated Ce(IV) compound (see figure 10.13).

11.6.1 Halides and Halide Complexes

The halides are obtained by a variety of standard methods; for example

$$ThO_2 \xrightarrow[400°]{CCl_2F_2} ThF_4$$

$$ThCl_4.aq \xrightarrow{SOCl_2} ThCl_4$$

$$Th \xrightarrow[700°]{Br_2} ThBr_4$$

$$Th \xrightarrow[400°]{I_2} ThI_4$$

All four halides have eight-co-ordinated Th^{4+} ions. Curiously, the smallest and largest ions, in ThF_4 and ThI_4, are disposed in a square antiprism, while the chloride and bromide (in β-$ThBr_4$) have dodecahedral co-ordination. The halides are slightly volatile: thus ThI_4 has m.p. 566° and b.p. 837°. In the vapour ThF_4 and $ThCl_4$ have Th–F = 214 pm and Th–Cl = 258 pm. In crystalline $ThCl_4$, the Th–Cl distances are four at 272 pm and four at 290 pm, the higher co-ordination number leading to the greater bond distance. The tetrafluoride is insoluble in water but forms a hydrate $ThF_4.2.5H_2O$, while the other halides are soluble. Oxyhalides $ThOX_2$ (X = F, Cl, Br and I) are known; for example

$$ThO_2 \ + \ ThBr_4 \ \xrightarrow{\ 500° \ } \ 2ThOBr_2$$

They are structurally similar to the corresponding protactinium and uranium compounds.

Thorium forms a wide variety of halide complexes. Besides fluoride complexes of 'simple' stoichiometry such as K_5ThF_9 (dodecahedral eight-co-ordination), Na_4ThF_8, Na_3ThF_7 and Na_2ThF_6, complexes such as $Na_7Th_6F_{31}$ (in which the Th is eight-co-ordinated to a square antiprism of fluoride ions) are also formed. Chloride, bromide and iodide complexes are exemplified by Rb_2ThCl_6, $RbThCl_5$, $(C_5H_5NH)_2ThBr_6$ and $(NBu_4)_2ThI_6$. In the great majority of cases these complexes are isomorphous with corresponding U(IV) complexes (*q.v.*) and are prepared similarly. The detailed structures are in many cases unknown, but $(NH_4)_4ThF_8$ and $(NH_4)_3ThF_7$ both have nine-co-ordinate thorium, the latter adopting the tricapped trigonal prismatic geometry.

11.6.2 Other Complexes

The halides readily form adducts with ligands that typically show affinity with actinides, for example CH_3CONMe_2, Me_2SO and Ph_3PO. Again, these complexes are generally similar to the uranium(IV) complexes but there are sometimes differences in stoichiometry. Thus thorium forms $ThCl_4(Me_2SO)_5$ while uranium, besides forming $UCl_4(Me_2SO)_5$ will form $UCl_4(Me_2SO)_7$ also; both metals also form $MCl_4(Me_2SO)_3$. These complexes are formed from the action of acetone–Me_2SO mixtures on the tetrahalide. Other complexes are $ThBr_4(CH_3CONMe_2)_5$, $ThBr_4(Ph_3PO)_3$ and $ThBr_4(Ph_3PO)_2$.

A fairly general method for conversion of lanthanide and actinide chlorides into thiocyanates is by treatment with a solution of KNCS in acetone (in which it is fairly soluble), KCl being precipitated. The simple thiocyanates are largely unknown, but many complexes can be obtained. These are often of high co-ordination number since the NCS^- ion, bonded to the metal through the nitrogen atom, occupies little space in the co-ordination sphere. In the case of thorium the complexes $Rb_4[Th(NCS)_8].2H_2O$, $Rb[Th(NCS)_5(H_2O)_3].[Th(NCS)_4(Ph_3PO)_4]$ and $[Th(NCS)_4(CH_3CONMe_2)_4]$ can be obtained.

Thorium forms a number of complexes with anionic chelate ligands. The acetylacetonate $[Th(acac)_4]$, m.p. 171°, has an eight-co-ordinated D_2 square antiprismatic co-ordination polyhedron and is quite volatile and soluble in benzene. A wide variety of similar thorium β-diketonates has been prepared. They form adducts with other ligands; for example, the tropolonate forms a nine-co-ordinated dimethylformamide adduct $[Th\ trop_4(HCONMe_2)]$ (trop = $\bar{C}H:CH.CO.\bar{C}H.CO.CH:\bar{C}H$). In

$K_4Th(C_2O_4)_4.(H_2O)_4$ oxalate anions are quadridentate giving a polymeric structure in which thorium ions are ten-co-ordinated in a bicapped square antiprismatic arrangement. Thorium is twelve-co-ordinate in $[Th(NO_3)_6]^{2-}$.

When in the form of a chelate anionic ligand, sulphur is capable of forming complexes with lanthanides and actinides although it does not otherwise do so. In the case of thorium a diethyldithiocarbamate $[Th(S_2CNEt_2)_4]$, with distorted dodecahedral eight-co-ordination, is formed.

11.6.3 Lower-valent Thorium Compounds

These are limited to the di-iodide and tri-iodide. Both of these compounds are prepared in a similar manner.

$$3ThI_4 + Th \xrightarrow[\text{Ta container}]{600-700^\circ} 4ThI_3$$

$$ThI_4 + Th \xrightarrow[\text{Ta container}]{700-850^\circ} 2ThI_2$$

ThI_3 is black while ThI_2 is golden. Both evolve hydrogen on treatment with water. ThI_2 has a polymeric structure in which the Th is six-co-ordinate, both trigonal prismatic and trigonal antiprismatic co-ordination polyhedra being present. Since the Th–I distances are 322 and 320 pm (320 pm in ThI_4) the thorium is presumably Th^{4+}. This idea is supported by the diamagnetism and high electrical conductivity of the solid. The solid should thus be regarded as a 'metal' made up of nTh^{4+} and $2nI^-$ ions with $2n$ electrons in conduction bands formed by overlap of the Th^{4+} and I^- orbitals, perhaps 5f and 5d orbitals, respectively. ThI_3 is probably similar.

11.7 Protactinium

Protactinium is a metal that is difficult to study. ^{231}Pa has a half-life of 32 340 years and hence must be handled in a glove-box. Furthermore, the solution chemistry is complicated by the ready hydrolysis of the most stable oxidation state, Pa(V). The metal, obtained by barium reduction of the tetrafluoride at 1300–1400°, has m.p. 1565° and density 15.4 g cm^{-3}; it adopts a tetragonal structure. The silvery metal rapidly forms a dark tarnish.

The most stable oxidation state is Pa(V) but Pa(IV) is readily available, for example by zinc and acid reduction, so that a wide variety of Pa(IV) compounds have been made. In aqueous solution, the standard potentials are

$$Pa \xrightarrow{0.9 \text{ V}} Pa^{4+}(aq) \xrightarrow{0.1 \text{ V}} Pa^{5+}(aq)$$

11.7.1 Halides and Oxyhalides

In the solid state all the halides PaX_5 and PaX_4 (X = F, Cl, Br, I) are known, $PaCl_5$ having been prepared by von Grosse in 1934.

The brown tetrafluoride is conveniently made from the pentoxide

$$Pa_2O_5 + H_2(\text{excess}) + 8HF \xrightarrow{600^\circ} 2PaF_4 + 5H_2O$$

The pentafluoride can be obtained by fluorination of PaF_4 at 700° and is a colourless, relatively involatile, solid.

The pentachloride is a fairly volatile yellow solid (sublimes at 200° in vacuo), which is prepared by the action of thionyl chloride vapour at 350–500° on the dried hydrous oxide. Its structure presents an example of the rather unusual seven-coordination, comprising pentagonal bipyramids with chains of $PaCl_2Pa$ bridges (see figure 11.9). It is reduced to the tetrachloride, which has an eight-co-ordinate structure similar to UCl_4, by hydrogen at 400°.

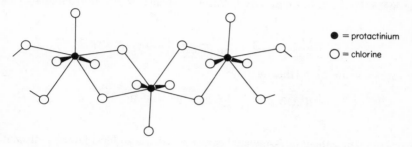

● = protactinium

○ = chlorine

Figure 11.9 The chain structure adopted by $PaCl_5$

The bromides may be prepared as follows

$$Pa_2O_5 \xrightarrow{\text{C+Br}_2 \text{ at } 700°} PaBr_5$$

$$PaBr_5 \xrightarrow{\text{Al at } 400°} PaBr_4$$

The dark red pentabromide is volatile in a vacuum at 300° and exists in two crystalline modifications. Of these, the β-form has the dimeric six-co-ordinate UCl_5 structure. It may be prepared from the pentoxide by heating with aluminium bromide at 400° in a vacuum. Reduction with aluminium powder at 400° gives the tetrabromide, which has the eight-co-ordinate UCl_4 structure (Pa–Br; 4 x 283 pm and 4 x 301 pm).

The black pentaiodide, obtained in a similar way to the bromide, or by action of SiI_4 on $PaBr_5$ at 180° in vacuum, is volatile at 400° in vacuum. It is reduced to PaI_4 by aluminium at 400°. The dark green PaI_4 is converted into a black substance believed to be the tri-iodide after one week in vacuum at 350°.

The protactinium penta- and tetrahalides are readily attacked by water to give hydrates and oxohalides; the ease of hydrolysis increases in the order F ⟨ Cl ⟨ Br ⟨ I.

Protactinium forms many oxohalides. In this it resembles both its d-transition analogues niobium and tantalum and its f-transition neighbour uranium. Two examples follow. Protactinium(IV) oxydichloride $PaOCl_2$ is made by the action of antimony trioxide on $PaCl_4$ in a vacuum at 400°. It has a polymeric structure involving three crystallographically distinct Pa^{4+} ions, respectively co-ordinated to $3\,O + 5\,Cl$, $4\,O + 5\,Cl$ and $3\,O + 4\,Cl$. Pa—Cl varies from 274–308 pm and Pa—O from 219–238 pm. There are no discrete PaO^{2+} groups; there is evidence for protactinium bonding as $Pa=O^{3+}$ to a terminal oxygen atom in only very few instances; in this it shows a difference from uranium. $PaOCl_2$ disproportionates on heating, as do other actinide oxydihalides.

$$2PaOCl_2 \xrightarrow[\text{vacuum}]{550^\circ} PaCl_4 + PaO_2$$

Protactinium(V) oxytribromide may be made by displacement of bromine by oxygen.

$$PaBr_5 + \tfrac{1}{2}O_2 \xrightarrow[\text{sealed tube}]{350^\circ} PaOBr_3 + Br_2$$

It disproportionates on further heating.

$$2PaOBr_3 \xrightarrow[\text{vacuum}]{500^\circ} PaO_2Br + PaBr_5$$

Its structure, determined by X-ray analysis, is polymeric and has seven-co-ordinate Pa^{5+} (to $3\,O + 4\,Br$). The co-ordination polyhedron approximates to a pentagonal bipyramid with Pa—O = 206–214 pm and Pa—Br = 269–302 pm.

11.7.2 Halide Complexes
The rather large, highly charged protactinium ions ($Pa^{5+} \approx 70$ pm; $Pa^{4+} \approx 85$ pm) very readily form halide complexes with co-ordination numbers of up to 9. Complex ions of stoichiometry PaF_8^{3-}, PaF_7^{2-}, PaF_6^{-}, $PaCl_8^{3-}$ and PaX_6^{-} ($X = Cl, Br, I$) are known for Pa(V), while Pa(IV) has PaF_8^{4-}, PaF_7^{3-}, PaF_5^{-} and PaX_6^{2-} ($X = Cl$, Br, I). Although structural data are limited, the stoichiometries probably reflect a fall in co-ordination number as the ionic size of the halide increases.

The complex fluorides are noteworthy for being stable in aqueous solution. All three species may be precipitated from aqueous hydrofluoric acid, the product depending on the detailed conditions, for example

$$2KF + PaF_5 \xrightarrow[\text{add acetone}]{17M\ HF} K_2PaF_7$$

They are all air-stable colourless solids. X-ray structural determinations show that Pa^{5+} is eight-co-ordinate in $RbPaF_6$ and in Na_3PaF_8. The co-ordination polyhedron in the latter compound is remarkable in being a nearly perfect cube; this polyhedron is rare, on account of the comparatively large ligand–ligand repulsions it necessitates. K_2PaF_7 has nine-co-ordinated Pa^{5+}.

The chloro, bromo and iodo-complexes are not made in aqueous solution owing to hydrolysis. The chloro-complexes are obtained from thionyl chloride solution

$$PaCl_5 + NMe_4Cl \xrightarrow{SOCl_2} (NMe_4)PaCl_6$$

$$PaCl_5 + 3NMe_4Cl \xrightarrow{SOCl_2} (NMe_4)_3PaCl_8$$

The chloro-complexes of Pa^{4+} and the bromo and iodo-complexes of both oxidation states may be obtained from acetonitrile solution (the Pa^{4+} complexes must be protected from aerial oxidation).

11.7.3 Complexes with Uncharged Ligands

Protactinium forms complexes with nitrogen and oxygen donors which are well known for their formation of stable complexes with other actinides in +4 or +5 oxidation states, for example $(CH_3)_2N.CO.CH_3$, CH_3CN, $(CH_3)_2SO$ and Ph_3PO. These adducts may usually be prepared by mixing the components in polar organic solvents such as methylene dichloride, acetone, or acetonitrile itself. Stoichiometries that have been obtained are $PaX_5(OPPh_3)_n$ ($X = Cl, Br; n = 1, 2$), $PaOCl_3(OPPh_3)$, $PaCl_3(OEt)_2(OPPh_3)$, $PaCl_4$ dma_3, $PaBr_4$ dma_5, $PaCl_4(OPPh_3)_2$, $PaBr_5(CH_3CN)_3$, $PaCl_4(CH_3CN)_4$, $PaCl_4(Me_2SO)_5$ and $PaCl_4(Me_2SO)_3$.

Little structural information is available concerning these complexes, but some points of interest arise from the infrared spectra. The P=O stretching frequency in the $Ph_3PO–Pa(V)$ complexes is reduced from its position in the free ligand (1195 cm^{-1}) to between 1075 and 960 cm^{-1}. This very large reduction probably corresponds to withdrawal of the P=O π-bonding electrons towards the formally pentapositive cation and overrides the increase in $v(P=O)$ expected from kinematic coupling with the Pa–O frequency. A similar shift of about 100 cm^{-1} is observed in the S=O frequencies of $Pa(IV) – Me_2SO$ complexes. In the complexes $PaOCl_3(OPPh_3)_2$ and $(NEt_4)_2PaOCl_5$ a band at 830 cm^{-1} is assigned to a Pa=O stretching mode; these compounds would be an example of PaO^{3+}, analogous to TaO^{3+}.

11.7.4 Solution Chemistry and Oxy-salts

In solution Pa(V) is extensively hydrolysed, a hydrous pentoxide being precipitated from solutions not containing complexing agents of which the most effective is fluoride ion. Very dilute solutions (10^{-5} M) in aqueous HNO_3 may also be prepared, but in fuming nitric acid 0.5 M solutions are obtained, which on evaporation give hydrated $PaO(NO_3)_3$. Hexanitrato-complexes, for example $CsPa(NO_3)_6$, are obtained by treating the corresponding $PaCl_6^-$ compound with liquid nitrogen pentoxide. From sulphuric acid solution, a sulphate $PaO(SO_4H)_3$ is obtained.

11.8 Uranium

This widely studied metal occurs in the oxidation states III, IV, V and VI. The general nature of its chemistry is determined by two factors. The first, of course, is the relative magnitudes of the standard oxidation potentials in aqueous solution and the second is the peculiar stability of the uranyl ion UO_2^{2+}

$$U \xrightarrow{1.80 \text{ V}} U^{3+} \xrightarrow{0.631 \text{ V}} U^{4+} \xrightarrow{-0.58 \text{ V}} UO_2^+ \xrightarrow{-0.063 \text{ V}} UO_2^{2+}$$

These standard potentials (1M $HClO_4$) indicate that uranium is a fairly electropositive metal with oxidation states of +4 and +6 which are stable in water. The +3 state is a highly reducing species which will rapidly reduce water to hydrogen; the

+5 state is liable to rapid disproportionation into an equimolar mixture of +4 and +6 because the former is much more stable than U(V) and the latter is about equally as stable as U(V). For nonaqueous systems similar results are obtained, although U(III) compounds are often perfectly stable in the absence of air and U(V) compounds such as the alkoxides are known that are not subject to disproportionation.

Uranium metal is very dense (19.0 g cm^{-3}) and is fairly reactive. It forms a non-protective tarnish in air and is attacked slowly by hot water. It dissolves easily in hydrochloric or nitric acids, but more slowly in sulphuric. Physically, it consists of two isotopes, mainly ^{238}U, with 0.72 per cent ^{235}U and traces of ^{234}U formed by α, β, β disintegration of ^{238}U.

11.8.1 Uranium(VI)

The most stable state of uranium is U(VI) as the uranyl ion $UO_2{}^{2+}$. A compound of this ion is the end product of the exposure to moist air of any other type of uranium compound and is the form in which uranium compounds are most often encountered, for example as uranyl nitrate $UO_2(NO_3)_2(H_2O)_2$. The $UO_2{}^{2+}$ ion is linear or very nearly so. There is probably considerable covalent character in the U—O bonds. The presence of this is supported by the bond length of about 175–190 pm. Thus, among many possible examples, in uranyl fluoride UO_2F_2 each U is co-ordinated to 2 O each at 191 pm and to 6F at 250 pm (ionic radii: $O^{2-} = 140$ pm; $F^- = 136$ pm), while in $BaUO_4$ there are 2 O at 189, 2 O at 219 and 2 O at 221 pm. Simple ionic bonding cannot easily explain why in the vast majority of (but not all) compounds of this type there are two short U—O links, shorter than the sum of the ionic radii of U^{6+} and O^{2-} as obtained from the observed uranium–ligand distances other than those in the UO_2 group. Thus the mixed oxide just mentioned can be written $Ba(UO_2)O_2$ just as α-UO_3, with a 2+6 co-ordination similar to UO_2F_2, can be considered to be $(UO_2)O$. It is an orange substance and may be prepared by heating the hydrated peroxide (itself prepared from aqueous solution) to $400°$. Four other modifications of UO_3 are known, one of which, δ-UO_3, has the ReO_3 structure with strictly octahedral co-ordination. Apart from σ covalent character in the $UO_2{}^{2+}$ entity there is very probably a large π contribution involving oxygen p orbitals and a combination of uranium 5f and 6d orbitals. Once the pair of strong collinear U—O bonds have been formed, formation of any further similarly short bonds to oxygen in an octahedral situation would be hindered by UO_2–O repulsions, $O_{(UO_2)}$—O then being less than the sum of the van der Waals radii (see figure 11.10).

Table 11.2 Number of ligand atoms in equatorial plane

Four	Five	Six
β-$SrUO_4$	$(NH_4)_3UO_2F_5$	$UO_2(NO_3)_2(H_2O)_2$
$MgUO_4$	$UO_2(S_2CNR_2)_2(Ph_3PO)$	$Na[UO_2(CH_3COO)_3]$
$CoUO_4$	$(NH_4)_2(UO_2)_2(C_2O_4)_3$	$UO_2(NO_3)_2(OPPh_3)_2$
$Cs_2UO_2Cl_4$		$Na_4[UO_2(O_2)_3]$
		$(Me_4N)[UO_2(S_2CNEt_2)_3]$

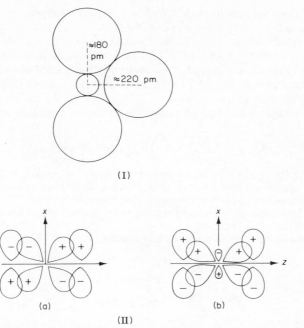

(I)

(II)

Figure 11.10 (I) shows the inability of a third O^{2-} ion to approach a $[UO_2]^{2+}$ group closely; (II) orbital overlaps that may be invoked to explain π bonding in uranyl complexes: (a) d_{xz}–p_x, (b) f_{xz^2}–p_x

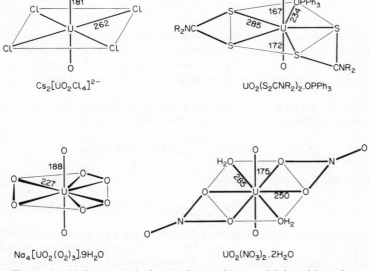

Figure 11.11 Some typical uranyl complexes, with bond lengths

Uranyl compounds may show 2+4, 2+5 or 2+6 co-ordination, with the set of 4, 5 or 6 ligands in or near a plane normal to the $O-U-O$ axis. These equatorial ligands may be uncharged, such as H_2O, Ph_3PO, Ph_3AsO, $(BuO)_3PO$ or C_5H_5N or anionic, such as NO_3^-, CH_3COO^-, F^-, Cl^- or O^{2-} (nitrate and acetate co-ordinate in a bidentate manner). Thus the co-ordination geometry is either octahedral (with one short axis), or pentagonal bipyramidal, or most commonly hexagonal bipyramidal (see figure 11.11 and table 11.2).

One novel compound worthy of mention is the peroxycomplex, prepared as follows

$$UO_4^{2-} + H_2O_2 \text{ (excess)} \xrightarrow[NaOH]{} Na_4[UO_2(O_2)_3]$$

It is of interest that n.m.r. studies of uranyl salts in acetone–water at low temperatures imply a maximum hydration number of four. Perchlorate ions do not compete for co-ordination but nitrate, chloride and bromide ions do. This conclusion is, of course, in accord with the solid-state structures determined for uranyl complexes.

The uranyl halides may be prepared in a number of ways; for example

$$UCl_4 \xrightarrow[350°]{O_2} UO_2Cl_2 + Cl_2$$

$$UBr_4 \xrightarrow[150°]{O_2} UO_2Br_2 + Br_2$$

$$UO_2Cl_2 \xrightarrow{HF \text{ (liq)}} UO_2F_2 + HCl$$

Uranyl chloride is a bright yellow deliquescent solid, m.p. 578°. It decomposes on heating

$$UO_2Cl_2 \xrightarrow{300°, \text{ vacuum}} UO_2 + Cl_2$$

Uranyl chloride has an interesting pentagonal bipyramidal structure in which linear UO_2 ($U-O = 173, 178$ pm) is co-ordinated with a pentagonal array of four Cl (273, 275 pm) and one O (252 pm). The Cl atoms are all linked to other U atoms, forming chains. The unique O atom forms part of a UO_2 group belonging to another chain.

In addition to the anhydrous material, a number of hydrates, for example $UO_2Cl_2.3H_2O$ are known. The compound will add chloride ion

$$UO_2Cl_2 + 2Me_4NCl \xrightarrow{4M \text{ HCl}} (Me_4N)_2UO_2Cl_4$$

A number of other ligands are also added

$$UO_2Cl_2 + 2L \rightarrow UO_2Cl_2L_2 \quad (L = Ph_3PO, Ph_3AsO, Ph_2SO, py)$$

Uranyl fluoride has a sheet structure incorporating a puckered hexagonal co-ordination of fluoride ions with uranyl ions, each fluoride being co-ordinated to three uranium atoms (see figure 11.12). It forms di- and trihydrates and adducts with ammonia. It can be reduced to UO_2 by hydrogen at 450°.

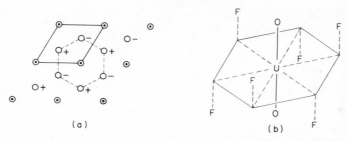

Figure 11.12 (a) The unit cell of UO_2F_2 in projection (after W.J. Zachariasen, *Acta crystallogr.*, **1** (1948), 279) ⊙ represents a UO_2 group perpendicular to the plane of the paper; ○ denotes a fluorine atom (+,− imply above/below the plane of the paper by 61 pm; (b) a view of the UO_2F_6 polyhedron showing the puckered in-plane co-ordination

The U−O infrared stretching frequency in uranyl compounds occurs at around 920–980 cm^{-1} and the O−U−O bending mode at around 265 cm^{-1}. The former value is similar to that for the 'osmyl' group $OsO_2{}^{2+}$. Many uranyl compounds show a bright green fluorescence when irradiated with near-ultraviolet light, and vibrational fine structure arising from O−U−O stretching vibrations is superposed on the emitted radiation.

Uranium (VI) also forms a number of compounds without the $UO_2{}^{2+}$ structure. Among these are the hexahalides UF_6 and UCl_6. The former can be made by the action of fluorine on almost any simple uranium compound but the following conditions are of preparative utility

$$UF_4 + F_2 \xrightarrow{220°} UF_6$$

$$UO_2F_2 + 2F_2 \xrightarrow{270°} UF_6 + O_2$$

An interesting reaction that uses oxygen as the oxidant is

$$2UF_4 + O_2 \xrightarrow{800°} UF_6 + UO_2F_2$$

The hexafluoride has been extensively investigated because of its importance in the separation of ^{235}U from ^{238}U by gaseous thermal diffusion. It is a colourless solid, m.p. 64° (under pressure), which sublimes at 57°. It is instantly hydrolysed by moisture. In the vapour and liquid states the molecule is octahedral as indicated by vibrational spectra and electron diffraction. The U−F distance is 199.4 pm, and the symmetrical stretching vibration (A_{1g}, Raman) is at 667 cm^{-1}. X-ray analysis shows the solid to be distorted from octahedral symmetry (F−Ū−F = 86.5°–92.9°; U−F = 188–228 pm).

The hexafluoride is converted into UO_2F_2 or UOF_4 by hydrolysis. It can be reduced to UF_5 by HBr at 80°; to $NO^+UF_6{}^-$ by nitric oxide; and to UF_4 by PF_3. However, $AlCl_3$ converts it into the hexachloride UCl_6, a black-green solid m.p. 177°, which is slightly volatile, soluble in carbon tetrachloride and hydrolysed by water to uranyl chloride.

The hexafluoride forms complexes with fluoride ion, when treated with sodium fluoride in perfluoroheptane. Both Na_2UF_8 and $NaUF_7$ may be obtained.

11.8.2 Uranium (V)

This ion almost certainly exists in aqueous solution as the species UO_2^+, similar to the uranyl ion. The reversibility of the polarographic oxidation to UO_2^{2+} is evidence that the stoichiometry of the two species is identical and further suggestive evidence is provided by the chemistry of neptunium and plutonium where the +5 oxidation state is more stable and the species observed in the solid state by X-ray crystallography are NpO_2^+ and PuO_2^+. Millimolar solutions of UO_2^+ ion can be obtained by electrolytic reduction of UO_2^{2+} and are reasonably stable to disproportionation within the pH range 2–4. At equilibrium, however, disproportionation is virtually complete

$$2UO_2^+ + 4H_3O^+ \rightarrow UO_2^{2+} + U^{4+} + 6H_2O; \quad K = 1.7 \times 10^6$$

There is no proven example of the UO_2^+ ion in the solid state.

Uranium pentoxide U_2O_5 may be obtained from a sulphuric acid solution of UO_3. It is oxidised by air. A variety of fairly stable U(V) mixed oxides may be obtained by heating a mixture of a U(IV) mixed oxide with UO_3, or a U(VI) oxide with UO_2

$$Na_2UO_4 + UO_2 \xrightarrow{700^\circ} 2NaUO_3$$

$$BaUO_3 + UO_3 \xrightarrow{550^\circ} Ba(UO_3)_2$$

A wide variety of halides and halide complexes of U(V) have been isolated. Thus all MX_5 and MX_6^- are known (X = F, Cl, Br, I) except MI_5. Some preparative methods are given below

$$UF_6 + HBr \xrightarrow{65^\circ} UF_5 + HF + \tfrac{1}{2}Br_2$$

$$UO_3 \xrightarrow{SiCl_4, 500^\circ} UCl_5$$

$$UO_3 + 3COBr_2 \xrightarrow{120^\circ} UBr_5 + 3CO_2 + \tfrac{1}{2}Br_2$$

$$(Ph_4As)_2UCl_6 + \tfrac{1}{2}Cl_2 \xrightarrow[80-90^\circ]{CH_3NO_2} (Ph_4As)UCl_6(yellow) + Ph_4AsCl$$

$$UF_5 + CsF \xrightarrow{aq.HF} CsUF_6 \text{ (blue)}$$

Uranium pentafluoride has two crystal modifications, but the structures of both the α and β form are unproven. The red-brown pentachloride is dimeric both in the solid state, where two UCl_6 octahedra share an edge, and in carbon tetrachloride solution. The pentabromide is again dimorphic, the β form having a dimeric structure similar to UCl_5.

All the pentahalides are somewhat unstable, either to disproportionation or to dissociation.

$$2UF_5 + 2H_2O \rightarrow UF_4 + UO_2F_2 + 4HF$$

$$2UCl_5 \xrightarrow{\text{R.T.}} UCl_4 + UCl_6$$

$$2UCl_5 \xrightarrow{\text{vacuum}} 2UCl_4 + Cl_2$$

The oxohalides also exist

$$Et_4N^+UX_6^- \xrightarrow[Et_4N^+]{Me_2CO} (Et_4N)_2(UOX_5) \qquad\qquad (X = F, Cl, Br)$$

These are coloured (X = F, pink; X = Cl, blue; X = Br, green). The phosphine oxide complexes of UCl_5 and UBr_5 have been made

$$UX_6^- + Ph_3PO \rightarrow UX_5(Ph_3PO) + X^- \qquad (X = Cl, Br)$$

while adducts UCl_5L (L = Ph_3P, bipy, pyrazine) and UCl_5L_2 (L = phen, py) have also been prepared, as has a novel compound $UF_5.XeF_6$, of as yet unknown structure. There is little structural information on any U(V) complexes but they are probably octahedral.

The largest series of uranium(V) compounds consists, a little surprisingly, of the alkoxides $U(OR)_5$. These are thermally stable and are also stable with respect to disproportionation into U(IV) and U(VI) but like all metal alkoxides are rapidly hydrolysed by moisture. They may be prepared by several routes, for example

$$UCl_5 + 5ROH + 5NH_3 \rightarrow U(OR)_5 + 5NH_4Cl$$

(a) $$UCl_4 + 4EtOH + 4NH_3 \rightarrow U(OEt)_4 + 4NH_4Cl$$

(b) $$U(OEt)_4 \xrightarrow{Br_2} UBr(OEt)_4 \xrightarrow{NaOEt} U(OEt)_5$$

$$U(OEt)_5 + 5ROH \rightarrow U(OR)_5 + 5EtOH$$

In the last reaction ROH is less volatile than EtOH, which is removed by distillation, thus driving the reaction to completion. These compounds tend to polymerise through

$$\begin{array}{c} R \\ \downarrow \\ U - O \rightarrow U \end{array}$$

bridging, the degree of polymerisation depending on the bulk of the group R. Thus the average molecularity of $U(OMe)_5$ in benzene is 3.01; for $U(OEt)_5$ the number is 1.90; $U(OCEt_3)_5$ is monomeric. The *n*-butyl compound is a brown liquid, b.p. 193° at 0.1 Pa (10^{-3} mm Hg). Adducts such as $U(OR)_5py$, salts $NaU(OR)_6$ and mixed compounds such as $UCl(OR)_4$ and $U(OR)_4(acac)$ are also known.

11.8.3 Uranium(IV)

The apple-green hydrated ion may be obtained by reduction of uranyl ion using zinc and sulphuric acid. It is stable in the absence of air but is slowly reoxidised on standing in air. There is some tendency to hydrolysis, but this is very slight in acid solution.

The dioxide UO_2 has the eight-co-ordinate fluorite structure and is prepared by hydrogen reduction of UO_3 or U_3O_8 at $300-600°$. It reoxidises in air, especially if finely divided. The uranium–oxygen system is very complex and has been exten sively studied. Besides the stoichiometric oxides UO (a semimetallic interstitial compound), UO_2 and UO_3, there are at least four nonstoichiometric phases. These are U_3O_8 (the stable oxide obtained by heating, for example, uranyl nitrate in air), U_4O_9 and two tetragonal phases of slightly variable composition within the range $UO_{2.25} - UO_{2.40}$. The U_4O_9 structure is related to that of UO_2 by inclusion of in-terstitial oxygen atoms.

The uranium(IV) halides UX_4 X = F, Cl, Br, I) are all known. Uranium tetra-fluoride is a green solid, m.p. $1036°$, which is stable in air and insoluble in water. In the crystal structure U is co-ordinated to eight F forming an irregular square anti-prism with U–F = 225 to 232 pm. It may be made by the action of excess gaseous HF on heated UO_2.

Uranium tetrachloride may be made in several ways, but the most convenient is by heating U_3O_8 with excess liquid hexachloropropene, the product being purified by sublimation. It is a dark green solid, m.p. $590°$, which is deliquescent and is eventually oxidised by air to UO_2Cl_2. It is soluble in water or acetone. The crys-tal structure contains U co-ordinated in a regular dodecahedron to eight Cl at 286.9 and 263.8 pm (neutron-diffraction determination).

Brown uranium tetrabromide, m.p. $519°$, is a deliquescent solid soluble in water or acetone. It is made by the action of a Br_2/He mixture on the metal at $650°$. In the crystal, uranium has pentagonal byramidal co-ordination to 7Br. The black tetraiodide slowly loses iodine at room temperature. It may be made by heating the metal with iodine vapour at $500°$ and 20 kPa (150 mm Hg), the UI_4 being conden-sed out at $300°$.

The uranium tetrahalides, and uranium(IV) compounds in general, form a variety of complexes that show the usual actinide attributes of high co-ordination number and variable geometry. UCl_4 and UBr_4 form complexes with R_3PO, Me_2SO, CH_3CONMe_2(dma) and pyridine. The stoichiometries vary between UX_4L and $UX_4.7L$ and structural information is limited. Some specific examples are UX_4(dma)$_{2.5}$ (X = Cl, Br); UBr_4(dma)$_5$; UCl_4(Me$_2$SO)$_x$ (x = 3, 5, 7); UX_4(Ph$_3$PO)$_2$ (X = Cl, Br) and UCl_4py$_2$. Where seven donor molecules are added as in UCl_4(Me$_2$SO)$_7$, a number of these (probably two) are almost certainly not co-ordinated to uranium but are held in the crystal structure by weak dipolar or van der Waals forces. The evidence for this comes from two infrared bands at 1047 and 942 cm^{-1}, the former almost exactly corresponding with the value for the free ligand (1045 cm^{-1}) while the latter shows a reduction of 103 cm^{-1} characteristic of R_2SO (and R_3PO) complexes, which possibly arises from polarisation of the d-p π-bonding electrons towards oxygen, thus weakening the S–O π bond. Besides these adducts of UCl_4 and UBr_4, similar thiocyanate adducts such as [U(NCS)$_4$(Ph$_3$PO)$_4$] and [U(NCS)$_4${OP(NMe$_2$)$_3$}$_4$] have been prepared. One compound studied by diffrac-tion methods is UCl_4{OP(NMe$_2$)$_3$}$_2$, which has a *trans* octahedral structure. Anionic complexes UX_6^{2-} (X = F, Cl, Br, I), UF_7^{3-}, UF_8^{4-} and $U(NCS)_8^{4-}$ and also formed. These, like the adducts with uncharged ligands, are often prepared by mixing the components in nonaqueous solvents; for example

$$2Ph_4AsI + UI_4 \xrightarrow{CH_3CN} (Ph_4As)_2[UI_6]$$

Other methods are sometimes also applicable

$$2CsCl + UCl_4 \xrightarrow{aq.\ HCl} Cs_2UCl_6$$

$$4NH_4F + UF_4 \xrightarrow[\text{sealed tube}]{100^\circ} (NH_4)_4UF_8 \xrightarrow{heat} (NH_4)_2UF_6$$

$$8KNCS + UCl_4 \xrightarrow{acetone} K_4U(NCS)_8 + 4KCl$$

It is probable that co-ordination numbers of six and eight predominate in the complexes mentioned so far (in $(NEt)_4M(NCS)_8$ (M = Th, Pa, U, Np, Pu) the co-ordination is cubic (U–N = 238 pm) while Rb_2UF_6 involves dodecahedral eight-co-ordination). This is probably also true of complexes with chelate ligands. Uranium(IV) complexes with H_4(edta) to give $U(edta)(H_2O)_2$ and with a variety of β-diketones to give the tetrakis(β-diketonate), for example $[U(acac)_4]$. This latter compound is isomorphous with $[Th(acac)_4]$, having square antiprismatic co-ordination.

Examples of complexes with sulphur and phosphorus bound to uranium are rare; many 'phosphine complexes' are now known to be phosphine oxide complexes, due to oxidation of the ligand. However, there is evidence for $UCl_4(PPh_3)_2$ and $(UCl_4)_2$(diphos), while dithiocarbamate complexes have been prepared as follows

$$UCl_4 + 4LiS_2CNEt_2 \xrightarrow{EtOH} U(S_2CNEt_2)_4 + 4LiCl$$

These are almost certainly eight co-ordinate. The complexes UCl_4L_2 (L = bipy, phen), made by reaction between the components in dimethoxyethane, are probably eight co-ordinate also. Reaction of UCl_4 with nitrogen in the presence of silicon or aluminium affords UNCl, which has the PbClF structure.

$$UCl_4 + N_2 + Si\ (or\ Al) \rightarrow UNCl + SiCl_4\ (or\ AlCl_3)$$

11.8.4 Uranium (III)

The hydrated ion is very unstable to oxidation by air or water; thus

$$U^{3+} + H_2O \rightarrow U^{4+} + OH^- + \tfrac{1}{2}H_2$$

It is, however, reasonably stable (air being excluded) as the solid hydrated sulphate $U_2(SO_4)_3.xH_2O$ (x = 8, 5, 2), as double sulphates, and double chlorides $MUCl_4.xH_2O$ (M =Rb, NH$_4$; x = 5, 6; M = K, x = 5). Hydrates $UF_3.3H_2O$ and $UBr_3.6H_2O$ have also been reported.

A few complexes are known, such as $UCl_3.7NH_3$ and $UBr_3.6NH_3$, $U(MeCN)X_3$ (X = Cl, Br) and UCl_3(phenazone)$_6$, but work on uranium(III) complexes is hampered by the ready reduction of many of the ligands, such as Ph_3PO or bipyridyl, which would otherwise be expected to co-ordinate with this very strongly reducing species.

Uranium hydride UH_3 is a black powder obtained by the reaction of the metal with hydrogen at 300°. It tends to be slightly hydrogen-deficient unless special precautions are taken in the preparation to avoid this. Its reaction with gaseous HCl affords a convenient method of preparation of UCl_3; some other reactions are also of interest

$$UH_3 + 3HCl \rightarrow UCl_3 + 3H_2$$

$$UH_3 + NH_3 \rightarrow UN + 3H_2$$

$$2UH_3 + 4H_2O \rightarrow 2UO_2 + 5H_2$$

$$2UH_3 + 4H_2S \rightarrow 2US_2 + 5H_2$$

11.9 Neptunium, Plutonium and Americium

These three elements will be dealt with together. Except for the important difference between the stabilities of their oxidation states, in the sense that lower oxidation states become more stable in the sequence Np, Pu, Am, the metals form very similar compounds. Their compounds are also very similar to those of uranium, which has, however, been considered separately because more is known about it than about the three succeeding elements, on account of their radioactivity and comparatively limited availability.

The aqueous chemistry is summarised by the list of oxidation potentials (volts) in 1M acid (except the VI–VII potential which is in 1M alkali).

$$Np \xrightarrow{1.83} Np^{3+} \xrightarrow{-0.155} Np^{4+} \xrightarrow{-0.739} NpO^{2+} \xrightarrow{-1.137} NpO_2^{2+};$$
$$NpO_4^{2-} \xrightarrow{-0.582} NpO_5^{3-}$$

$$Pu \xrightarrow{2.03} Pu^{3+} \xrightarrow{-0.98} Pu^{4+} \xrightarrow{-1.17} PuO_2^{+} \xrightarrow{-0.91} PuO_2^{2+}; PuO_4^{2-} \rightarrow PuO_5^{3-}$$

$$Am \xrightarrow{2.36} Am^{3+} \xrightarrow{\approx -2.8} Am^{4+} \xrightarrow{\approx -0.7} AmO_2^{+} \xrightarrow{-1.60} AmO_2^{2+}$$

The metals are clearly quite electropositive; compare the potentials $Mg-Mg^{2+}$(aq), 2.37 V and $Al-Al^{3+}$(aq), 1.66 V. Thus metallic plutonium reacts slowly with water but easily with dilute or concentrated hydrochloric acid. It is passive in dilute or concentrated nitric acid and dissolves slowly in dilute sulphuric acid. Curiously, the $M-M^{3+}$(aq) potentials increase in the order U⟨Np⟨Pu⟨Am. This increase of electropositive character along part of the actinide series is without parallel in the lanthanide series, where the standard oxidation potentials for $M-M^{3+}$aq. all fall with increasing atomic number. The lower oxidation states become more stable along the series U→Am; thus U^{3+} reduces water while Am^{3+} can only be oxidised with difficulty. Neptunium is the first actinide with seven outer electrons and can be oxidised to Np(VII). Plutonium(VII) is attainable but less stable.

11.9.1 Oxidation States
Plutonium is a unique element in that four of its oxidation states, III, IV, V and VI, are each separated from the adjacent state by oxidation potentials of about equal

value, namely about 1 volt. This means that if ions of, say, Pu^{4+} are introduced into water, the simultaneous formal processes

$$Pu^{4+} + e \rightarrow Pu^{3+}$$

$$Pu^{4+} + 2H_2O \rightarrow PuO_2^+ + 4H^+ + e$$

will take place with little overall free-energy change and the final result will be the simultaneous presence of Pu^{3+}, Pu^{4+} and PuO_2^+ with the latter species in turn producing Pu^{4+} and PuO_2^{2+} (see figure 11.13). A further complication is the slower re-

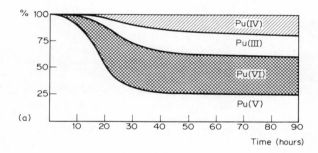

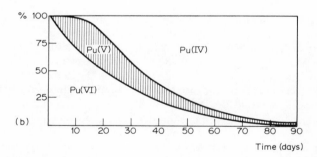

Figure 11.13 (a) Disproportionation of Pu(V) in 0.1N HNO_3 + 0.2N $NaNO_3$; (b) Self-reduction of Pu(VI) due to its α-emission, in 5N $NaNO_3$ (after P.I. Artiukhin, V.I. Medcedovskii and A.D. Gel'man, *Radiokhimiya*, **1** (1959), 131 and *Zh. neorg. Khim.*, **4** (1959), 1324)

duction of PuO_2^{2+} and PuO_2^+ to Pu^{4+} by its own energetic α radiation, which produces reducing species from interaction with the aqueous medium. Specific oxidations or reductions of the various Pu species can, however, be carried out by ceric ion or iodide ion respectively, or other reagents.

The behaviour of aqueous solutions of neptunium or americium is simpler. Thus Np^{3+}, while it may be obtained by dissolution of the trichloride, is oxidised by air to Np^{4+} but is stable under nitrogen. Np^{4+} can be oxidised to NpO_2^+ by mild reagents and the latter is normally stable against disproportionation owing to the comparatively difficult oxidation to NpO_2^{2+}, but in acid solutions it disproportionates, as the step

$$NpO_2^+ + 4H^+ + e \rightarrow Np^{4+} + 2H_2O$$

will then be favoured. NpO_2^{2+} is similar to UO_2^{2+} but is very easily reduced. Under alkaline conditions a Np(VII) species is produced in solution by electrolytic oxidation

$$NpO_4^{2-} + 2OH^- \xrightarrow{1M\ OH^-} NpO_5^{3-} + H_2O + e;\ E_o = -0.582\ V$$

The exact nature of the aqueous species involved is uncertain; however, hydrated salts of NpO_5^{3-} may be precipitated by addition of large cations such as Sr^{2+}, Ba^{2+} or $[Co(NH_3)_6]^{3+}$. Np(VII) is stable at pH 7. A corresponding but less stable species of Pu may be made similarly; Pu(VII) in solution at pH 7 disappears in three minutes.

Americium(III) is the most stable species of that metal and is oxidised with difficulty to Am^{4+}, which normally disproportionates rapidly into $Am^{3+} + AmO_2^+$. However, hypochlorite oxidation in alkaline solution followed by addition of fluoride ion gives a stable solution of Am(IV) from which $(NH_4)_4AmF_8$ can be crystallised. The oxidation $Am^{3+} \rightarrow AmO_2^{2+}$ can be effected by peroxydisulphate ion.

11.9.2 Oxides

For all three elements the dioxide, of fluorite structure, is the stable oxide, obtained by strong heating of hydroxides or nitrates in air. This is perhaps unexpected for Am, which has such a stable trivalent state in solution. None of the dioxides will react with further oxygen. An oxide Np_3O_8, isomorphous with U_3O_8, can, however, be obtained by heating $NpO_2(NO_3)_2$ in air to about 400°. Lower oxides Pu_2O_3 and Am_2O_3 are obtained from the dioxides by reduction with plutonium at 1500° and with hydrogen, respectively. Prepared at 800° the Am_2O_3 had the rare earth 'A' structure; the 'C' form was obtained at 600°.

Mixed oxides   The formation of mixed oxides presents a ready way of stabilising Np(VII) and Pu(VII)

$$5Li_2O + 2NpO_2 \xrightarrow[400°]{O_2} 2Li_5NpO_6$$

The reaction with PuO_2 proceeds similarly. These compounds contain isolated octahedral $[MO_6]^{5-}$ anions (but with the oxygen atoms also co-ordinated to Li^+ ions). The oxidation state of the metal has in the case of neptunium been unambiguously demonstrated to be VII by Mössbauer spectroscopy, an isomer shift of -68.4 mm s^{-1} being obtained (compare Np(VI), 48.4 mm s^{-1}). The neptunium compound is diamagnetic (f^0) while the plutonium analogue (f^1) has a low moment ($\mu = 0.34$ B.M.). Both compounds dissolve in water or alkali to give solutions of the heptavalent metal. There are a considerable number of other mixed oxides with the metals in lower oxidation states, but these will not be considered here except to say that they are usually made by direct combination of component oxides; examples are Na_6NpO_6, Na_6PuO_6, Li_6AmO_6 and Li_3MO_4 (M = U, Np, Pu, Am).

11.9.3 Halides

Black AmX_2 (X = Cl, Br, I) may be prepared from HgX_2 and the metal at 300–400°. The trihalides are all known and have been dealt with previously. The tetrafluorides are also all known; green NpF_4, brown PuF_4 and tan AmF_4 can be made respectively by the action of HF and oxygen on NpO_2 or on PuO_2 and by fluorination of AmO_2, all at 500°. They are all isostructural with ThF_4. Many of the other possible halides, however, are conspicious by their absence as the following list shows.

		$AmCl_2$
		$AmBr_2$
		AmI_2
All MX_3 known		
NpF_4	PuF_4[†]	AmF_4
$NpCl_4$	$PuCl_4$[‡]	
$NpBr_4$		
NpF_5		
NpF_6	PuF_6	

[†] Pu_4F_{17} also exists; [‡] Gas phase only.

This array is curious in some respects. For example, Np^{4+} is the most stable state of neptunium in aqueous media but gives no NpI_4 (compare Cu^{2+}), and while the V and VI states are represented only by fluorides, the higher states Np(VII) and Pu(VII) are attainable in water.

Neptunium tetrachloride, made by the action of CCl_4 vapour on the dioxide, is isostructural with UCl_4. It is readily reduced to $NpCl_3$ by hydrogen. Plutonium tetrachloride has been detected in gas-phase equilibrium at 928°.

$$2PuCl_3 + Cl_2 \rightleftharpoons 2PuCl_4$$

It has not been isolated. Neptunium tetrabromide is obtained by direct action of bromine on metallic neptunium at 425°; it too is isostructural with its uranium analogue.

Neptunium and plutonium hexafluorides, made by the action of fluorine on the tetrafluorides, are volatile solids, orange and red-brown respectively, m.p.s 55° and 52°. NpF_6 also boils at 55°, PuF_6 at 62°. They have octahedral monomeric molecules with Np–F = 198.1 pm and Pu–F = 197.1 pm (electron diffraction). Both the heats of formation and the symmetrical stretching frequency fall in the sequence $UF_6 \rightarrow NpF_6 \rightarrow PuF_6$, the values being 2140, 1940 and 1750 kJ $mole^{-1}$ (511, 463 and 418 kcal $mole^{-1}$) and 667, 654 and 628 cm^{-1}. This accords with the decreasing stability of the VI oxidation state along the sequence. Neptunium hexafluoride can be reduced by iodine in iodine pentafluoride at room temperature to the pentafluoride, which is isostructural with β-UF_5.

11.9.4 Complex Halides and Oxyhalides

A number of these compounds can be obtained. They are similar in structure to the corresponding uranium compounds. Thus the oxyfluorides MO_2F_2 (M = Np, Pu, Am) are all obtainable; for example, NpO_2F_2 results from the action of HF on NpO_3 at 300°. AmO_2F_2 (made from the action of HF and F_2 on $NaAmO_2(CH_3COO)_3$ at −190°) is known although AmO_3 and AmF_6 are not; doubtless the stability of the $MO_2{}^{2+}$ entity is an important factor here (all these oxyfluorides are isostructural with UO_2F_2).

Neither NpF_6 or PuF_6 give hexavalent complex fluorides on treatment with alkali fluorides. Lower oxidation state complex fluorides are formed instead.

$$NpF_6 \; + \; 3NaF \; \xrightarrow{\;200°\;} \; Na_3NpF_8 \; + \; \tfrac{1}{2}F_2$$

$$PuF_6 \; + \; 2CsF \; \xrightarrow{\;300°\;} \; F_2 \; + \; Cs_2PuF_6 \; \xrightarrow{\;F_2; \, \rangle \, 300°\;} \; CsPuF_6$$

Na_3NpF_8 is isostructural with Na_3PaF_8 and Na_3UF_8 and thus has cubic eight-coordination of Np by F. $CsPuF_6$ is isostructural with $CsUF_6$ and with $CsNpF_6$, which is obtained similarly. These have a nearly octahedral six-co-ordination of U, Np or Pu by F (see figure 11.14). Seven-co-ordination appears in the context of fluoride complexes of U, Np and Pu in the +5 oxidation state; thus Rb_2MF_7, which have an irregular trigonally capped square-based polyhedron, have been isolated by fluorination of a rubidium–actinide(IV) fluoride mixture.

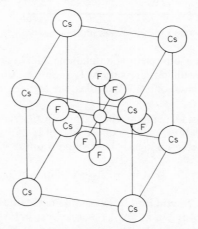

Figure 11.14 The unit cell of $CsMF_6$ (M = U, Np, Pu) (after M. Reisfield and C.A. Crosby, *Inorg. Chem.*, **4** (1965), 65)

A variety of fluoro-M(IV) complexes are available. Thus for U, Np, Pu and Am the stoichiometries $LiMF_5$, Na_2MF_6, $(NH_4)_4MF_8$ (nine-co-ordinate M), and the unusual $Na_7M_6F_{31}$ are known in all four cases. A great deal of careful structural work has been done on the actinide (IV) fluoro-complexes, of which there are literally dozens, and indeed on other actinide halide complexes also, but only a small selection can be noted here. A number of chloro-complexes such as Cs_2MCl_6 and $(NEt_4)_2MCl_6$ (M = U, Np, Pu; not Am) can be obtained from aqueous solution if the concentration of chloride ion is fairly high. They all have octahedral co-ordination of the actinide. Bromides $(NEt_4)_2MBr_6$ (M = U, Np, Pu) may be obtained from an ethanol–acetone–HBr solution. Complexes of the trivalent metals, which are poorer complex-formers than higher oxidation states, are fewer but chloro-complexes of Pu and Am are known, for example Cs_3PuCl_6, obtained from its components in 4M aqueous HCl and Cs_3AmCl_6 from ethanolic HCl.

11.9.5 Other Complexes of Neptunium, Plutonium and Americium
All these metals form analogues of the uranyl ion, namely the neptunyl, plutonyl and americyl ions, which form complex compounds similar to those of uranium. These have been comparatively little investigated. However, $NpO_2{}^{2+}$ and $PuO_2{}^{2+}$ are known to be extracted from aqueous nitrate solution by tributylphosphate as $MO_2(NO_3)_2\{OP(OBu)_3\}_2$ similarly to uranium. In the lower oxidation states, β-Np(acac)$_4$ has square antiprismatic D_{2d} co-ordination, as shown by an X-ray determination. [Pu(acac)$_4$] is a brown volatile solid and [Pu(oxine)$_4$] has also been prepared; with $PuO_2{}^{2+}$, 8-hydroxyquinoline gives [PuO$_2$(oxine)$_2$(H.oxine)] similarly to the uranium compound where the third oxine is present in the form of a zwitterion

co-ordinated by oxygen only, giving a pentagonal bipyramidal arrangement.

Several adducts of Np and Pu tetrahalides with uncharged donor molecules have been reported; they are generally rather similar to the corresponding Th, Pa and U compounds but there are some differences in stoichiometry. The complexes are obtained from nonaqueous solutions, for example, acetone or a solution of the donor itself. They include $MCl_4.2.5(MeCONMe_2)$, $MCl_4(Me_2SO)_7$ and $MCl_4(Me_2SO)_3$ (M = Np, Pu), $NpCl_4(Me_2SO)_5$, $MBr_4(OPPh_3)_2$ and $MBr_4\{OP(NMe_2)_3\}_2$ (M = Np, Pu) and $Np(NCS)_4(OPPh_3)_2$. $Pu(NO_3)_4(Ph_3AO)_2$ (A = P, As) are also known – the extraction of Np, Pu and Am depends on the formation of such complexes as these and $Am(NO_3)_3.3(R_3PO)$. Some alkoxides such as $Np(OR)_4$ exist: in contrast to uranium, $Np(OR)_5$ does not form by disproportionation, but in the presence of excess sodium ethoxide, the complex of pentavalent neptunium $NpBr(OEt)_4$ obtains. The plutonium compound $Pu(OCHMe_2)_4$ is also known.

Despite the relative lack of affinity of sulphur ligands for lanthanides and actinides, the chelate diethyldithiocarbamate anion does form complexes with trivalent and tetravalent actinides, namely $M(Et_2NCS_2)_4$, $M(Et_2NCS_2)_3$ and $[M(Et_2NCS_2)_4]^-$ (M = Np and Pu).

11.10 Curium and the Succeeding Actinides

These, in order of atomic number, are curium, berkelium, californium, einsteinium, fermium, mendelevium, nobelium and lawrencium. In this section of the actinide series the chemical behaviour reverts to that of the lanthanides. Thus the trivalent state is generally the most stable, with deviations that favour the stability of the +2 or +4 states when these would attain a filled or half-filled f-shell. Thus the ions

$Bk^{4+}(f^7)$ and $No^{2+}(f^{14})$ would be expected to be stable, as to a lesser extent would be $Cf^{4+}(f^8)$ and $Md^{2+}(f^{13})$. The MO_2^{2+} and MO_2^+ species are unknown in this part of the series, and even after allowance is made for the severe limitations on preparative work arising from radioactivity and lack of availability, this absence is probably genuine.

11.10.1 Solid Compounds
A surprisingly large number of crystalline compounds of these metals have been characterised. They are mainly halides, oxides, sulphides, oxohalides and halide complexes. As described previously, all the trihalides are known through to californium and $EsCl_3$ has also been prepared. Their structures are analogous to the lanthanide trihalides and the methods of preparation are conventional; an account of these trihalides has already been given. $EsBr_2$ has been obtained by reduction of $EsBr_3$ in H_2 at $650°$.

The tetrafluorides of Cm, Bk and Cf have all been prepared, that of curium by fluorination of $^{244}CmF_3$ and those of berkelium and californium by fluorination of the oxides. No other tetrahalides of curium and its succeeding elements have been reported, but the complex ion $[BkCl_6]^{2-}$ is quite stable and its dicaesium salt has been isolated. The experimental work in this area of chemistry, which is carried out on a very small scale, is often very elegant and as an illustration some details of the preparation of the last-mentioned two compounds will now be given.

11.10.2 Berkelium Tetrafluoride
In preliminary experiments $^{249}Bk^{3+}$ was purified by solvent extraction on the microgram scale using di-2-ethylhexylphosphoric acid as extractant. A single bead, about 1 mm diameter, of sulphonated polystyrene cation-exchange resin was next used to absorb all the Bk^{3+} ions from solution. The bead was then dried and heated to $400°$ in a sapphire crucible (sapphire, Al_2O_3, resists fluorine). The residue was treated with He (atmospheric pressure) and successive small quantities of F_2 (3 kPa). Finally, treatment with F_2 (200 kPa) for 16 hours gave microcrystalline BkF_4, which was introduced into a tapered capillary tube and characterised by its X-ray powder photograph, which showed it to be isostructural with UF_4.

11.10.3 Dicaesium Hexachloroberkelate (IV)
In preliminary experiments the Bk^{3+} was purified using a cation-exchange column with hydroxyisobutyric acid as eluant. A solution was prepared of 5 µg Bk^{3+} in 5 mm^3 of $2M$ H_2SO_4, which was introduced into a fine tapered capillary. This solution was treated with 5 mm^3 of $NaBrO_3$ in $2M$ H_2SO_4 for three minutes at $90°$ to oxidise the berkelium to Bk^{4+}, which was precipitated as yellow-green $Bk(OH)_4$ by NH_4OH. After centrifugation and washing, the tube was cooled to $-23°$ and the contents treated with 20 per cent excess $0.1M$ CsCl in concentrated aqueous HCl containing chlorine, and the mixture was saturated with gaseous HCl. A red-orange precipitate was produced (50 per cent yield) which after centrifugation and washing with glacial acetic acid and with ether, was characterised as $Cs_2[BkCl_6]$ by X-ray powder photography and by analysis. The analyses were performed by means of very small-scale potentiometric titrations for Bk^{4+} and Cl^- using standard solutions of Fe^{2+} and Ag^+, respectively. The $[BkCl_6]^{2-}$ ion is octahedral with Bk–Cl = 255 pm. Other

halide complexes that have been isolated include $Cs_2NaBk^{III}Cl_6$ and $Na_7M_6{}^{IV}F_{31}$ (M = Cm, Cf). The oxides M_2O_3 (M = Cm, Bk, Cf, Es) and MO_2 (M = Cm, Bk, Cf) have also been characterised as have the oxochlorides MOCl (M = Bk, Cf, Es).

11.10.4 Solution Studies

In aqueous solution curium and the succeeding elements all form the M^{3+} ion which is for all of them the most stable state, except for nobelium where the f^{14} ion No^{2+} is more stable. Another ion not in the tripositive state is Cm^{4+}, which is stabilised in solution by fluoride ion but otherwise has an oxidation potential for the aqueous reaction $Cm^{3+} \rightarrow Cm^{4+} + e$ estimated at rather more negative than -2.8 V. Cm^{3+} is an f^7 ion and the fact that it attains the +4 state at all is a residue of the behaviour of the earlier actinides, which attain stable higher oxidation states. The next actinide, berkelium, has a Bk^{4+} aqueous ion which is considerably more stable than Cm^{4+}, the potential for the aqueous reaction $Bk^{3+} \rightarrow Bk^{4+} + e$ being estimated at -1.6 V. In this case the Bk^{4+} ion is stabilised by its f^7 configuration; the succeeding elements do not form tetravalent ions or compounds.

There is a considerable tendency towards divalency at the heavier end of the series. Thus the f^{13} and f^{14} ions Md^{2+} and No^{2+} are stable in aqueous solution. The standard oxidation potential for $Md^{2+}(aq.) \rightarrow Md^{3+}(aq.) + e$ was found to be about $+0.2$ V. Thus Md^{3+}, although more stable than Md^{2+}, was reduced by only moderately strong reducing agents such as V^{2+} and Eu^{2+}. In the case of nobelium the f^{14} configuration of No^{2+} makes it more stable than No^{3+}. To conclude this section a short account of the elegant work with No^{2+}/No^{3+} will be given.

A target, mainly ^{244}Pu, was bombarded with energetic ^{16}O ions. The resulting shower of ^{255}No atoms, first caught in helium and then transferred to Pt foil was examined in one of several ways. The nobelium atoms numbered about 100 for each experiment and were assayed by α-counting. Since ^{255}No has $t_{1/2} = 3$ min., rapid procedures were necessary. Elution sequences were obtained using a 50 x 12 mm cation-exchange column and 0.3M and 1.9M ammonium α-hydroxyisobutyrate as elutants. The nobelium was not eluted with the 0.3M solution, though tracer quantities of Es^{3+} were, but was eluted by the 1.9M solution just before tracer quantities of Sr^{2+}. In coprecipitation experiments with LaF_3 and BaF_2, nearly all the nobelium came down with the BaF_2; Am^{3+} tracer mainly came down with the LaF_3, while Ra^{2+} tracer mainly came down with the BaF_2. However, in experiments where Ce^{4+} oxidant was added, rather more than half the nobelium came down with the LaF_3. Elution experiments were performed in which nobelium, Es^{3+} and Am^{3+} were eluted with 6M H_2So_4 from a 10 x 2 mm column packed with $SrSO_4$. When Ce^{4+} ion was added, the order of elution was No = Es^{3+}, Am^{3+}. When Zr^{4+} (non-oxidant; added as a blank) was used instead of Ce^{4+}, the order was Es^{3+}, Am^{3+} No. Thus in the second case, the larger No^{2+} ion is eluted after the smaller Es^{3+} and Am^{3+} ions.

Some of the actinides also show lattice-stabilised dipositive oxidation states. Thus both Am^{2+} and Es^{2+} have been obtained in CaF_2 host lattices; this type of behaviour is also observed throughout the lanthanide series. Finally, polarographic experiments using $Cf(ClO_4)_3$ in acetonitrile have clearly shown a reduction wave corresponding to $Cf^{3+} \rightarrow Cf^{2+}$ at about the same potential as the $Sm^{3+} \rightarrow Sm^{2+}$ reduction.

11.11 Organo-compounds of the Actinides

The actinides form a number of interesting organometallics, especially with aromatic anionic ligands such as $C_5H_5{}^-$. A series of triscyclopentadienyls $M(C_5H_5)_3$ may be prepared from the trichlorides and beryllium cyclopentadienyl (or in the case of uranium, potassium cyclopentadienyl)

$$2MCl_3 + 3Be(C_5H_5)_2 \xrightarrow{65°} 2M(C_5H_5)_3 + 3BeCl_2$$

where M = Pu, Am, Cm, Bk, Cf.

$$UCl_3 + KC_5H_5 \xrightarrow{\text{benzene}} U(C_5H_5)_3 + 3KCl$$

These compounds are stable to heat ($Am(C_5H_5)_3$; decomp. 330°) but are easily hydrolysed and decompose in air. Like the corresponding lanthanide compounds, they readily form adducts, including examples where the added ligand (for example, *cyclo*-$C_6H_{11}NC$) is usually associated with covalently bonded complexes.

The tetravalent actinides also form cyclopentadienyls. Thus compounds $M(C_5H_5)_4$ may be prepared by conventional methods

$$MCl_4 + 4KC_5H_5 \xrightarrow{C_6H_6} M(C_5H_5)_4 + 4KCl$$

where M = Th, U, Np.

The corresponding protactinium compound has also been prepared from $Be(C_5H_5)_2$. These are monomeric compounds in solution, with zero dipole moment, and the uranium compound has tetrahedrally disposed π-cyclopentadienyl groups.

A deficiency of potassium cyclopentadienyl in the preparation leads to the monochloro compounds

$$MCl_4 + 3KC_5H_5 \rightarrow M(C_5H_5)_3Cl + 3KCl$$

where M = Th, U.

$U(C_5H_5)_3Cl$ has been shown by X-ray diffraction to have an approximately tetrahedral disposition of the four ligands (see figure 11.15). The three C_5H_5 groups are bonded such that all U–C distances are approximately equal, as they are in $U(C_5H_5)_4$. Their description as π-bonded C_5H_5 groups, though geometrically correct, should not be taken as implying that they are necessarily covalently bonded to the metal. The description as *pentahapto*cyclopentadienyl(h^5) is better in that it has no implications as to the nature of the bonding. The chlorine in $U(C_5H_5)_3Cl$

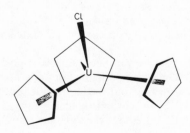

Figure 11.15 The structure of $U(h^5\text{-}C_5H_5)_3Cl$

and the thorium analogue can be replaced by other groups such as F, Br, I, OH, OEt or BH_4. The action of lithium alkyls or aryls on $U(C_5H_5)_3Cl$ gives $U(C_5H_5)_3R$, where R = Me, Bu, Ph, CCPh or C_6F_5. When R = CCPh, X-ray analysis shows the phenylacetylide group to be σ-bonded. It is remarkable that while the $M(C_5H_5)_3$ compounds are hydrolysed by water

$$M(C_5H_5)_3 + 3H_2O \rightarrow M(OH)_3 + 3C_5H_6$$

$U(C_5H_5)_3Cl$ gives a stable $[U(C_5H_5)_3]^+$ species in water, which can be precipitated as, for example, its $[PtCl_6]^{2-}$ salt. This suggests that the bonding of the C_5H_5 groups in the U(IV) compound is stronger than in U(III) and may be qualitatively different, being largely covalent as opposed to largely electrostatic in the U(III) compound. Both $M(C_5H_5)_3$ and $M(C_5H_5)_3Cl$ are volatile in vacuum at about 200° but $M(C_5H_5)_4$ are relatively involatile. An interesting adduct, presumably involving a fluoride bridge, may be made as follows

$$U(C_5H_5)_3F + U(C_5H_5)_3 \cdot THF \xrightarrow{\text{benzene}} U(C_5H_5)_3F \cdot U(C_5H_5)_3$$

The proton n.m.r. spectra of some of these compounds show large paramagnetic shifts, though not so large as those of some of the corresponding lanthanide compounds. Thus $U(C_5H_5)_4$ shows a shift of 20.4 p.p.m. and $Np(C_5H_5)_3Cl$ one of 27.4 p.p.m. The unstable $U(C_5H_5)_2Cl_2$ has also been reported; it is a salt.

$$UCl_4 \xrightarrow{C_5H_5Tl} [U(C_5H_5)_3]_2[UCl_6]$$

Other conjugated systems have also been used to obtain similar complexes. Thus the indenyl $U(C_9H_7)_3Cl$ is known; it has a structure similar to the C_5H_5 analogue. $Th(C_9H_7)_3Cl$ also exists, while there is some evidence for $M(C_9H_7)_4$ and $M(C_9H_7)_2Cl_2$. The tetra-allyls of uranium and thorium are also known but are thermally unstable, the former decomposing at $-30°$

$$4C_3H_5MgBr + ThCl_4 \rightarrow Th(C_3H_5)_4 + 4MgCl_2$$

$Th(C_3H_5)_4$ has a proton n.m.r. spectrum that agrees with the formulation as a π complex (3.97, 6.46, 7.61 τ at 10°; 3.95, 7.09 τ at 80° due to scrambling of the CH_2 protons' resonances).

A very interesting group of compounds is formed by cyclo-octatetraene. The first of these to be synthesised was bis(cyclo-octatetraenyl)uranium(IV), now commonly known as uranocene

$$2K_2C_8H_8 + UCl_4 \xrightarrow{THF, 0°} U(C_8H_8)_2 + 4KCl$$

Uranocene is a green crystalline substance, volatile at 180° in a moderate vacuum. It is stable to water, acetic acid and aqueous sodium hydroxide but inflames in air. The structure (X-ray analysis) has the eclipsed D_{8h} symmetry (see figure 11.16) and the U–C distance is 265 pm, fairly close to the value of 274 pm in $U(C_5H_5)_3Cl$, where the $U-C_5H_5$ link is also unattacked by water. Similar compounds $M(C_8H_8)_2$ (M = Th, Np, Pu), which are X-ray isomorphous and thus have similar

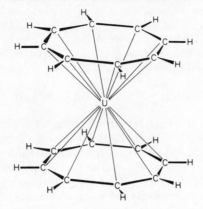

Figure 11.16 The structure of uranocene $U(C_8H_8)_2$

structures to $U(C_8H_8)_2$, but are decomposed by water, have also been made. The probability is that in these compounds the five energetically lowest $C_8H_8^{2-}$ molecular orbitals, which have respectively $\sigma, \pi(2)$, and $\delta(2)$ symmetry relative to the eightfold symmetry axis are stabilised by overlap with 6d and 5f orbitals, in a manner similar to the stabilisation of the $C_5H_5^-$ M.O.s by overlap with the iron 3d orbitals in ferrocene. Orbital reduction factors of about 0.8, which indicate considerable covalency, are necessary to explain the observed magnetic moments of $U(C_8H_8)_2$ (2.43 B.M.) and $Np(C_8H_8)_2$ (1.81 B.M.); theory predicts 3.20 and 3.27 B.M.

When UCl_4 is heated in benzene with aluminium and aluminium trichloride (conditions analogous to those that give $Cr(C_6H_6)_2$), a black compound $U(AlCl_4)_3C_6H_6$ is formed. X-ray analysis shows that the uranium atom is co-ordinated to three pairs of Cl from $AlCl_4^-$ ions and in a *hexahapto* manner to a benzene ring with U–C = 291 pm. The compound is fairly stable to heat but is decomposed by water, as might be expected.

We can finally mention an interesting report of a heteronuclear carbonyl involving uranium

$$4NaMn(CO)_5 + UCl_4 \rightarrow [U\{Mn(CO)_5\}_4] + 4NaCl$$

The orange product in unstable to air, but fairly stable thermally. Further investigations in this area should be of very considerable interest.

11.12 The Post-actinide Elements

Considerations of the periodic table suggests that, following lawrencium, element 103, a '6d' transition series should begin. Claims have been made for some of these elements, and we therefore discuss them briefly.

11.12.1 Element 104
There are few doubts concerning its existence, but considerable controversy surrounds (i) its name and (ii) which workers first identified it.

In 1966, workers from the Russian Nuclear Institute at Dubna reported that bombardment of ^{242}Pu with accelerated ^{22}Ne particles afforded a short-lived nucleus whose chemical properties closely resembled Hf (that is, a volatile chloride). This nucleus was said to be ^{260}104 ($T_{\frac{1}{2}} \approx 0.3$ s). Another isotope ^{259}104 ($T_{\frac{1}{2}} \approx 4.5$ s) was later reported, and the element named Kurchatovium (Ku). American workers at Berkeley reported bombardment of ^{249}Cf with ^{13}C and ^{12}C to yield ^{257}104 and ^{259}104, while bombardment of ^{248}Cm with ^{18}O gave ^{261}104 (like the other isotopes, decaying by α emission; $t_{\frac{1}{2}} \approx 70$ s). They proposed the name Rutherfordium (Rf) and disagreed with the early Russian work.

The current situation thus appears to be that element 104 can be prepared, has several isotopes and forms a chloride with similar properties to, but less volatile than. $HfCl_4$, while its solution behaviour is unlike di- and tervalent actinides.

11.12.2 Element 105
American workers in 1970 presented clear evidence for this element

$$^{249}Cf \xrightarrow{\,^{15}N\,} {}^{260}105 \xrightarrow[t_{\frac{1}{2}} = 1.6\ s]{-\alpha} {}^{256}Lr$$

They name it hahnium (Ha); their claim to originality is disputed by Russian workers — their route involved bombardment of ^{243}Am with ^{22}Ne. Isotopes ^{261}105 and ^{262}105 have also been reported, the latter having a half-life of 40 s, giving rise to the possibility of chemical experiments to assign its position in the periodic table.

11.12.3 Further Elements
Recently, both American and Russian workers claimed to have made element 106. The former bombarded a californium target with oxygen nuclei, and believe they obtained a 106 isotope of half-life 0.9 s. The latter bombarded a lead target with accelerated ^{54}Cr nuclei and identified an activity of half-life 10 ms with a different isotope of element 106. It seems possible that even heavier elements may be characterised shortly: element 110 (eka-Pt) might be especially interesting. Attempts to synthesise element 114 by bombarding ^{248}Cm with ^{40}Ar ions ()200 MeV) have been unsuccessful, however.

Bibliography

Chapter 1

1. E.M. Larsen. Zirconium and hafnium chemistry. *Prog. inorg. Chem.*, **13** (1970), 1.

Chapter 2

1. M.T. Pope and B.W. Dale. Isopolyvanadates, -niobates, and -tantalates. *Q. Rev. chem. Soc.*, **22** (1968), 527.
2. F. Fairbrother, *The Chemistry of Niobium and Tantalum*, Elsevier, Amsterdam (1967).

Chapter 3

1. P.C.H. Mitchell. Coordination compounds of molybdenum. *Chem. Rev.*, **1** (1966), 315.
2. P.C.H. Mitchell. Oxo-species of molybdenum V and VI. *Q. Rev. chem. Soc.*, **20** (1966), 103.
3. J.T. Spence. Biological aspects of molybdenum coordination chemistry. *Coord. Chem. Rev.*, 4 (1969), 475.
4. R.C. Bray and J.C. Swann. Molybdenum-containing enzymes. *Struct. Bond.*, **11** (1972), 107.
5. R.V. Parish. The inorganic chemistry of tungsten. *Adv. inorg. Chem. Radiochem.*, 9 (1966), 315.
6. P.G. Dickens and M.S. Whittingham. The tungsten bronzes and related compounds. *Q. Rev. chem. Soc.*, **22** (1968), 30.
7. D.L. Kepert. Isopolytungstates. *Prog. inorg. Chem.*, 4 (1962), 199.
8. D.L. Kepert, *The Earlier Transition Elements*, Academic Press (1972).

Chapter 4

1. R. Colton, *The Chemistry of Rhenium and Technetium*, Interscience, New York (1965).
2. R.D. Peacock, *The Chemistry of Technetium and Rhenium*, Elsevier, Amsterdam (1966).
3. J.E. Ferguson. Recent advances in the coordination chemistry of rhenium. *Coord. Chem. Rev.*, 1 (1966), 459.
4. K.V. Kotegov, O.N. Pavlow and V.P. Shvedov. Technetium, *Adv. inorg. Chem. Radiochem.*, **11** (1968), 2.

Chapter 5

1. W.P. Griffith, *The Chemistry of the Rarer Platinum Metals*, Interscience, New York (1967).
2. W.P. Griffith. Osmium and its compounds. *Q. Rev. chem. Soc.*, **19** (1965), 254
3. P.C. Ford. Properties and reactions of ruthenium(II) amine complexes. *Coord. Chem. Rev.*, **5** (1970), 75.

Chapter 6

1. W.P. Griffith, *The Chemistry of the Rarer Platinum Metals*, Interscience, New York (1967).
2. B.R. James. Reactions and catalytic properties of rhodium complexes in solution. *Coord. Chem. Rev.*, **1** (1966), 505.

Chapter 7

1. J.R. Miller. Recent advances in the stereochemistry of nickel, palladium and platinum. *Adv. inorg. Chem. Radiochem.*, **4** (1962), 133.
2. R.J. Cross. σ-Complexes of platinum(II) with hydrogen, carbon and other Group IV elements. *Organometal. Chem. Revs.*, **2A** (1967), 97.
3. F.R. Hartley. Starting materials for the preparation of organometallic complexes of platinum and palladium. *Organometal. Chem. Rev.*, **6A** (1970), 119.
4. J.S. Thayer. Organo-platinum(IV) compounds. *Organometal. Chem. Rev.*, **5A** (1970), 53
5. R. Ugo. The coordinative reactivity of phosphine complexes of platinum(0), palladium(0) and nickel(0). *Coord. Chem. Rev.*, **3** (1968), 319.
6. L.M. Venanzi. Phosphine complexes: σ vs. π-bonding and the nature of the *trans*-effect. *Chem. Br.*, **4** (1968), 162.
7. F. Basolo and R.G. Pearson. The *trans*-effect in metal complexes. *Prog. Inorg. Chem.*, **4** (1962), 381.
8. J.H. Nelson and H.B. Jonassen. Monoolefin and acetylene complexes of nickel, palladium and platinum. *Coord. Chem. Rev.*, **6** (1971), 27.
9. F.R. Hartley. Olefin and acetylene complexes of platinum and palladium. *Chem. Rev.*, **69** (1969), 799.
10. A.J. Thompson, R.J.P. Williams and J. Reslova. Compounds related to *cis*-Pt(NH$_3$)$_2$Cl$_2$ − an antitumor drug. *Struct. Bond.*, **11** (1972), 1.
11. F.R. Hartley, *The Chemistry of Platinum and Palladium*, Applied Science Publishers (1973).

Chapter 8

1. J.A. McMillan. Higher oxidation states of silver. *Chem. Rev.*, **62** (1962), 65.
2. D.C.M. Beverwijk, G.J.M. Van derKerk, A.J. Leusink and J.G. Noltes. Organo-silver chemistry. *Organometal. Chem. Rev.*, **5A** (1970), 215.
3. B. Armer and H. Schmidbaur. Organogold chemistry. *Angew. Chem. (int. edn)*, **9** (1970), 101.

Chapter 9

1. G. Wilkinson and F.A. Cotton. Cyclopentadienyl and arene metal complexes. *Prog. inorg. Chem.*, **1** (1959), 1.
2. E.O. Fischer and H.P. Fritz. Complexes of aromatic ring systems and metals. *Adv. inorg. Chem. Radiochem.*, **1** (1959), 56.

3. J.M. Birmingham. Synthesis of cyclopentadienyl metal compounds. *Adv. organometal Chem.*, **2** (1964), 365.
4. H.W. Quinn and J.H. Tsai. Olefin complexes of the transition metals. *Adv. inorg. Chem. Radiochem.*, **12** (1969), 217.
5. R.G. Guy and B.L. Shaw. Olefin, acetylene and π-allylic complexes of transition metals. *Adv. inorg. Chem. Radiochem.*, **4** (1962), 78.
6. M.R. Churchill. Transition-metal complexes of azulene and related ligands. *Prog. inorg. Chem.*, **11** (1970), 53.
7. P.M. Maitlis. Cyclobutadiene metal complexes. *Adv. organometal Chem.*, **5** (1967), 95.
8. E.W. Abel and F.G.A. Stone. The chemistry of transition-metal carbonyls: structural considerations. *Q. Rev. chem. Soc.*, **23** (1969), 325.
9. E.W. Abel and F.G.A. Stone. The chemistry of transition-metal carbonyls: synthesis and reactivity. *Q. Rev. chem. Soc.*, **24** (1970), 498.
10. W. Hieber. Metal carbonyls, forty years of research. *Adv. organometal Chem.*, **8** (1970), 1.
11. E.W. Abel and S.P. Tyfield. Metal carbonyl cations. *Adv. organometal. Chem.*, **8** (1970), 117.
12. P.S. Braterman. Spectra and bonding in metal carbonyls. *Struct. Bond.*, **10** (1972), 57.
13. L.M. Haines and M.B.H. Stiddard. Vibrational spectra of transition-metal carbonyl compounds. *Adv. inorg. Chem. Radiochem.*, **12** (1969), 53.
14. W.P. Griffith. Organometallic nitrosyls. *Adv. organometal. Chem.*, **7** (1968), 211.
15. M.F. Hawthorne. The chemistry of the polyhedral species derived from transition metals and carboranes. *Accts. chem. Res.*, **1** (1968), 281.
16. L.J. Todd. Transition-metal carborane complexes. *Adv. organometal. Chem.*, **8** (1970), 87.
17. G.W. Parshall and J.J. Mrowca. σ-Alkyl and σ-aryl derivatives of transition metals. *Adv. organometal. Chem.*, **7** (1968), 157.
18. R.S. Nyholm. Transition-metal complexes of some perfluoro-ligands. *Q. Rev. chem. Soc.*, **24** (1970), 1.
19. F.A. Cotton. Fluxional organometallic molecules. *Accts chem. Res.*, **1** (1968), 257.
20. E.L. Muetterties. Sterically nonrigid structures. *Accts chem. Res.*, **3** (1970), 766.
21. M.R. Churchill and R. Mason. Structural chemistry of organo-transition metal complexes — some recent developments. *Adv. organometal. Chem.*, **5** (1967), 93.
22. M.L. Maddox, S.L. Stafford and H.D. Kaesz. Applications of n.m.r. to the study of organometallic compounds. *Adv. organometal. Chem.*, **3** (1965), 1.
23. M.I. Bruce. Mass spectra of organometallic compounds. *Adv. organometal. Chem.*, **6** (1968), 273.
24. J. Lewis and B.F.G. Johnson. Mass spectra of some organometallic molecules. *Accts chem. Res.*, **1** (1968), 245.

Chapter 10

1. T. Moeller, D.F. Martin, L.C. Thompson, R. Ferrus, G.R. Feistel and W.J. Randall. The coordination chemistry of yttrium and the rare earth metal ions. *Chem. Rev.*, **65** (1965), 1.
2. *Progress in the Science and Technology of the Rare Earths*, vol. 1 (1964); vol. 2 (1966); vol. 3 (1968), (ed. LeRoy Eyring), Pergamon, Oxford.
3. H.J.S. Gysling and M. Tsutsui. Organolanthanides and organoactinides. *Adv. organometal. Chem.*, **9** (1970), 361.

4. R.G. Hayes and J.M. Thomas. Organometallic compounds of the lanthanides and actinides. *Organometal. Chem. Rev.,* **7A** (1971), 1.
5. *Lanthanides and Actinides, M.T.P. International Review of Science,* Series One, volume 7 (ed. K.W. Pagnall), Butterworths, London (1971).
6. G.A. Melson and R.W. Stotz. *Coord. Chem. Rev.,* **7** (1971), 133.

Chapter 11

1. C. Keller, *The Chemistry of the Transuranium Elements,* Verlag Chemie, Weinheim (1971).
2. K.W. Bagnall. The co-ordination chemistry of the actinide halides. *Coord. Chem. Rev.,* **2** (1967), 145.
3. J. Selbin and J.D. Ortego. The chemistry of uranium(V). *Chem. Rev.,* **69** (1969), 657.
4. D. Brown. Some recent preparative chemistry of protactinium. *Adv. inorg. Chem. Radiochem.,* **12** (1969), 1.
5. J.M. Cleveland. Aqueous coordination chemistry of plutonium. *Coord. Chem. Rev.,* **5** (1970), 101.
6. E.H.P. Cordfunke, *The Chemistry of Uranium,* Elsevier, Amsterdam (1969). (see also bibliography to chapter 10).

General Interest

1. J.H. Canterford and R. Colton, *Halides of the Second and Third Row Transition Metals,* Wiley, New York (1968).
2. D. Brown, *Halides of the Lanthanides and Actinides,* Wiley, New York (1968). (Both these books deal extensively with the chemistry of both halides and halide complexes).
3. F.A. Cotton. Transition-metal compounds containing clusters of metal atoms. *Q. Rev. chem. Soc.,* **20** (1966), 389.
4. F.A. Cotton. Strong homonuclear metal–metal bonds. *Accts chem. Res.,* **2** (1969), 240.
5. J.E. Fergusson. Metal–metal bonded halogen compounds of the transition-metals. *Prep. inorg. react.,* **7** (1971), 93.
6. M.C. Baird. Metal–metal bonds in transition-metal complexes. *Prog. inorg. Chem.,* **9** (1968), 1.
7. B.M. Chadwick and A.G. Sharpe. Transition-metal cyanides and their complexes. *Adv. inorg. Chem. Radiochem.,* **8** (1966), 84.
8. D. Coucouvanis. The chemistry of the dithioacid and 1,1-dithiolate complexes. *Prog. inorg. Chem.,* **11** (1970), 233.
9. J.A. McCleverty. Metal-1,2-dithiolene and related complexes. *Prog. inorg. Chem.,* **10** (1968), 49.
10. G. Booth. Complexes of the transition metals with phosphines, arsines and stibines. *Adv. inorg. Chem. Radiochem.,* **6** (1964), 1.
11. C.A. McAuliffe, *Transition Metal Complexes of Phosphorus, Arsenic and Antimony Ligands,* Macmillan, London (1973).
12. B.F.G. Johnson and J.A. McCleverty. Nitric oxide complexes of transition-metals. *Prog. inorg. Chem.,* **7** (1966), 277.
13. W.P. Griffith. Transition-metal oxo-complexes. *Coord. Chem. Rev.,* **5** (1970), 459.
14. J.A. Connor and E.A.V. Ebsworth. Peroxy-compounds of transition-metals. *Adv. inorg. Chem. Radiochem.,* **6** (1964), 280.

15. E.I. Ochiai. Catalytic functions of metal ions and their complexes. *Coord. Chem. Rev.*, **3** (1968), 49.
16. J.P. Collman. Reactions of ligands coordinated with transition metals. *Transit. Metal Chem.*, **2** (1966), 2.
17. J.P. Collman and W.R. Roper. Oxidative addition reactions of d^8 compounds. *Adv. organometal. Chem.*, **7** (1968), 53.
18. J. Halpern. Oxidative addition reactions of transition-metal complexes. *Accts chem. Res.*, **3** (1970), 386.
19. E.L. Muetterties and C.M. Wright. Molecular polyhedra of high coordination numbers. *Q. Rev. chem. Soc.*, **21** (1967), 109.
20. R.V. Parish. Eight-coordination in the transition series. *Coord. chem. Rev.*, **1** (1966), 439.
21. S.J. Lippard. Eight-coordination chemistry. *Prog. inorg. Chem.*, **8** (1967), 109.
22. E.L. Muetterties and R.A. Schunn. Pentaco-ordination. *Q. Rev. chem. Soc.*, **20** (1966), 245.
23. *Transition Metal Hydrides* (ed. E.L. Muetterties), Dekker, New York (1971).
24. A.P. Ginsberg. Hydride complexes of the transition metals. *Transit. Metal Chem.*, **1** (1965), 143.
25. J. Chatt and G.J. Leigh. Nitrogen fixation. *Chem. Soc. Rev.*, **1** (1972), 121.

Index